ACADEMY OF SCIENCES OF THE CZECH REPUBLIC

Checklist of the Metazoan Parasites of Fishes of the Czech Republic and the Slovak Republic (1873–2000)

The publisher thanks the Academy of Sciences of the Czech Republic for a financial contribution that made the publication of this book possible.

CHECKLIST

of the Metazoan Parasites of Fishes of the Czech Republic and the Slovak Republic

FRANTIŠEK MORAVEC

(1873–2000)

Institute of Parasitology
Academy of Sciences of the Czech Republic
České Budějovice

ACADEMIA
PRAHA 2001

ISBN 80-200-0907-8

CONTENTS

INTRODUCTION

Metazoan parasites of fishes represent a large group of organisms that have adapted to the parasitic mode of life in members of this important group of aquatic vertebrates. They attack most body organs of fishes, parasitizing them as adults or as larvae. Some of them are known agents of fish diseases. They affect the health and the reproduction of their hosts, making them easy prey to predators and they decrease fish resistence to various infections and any unfavourable environmental conditions. In fish farming, parasites may lead to epidemics and mortality rates resulting in considerable economic loss. Some fish parasites are even transferable to man.

Investigations into the parasites of fishes in the territory of the present Czech and Slovak Republics (parts of the Austrian-Hungarian Empire from the seventeenth century until 1918 and of former Czechoslovakia between 1918 and 1992) have a long history. In the second half of the nineteenth century, the first studies on fish parasites were associated with an increased interest in aquatic organisms. During this period, inventories were attempted in the Czech Lands, this being undoubtedly influenced by a long tradition of the Czech fishpond cultivation (numerous South Bohemian ponds were founded during the Middle Ages, some of them in the eleventh century or even earlier).

The first paper dealing with the parasites of fishes is the comprehensive work of Prof. A. Frič, a well-known Czech zoologist at that time, entitled "On crustaceans of the Czech Lands" published in Czech in 1873. Frič's paper reported some crustaceans parasitic to fishes in addition to those free-living. This was soon followed by the first paper devoted to fish helminths by J. John (1877) "Helminths of Czech fishes". An important systematic – faunistic work reporting the results of investigations into fish parasites, carried out in the Elbe River at the end of the nineteenth century by Prof. Frič and his assistant A. Šrámek, was published by Šrámek (1901, 1903). His data were later included in Frič's (1908) book, "Czech fishes and their parasites".

In the course of the twentieth century, parasites of fishes were intensively studied both in the Czech countries (presently the Czech Republic) and Slovakia (presently the Slovak Republic). A large amount of information has accummulated so that at present this territory is among the best documented regions for fish parasites both in Europe and in the world. In 1970, Ergens and Lom in their book "Agents of parasitic diseases of fishes" attempted to catalogue the fish parasites occurring in the territory of former Czechoslovakia and their distribution; unfortunately, the authors mentioned only the main river drainage systems for the individual parasite species listed, without giving their exact localities. Therefore, it is not clear for some of the included species whether or not they were actually recorded in former Czechoslovakia and whether their distribution in individual main river basins was based on data from the literature or was verified either by the authors' unpublished findings or was only a presumed distribution. Moreover, now this book is somewhat out of date. Many papers dealing with parasites of fishes were published during the last thirty years, considerably extending the number of species of fish parasites as well as knowledge about their hosts and distribution.

At present Czech and Slovak ichthyoparasitological literature is rather extensive, including a large number of papers and books published in Czech and Slovak, but also in other languages. The papers are scattered in many journals, some of which are difficult to access. Since studies on fish parasites have become increasingly attractive to the young generation of parasitologists during the recent decade, it seems necessary and timely to make a checklist and bibliography of the work that has been done. Because many papers on fish parasites in the Czech and Slovak waters were written in the Czech and Slovak languages, the checklist is compiled in English in order to be useful for the introduction of these data to foreign fish parasitologists.

In spite of the present political independence of the Czech and Slovak Republics, the checklist covers data from both countries, because during the long-lasting existence of former Czechoslovakia (1918–1992), ichthyoparasitological investigations by some authors were carried out in both the Czech Lands (Bohemia, Moravia, Silesia) and Slovakia.

The checklist involves records from the period of 1873–2000 from published scientific literature. Unpublished records, such as those in students' theses, various technical reports, lectures or abstracts of papers from meetings, etc., have been excluded. Only the abstracts published in scientific journals have been included, provided that the published text clearly shows the original findings of the parasites from the territory in question. In a case of review papers, where the data were taken over from previous authors, these are not strictly cited in the text concerning the individual species of parasites in the checklist.

The territory under investigation includes four main river drainage systems: the Elbe (= Labe) R. system (L), the Oder (= Odra) R. system (O), the Visla R. system (V), and the Danube (= Dunaj) R. system (D); the Czech Republic includes the basins of the Rivers Elbe, Oder and Danube, whereas the Slovak Republic belongs to the basins of the Rivers Danube and Visla.

The checklist consists of seven sections: Introduction, Parasite – host list, Host – parasite list, List of metazoan parasites of fishes of the Czech Republic, List o metazoan parasites of fishes of the Slovak Republic, Literature, and Index.

In the PARASITE – HOST LIST, parasites are arranged by the following main groups (phyla or classes) with indication of the respective families: Coelenterata, Monogenea, Cestoda, Trematoda, Nematoda, Nematomorpha, Acanthocephala,

Hirudinea, Bivalvia, Copepoda, Branchiura, and Acariformes; larval forms are treated separately within individual groups. Although during recent years some authors take Myxosporea for metazoans, these parasites, mostly considered to be protozoans, have not been included. Species within the families are listed alphabetically. For each species of parasite contained herein, the following information is provided:

The current *scientific name*, including its authority (author(s) and year), is followed by the synonyms, but only those under which it was reported in the domestic ichthyoparasitological literature. The checklist is not a taxonomic work and, therefore, no attempt has been made to evaluate systematically the validity of the published records. However, attention is drawn to obvious errors and some tentative or uncertain identifications.

The *site* (or location) of occurrence of the parasite in/on its host(s) is given. Names designated by an asterisk (*) represent accidental findings or questionable host records. Numbers in parentheses after each host name correspond with the numbers assigned to the references establishing the particular parasite – host records. The *distribution* of the parasite is given separately for the Czech Republic (CR) and the Slovak Republic (SR), indicating the four main river drainage systems to which the localities belong (L – Labe (= Elbe) R., O – Odra (= Oder) R., V – Visla R., D – Dunaj (= Danube) R.).

The authors responsible for the *records* are listed in chronological order. The reference numbers correspond to those given after the appropriate host species. *Remarks* are given on the systematics, nomenclature and some specific topics such as the reassignment of misidentified material or tentative identifications.

In the HOST – PARASITE LIST, the fish orders and families are arranged systematically and their sequence follows the scheme of Eschmeyer (1998); fish species and subspecies in each family are arranged alphabetically. Only the fish species in which metazoan parasites were recorded in the Czech Republic or the Slovak Republic are included; those free of these parasites or the fish species that have not yet been examined for metazoan parasites are omitted. Fish names marked with an asterisk (*) are introduced fish species, not native to Central Europe. The parasites are listed according to their main taxonomic groups; the parasite names designated with an asterisk (*) represent accidental or uncertain findings.

Recent genetic studies of Ráb et al. (2000), Šlechtová et al. (2000) and Lusk et al. (2000) have shown that the previous records of *Cobitis taenia* from the Czech Republic represent, in fact, diploid-polyploid complexes of spined loach, involving *C. taenia*, *C. elongatoides* and *Cobitis* spp. In spite of this, species names of *Cobitis* are reported here as they were given in the respective original ichthyoparasitological papers.

The LISTS OF METAZOAN PARASITES OF FISHES OF THE CZECH REPUBLIC AND THE SLOVAK REPUBLIC are included to provide better and quicker orientation for the reader, showing how many and which species of these parasites occur in each of these two countries, making their comparison easier.

In the LITERATURE, the reader can find all references directly referred to in the previous text, but included are also some publications that do not contain the original parasite – host records or some that deal with various aspects of metazoan parasites in the Czech Republic and the Slovak Republic.

An INDEX to parasite and host scientific names is given to faciliate access to information in the parasite – host and host – parasite lists.

The checklist documents how much work was done by several generations of Czech and Slovak parasitologists to study the fauna of metazoan parasites of fishes in these countries during the last 128 years (a total of 344 species were recorded to date). However, the author is aware of the fact that there exist some additional data on these parasites in unpublished materials and that many new important records have not yet been published.

It is clear that a project of this sort is not easily accomplished without considerable help and advice from many colleagues. I am especially grateful to my co-workers from previous years Drs. R. Ergens and M. Gelnar (latter presently from the University of Brno) for their assistance with questions of monogenean nomenclature and for providing some of their papers to this project. For information and assistance in the preparation of the manuscript, I extend my sincere appreciation to Prof. V. Baruš, Institute of Vertebrate Biology, ASCR, Brno; Dr. T. Scholz, Institute of Parasitology, ASCR, České Budějovice; and Dr. M. Špakulová, Institute of Parasitology, SAS, Košice. Special thanks are due to present or earlier technicians of the Institute of Parasitology, ASCR, in České Budějovice, Mrs. I. Husáková, Mrs. I. Pouzarová, Mrs. J. Vokurková and Mr. T. Urzedovský, for searching for literary sources and their assistance with compiling the respective lists. My sincere thanks are due to Miss Heidi Splittgerber from the University of Colorado for revising the English.

The work on this project was supported by grant no. 524/00/0267 from the Grant Agency of the Czech Republic. I wish to thank the Comission of the Academy of Sciences of the Czech Republic for Publishing Scientific Literature for the financial support which made the publication of this work possible. I am most grateful for the excellent editorial and production work of Mr. Pavel Zátka from Academia who helped in many ways, especially in working out the format for the book.

PARASITE - HOST LIST

COELENTERATA

(exact systematic status not determined)

Polypodium hydriforme Ussov, 1885
Site: eggs.
Host: *Acipenser ruthenus* (1, 2)
Distribution: **SR:** D – River Dunaj (= Danube).
Records: 1. Rašín 1930b; 2. Rašín 1949b.

MONOGENEA

Fam. Dactylogyridae Bykhovsky, 1933

Acolpenteron sp.
Site: urinary bladder and urethers.
Host: *Cyprinus carpio* (1)
Distribution: **CR:** D – pond Jaroslavický near Jaroslavice (South Moravia).
Record: 1. Lucký and Dyk 1964.

Dactylogyrus achmerowi Gussev, 1955
Site: gills.
Host: *Cyprinus carpio* (1)
Distribution: **CR:** L – Vltava River.
Records: 1. Gelnar and Lux 1991; 2. Gelnar and Špakulová 1997.

Dactylogyrus alatus Linstow, 1878
Site: gills.
Hosts: *Abramis bjoerkna* (1, 2)
Alburnus alburnus (3, 4, 6, 7, 8)
Leucaspius delineatus (3)
Distribution: **CR:** L – ponds near Třeboň; D – River Morava; O – localitied not given; **SR:** D – Rivers Dunaj (= Danube) and Tisa, localities Čierna Voda and Senné (tributaries of Laborec R.).
Records: 1. Kašťák 1956d; 2. Kašťák 1957d; 3. Ergens 1962a; 4. Žitňan 1966a; 5. Žitňan 1968; 6. Ergens and Lom 1970; 7. Žitňan 1971c; 8. Gelnar et al. 1994.

Dactylogyrus amphibothrium Wagener, 1857
Site: gills.
Hosts: **Carassius carassius* (6)
**Cyprinus carpio* (6)
Gymnocephalus cernuus (1, 2, 3, 4, 5, 7, 8, 9, 10, 11, 12, 13, 14, 15, 16, 18, 19, 20)
Gymnocephalus schraetser (14, 21)
Perca fluviatilis (11, 14)
Distribution: **CR:** L – South-Bohemian ponds near Třeboň and Jindřichův Hradec; D – River Dyje and its pools, Kníničská and Vranovská water reservoirs; O – localities not specified; **SR:** D – Rivers Dunaj (= Danube), Hron, Laborec, Latorica and Tisa, Zemplínská Šírava water reservoir, localities Čierna Voda and Senné.
Records: 1. Vojtek et al. 1954; 2. Ergens 1956b; 3. Kašťák 1956d; 4. Kašťák 1957b; 5. Kašťák 1957d; 6. Lucký 1957b; 7. Lucký 1957d; 8. Vojtek 1959b; 9. Dyk 1961a; 10.

Ergens 1962a; 11. Lucký and Dyk 1964; 12. Žitňan 1964f; 13. Žitňan 1965a; 14. Žitňan 1965b; 15. Žitňan 1966d; 16. Přibyslavský and Lucký 1967; 17. Žitňan 1968c; 18. Ergens and Lom 1970; 19. Ergens et al. 1975; 20. Žitňan 1977; 21. Žitňan 1986.

Dactylogyrus anchoratus (Dujardin, 1845)
Site: gills.
Hosts: *Carassius auratus gibelio* (22, 29, 40)
Carassius carassius (3, 9, 15, 17, 21, 22, 27, 31)
Cyprinus carpio (1, 2, 3, 4, 5, 6, 7, 8, 9, 10, 11, 12, 13, 14, 15, 16, 17, 18, 19, 20, 21, 22, 23, 24, 26, 28, 30, 32, 33, 34, 35, 36, 38, 39)
Gobio gobio (14)
Leucaspius delineatus (14)
Distribution: **CR:** L – ponds near Třeboň and Jindřichův Hradec; D – Rivers Dyje, Morava and Rokytná, Kníničská water reservoir, South-Moravian ponds and pools, North-Moravian ponds; **SR:** D – Rivers Dunaj (= Danube), Hron, Laborec, Latorica and Tisa, Domaša, Orava, Ružín, Slňava and Zemplíská Šírava water reservoirs, locality Senné, Slovak ponds; V – River Dunajec.
Records: 1. Dyk 1954c; 2. Vojtek et al. 1954; 3. Lucký 1955b; 4. Ergens 1956b; 5. Kašťák 1956d; 6. Kašťák 1956f; 7. Kašťák 1957b; 8. Kašťák 1957d; 9. Lucký 1957d; 10. Lucký 1959a; 11. Dyk 1961a; 12. Dyk 1961b; 13. Lucký 1961; 14. Ergens 1962a; 15. Dyk and Lucký 1963; 16. Dyk 1964a; 17. Lucký and Dyk 1964; 18. Žitňan 1964a; 19. Žitňan 1964f; 20. Žitňan 1965a; 21. Žitňan 1965b; 22. Žitňan 1966a; 23. Žitňan 1966c; 24. Lucký and Jílek 1967; 25. Žitňan 1968c; 26. Ergens and Lom 1970; 27. Žitňan 1973b; 28. Tesarčík and Prouza 1974; 29. Žitňan 1974b; 30. Žitňan 1974c; 31. Ergens et al. 1975; 32. Žitňan 1977; 33. Žitňan 1978; 34. Žitňan 1979b; 35. Žitňan 1979c; 36. Žitňan 1979d; 37. Žitňan 1981a; 38. Žitňan 1982a; 39. Žitňan 1985; 40. Gelnar et al. 1994.

Dactylogyrus auriculatus (Nordmann, 1832)
Site: gills.
Hosts: *Abramis brama* (1, 2, 3, 4, 5, 6, 7, 8, 9, 10, 12, 13, 14, 15)
**Barbus barbus* (1)
**Cyprinus carpio* (1)
Distribution: **CR:** L – River Malše, Římovská water reservoir, ponds near Jindřichův Hradec, Bohemia; D – Rivers Dyje, Morava and Svratka, Kníničská and Vranovská water reservoirs; O – localities not specified; **SR:** D – Rivers Latorica and Tisa, surroundings of Komárno.
Records: 1. John 1877; 2. Vojtek et al. 1954; 3. Ergens 1956b; 4. Lucký 1959b; 5. Vojtek 1959b; 6. Vojtková 1959b; 7. Ergens 1962a; 8. Žitňan 1966a; 9. Žitňan 1966c; 10. Lucký and Jílek 1967; 11. Žitňan 1968c; 12. Ergens and Lom 1970; 13. Ergens et al. 1975; 14. Gelnar et al. 1990; 15. Gelnar et al. 1994; 16. Šindlerová et al. 1996.

Dactylogyrus bicornis Malewitzkaja, 1941
Site: gills.
Host: *Rhodeus sericeus* (1, 2, 3, 4, 6, 7, 8, 9)
Distribution: **CR:** L – localities not specified; D – Rivers Morava and Rokytná, South-Moravian ponds and pools; O – localities not specified; **SR:** D – Rivers Dunaj (= Danube) and Tisa.
Records: 1. Lucký 1957b; 2. Lucký 1959a; 3. Lucký and Dyk 1964; 4. Žitňan 1966a; 5. Žitňan 1968c; 6. Ergens and Lom 1970; 7. Žitňan 1986; 8. Gelnar et al. 1994; 9. Baruš et al. 1997b.

Dactylogyrus borealis Nybelin, 1936
Site: gills.
Host: *Leuciscus cephalus* (14)
Phoxinus phoxinus (1, 2, 3, 4, 5, 6, 7, 8, 9, 10, 11, 12, 13)
Distribution: **CR:** L – localities not specified; D – Rivers Rokytná and Bystřice near Bystřice nad Pernštejnem; O – River Osoblaha, Hořina Brook; **SR:** D – Rivers Hornád and Hron, Dobšinská water reservoir; V – River Dunajec.
Records: 1. Vojtek 1958a; 2. Lucký 1959a; 3. Lucký 1959b; 4. Žitňan 1960c; 5. Lucký and Dyk 1964; 6. Ergens 1965a; 7. Žitňan

1965b; 8. Řehulka 1968; 9. Ergens and Lom 1970; 10. Žitňan 1970–1971; 11. Žitňan 1973a; 12. Žitňan 1974c; 13. Žitňan 1986; 14. Moravec 1995a.

Dactylogyrus carpathicus Zakhvatkin, 1951
Site: gills.
Hosts: *Barbus barbus* (1, 4, 5, 6, 7, 8, 9, 10, 11, 12, 13, 15, 17, 18, 19, 20)
Barbus petenyi (2, 3, 8, 10, 12, 13, 15, 16, 17, 20)
Distribution: **CR:** L – River Malše, Římovská water reservoir; D – Rivers Dyje and Rokytná, fish farm on Dyje R., Vranovská water reservoir; O – localities not specified; **SR:** D – Rivers Hornád, Hron, Laborec, Latorica, Tisa and Torysa and tributaries, Myslava Brook; V – River Dunajec.
Records: 1. Vojtek et al. 1954; 2. Kašťák 1956d; 3. Kašťák 1957d; 4. Lucký 1957b; 5. Lucký 1959a; 6. Žitňan 1960c; 7. Dyk 1961a; 8. Hradil 1964; 9. Lucký and Dyk 1964; 10. Žitňan 1964f; 11. Žitňan 1965a; 12. Žitňan 1965b; 13. Žitňan 1966a; 14. Žitňan 1968c; 15. Ergens and Lom 1970; 16. Žitňan 1970–1971; 17. Žitňan 1974c; 18. Gelnar et al. 1990; 19. Navrátil 1991; 20. Baruš et al. 1997b.

Dactylogyrus chondrostomi Malewitzkaja, 1941
Site: gills.
Host: *Chondrostoma nasus* (1, 2, 3, 4, 5, 7, 8, 9)
Distribution: **CR:** D – Rivers Svitava and Želetavka; O – localities not specified; **SR:** D – Rivers Hornád, Latorica and Tisa.
Records: 1. Lucký 1957b; 2. Ergens 1959a; 3. Žitňan 1960c; 4. Dyk 1961a; 5. Žitňan 1966a; 6. Žitňan 1968c; 7. Ergens and Lom 1970; 8. Ergens et al. 1975; 9. Baruš et al. 1997b.

Dactylogyrus chranilowi Bychowsky, 1931
Site: gills.
Hosts: *Abramis ballerus* (1, 3, 4, 6, 7)
Abramis brama (2)
Distribution: **SR:** D – Rivers Dunaj (= Danube), Latorica and Tisa, periodical pools of Váh R.
Records: 1. Vojtková 1959a; 2. Dyk 1961a; 3. Žitňan 1966a; 4. Žitňan 1967d; 5. Žitňan 1968c; 6. Ergens and Lom 1970; 7. Ergens et al. 1975.

Dactylogyrus cordus Nybelin, 1937
Site: gills.
Hosts: *Abramis brama* (1, 7)
Leuciscus cephalus (3, 4, 5)
Leuciscus leuciscus (2, 5, 6, 8)
Distribution: **CR:** L – ponds near Třeboň; D – Kníničská and Vranovská water reservoirs; O – River Osoblaha.
Records: 1. Dyk 1955a; 2. Ergens 1956a; 3. Ergens 1956b; 4. Dyk 1961a; 5. Ergens 1962a; 6. Ergens 1965a; 7. Lucký and Jílek 1967; 8. Ergens and Lom 1970.

Dactylogyrus cornoides Gläser et Gussev, 1971
Site: gills.
Hosts: *Abramis bjoerkna* (1, 2, 3, 4)
Abramis sapa (4)
Vimba vimba (1, 3)
Distribution: **CR:** L – River Malše, Římovská water reservoir; D – River Morava; **SR:** D – River Latorica.
Records: 1. Ergens and Lom 1970; 2. Ergens et al. 1975; 3. Gelnar et al. 1990; 4. Gelnar et al. 1994.

Dactylogyrus cornu Linstow, 1878
Site: gills.
Hosts: *Abramis ballerus* (19)
Abramis bjoerkna (6, 7, 9, 12, 14, 15, 17, 18, 19, 21, 22, 23, 24, 25)
Abramis brama (6, 7, 13, 14, 15, 18)
Aspius aspius (3, 5)
Leuciscus cephalus (3, 4, 5)
Rutilus rutilus (1, 2, 6, 7, 8, 10, 11, 15, 16, 18, 19)
Vimba vimba (8, 18, 19, 21)
Distribution: **CR:** L – River Malše, Římovská water reservoir, ponds near Třeboň; D – Rivers Dyje, Rokytná and Svratka, South-Moravian ponds and pools;

O – localities not specified; **SR:** D – Rivers Dunaj (= Danube), Hron, Laborec, Latorica, Morava and Tisa, localities Čierna Voda and Senné.
Records: 1. Vojtek et al. 1954; 2. Ergens 1956b; 3. Kašťák 1956d; 4. Kašťák 1957b; 5. Kašťák 1957d; 6. Lucký 1957b; 7. Lucký 1957d; 8. Lucký 1959a; 9. Vojtek 1959b; 10. Vojtková 1959b; 11. Dyk 1961a; 12. Ergens 1962a; 13. Pacák 1962; 14. Dyk and Lucký 1963; 15. Lucký and Dyk 1964; 16. Žitňan 1964f; 17. Žitňan 1965a; 18. Žitňan 1965b; 19. Žitňan 1966a; 20. Žitňan 1968c; 21. Ergens and Lom 1970; 22. Žitňan 1971c; 23. Ergens et al. 1975; 24. Gelnar et al. 1990; 25. Gelnar et al. 1994.

Dactylogyrus crassus Kulwiec, 1927
Site: gills.
Hosts: *Carassius carassius* (1, 2)
Distribution: **CR:** D – periodical pools of Dyje River, lake Černé jezero (Svratka R. drainage system).
Records: 1. Dyk 1955a; 2. Lucký 1957d.
Remark: According to Gussev (1985), this species has been frequently confused with *D. vastator*.

Dactylogyrus crucifer Wagener, 1857
Site: gills.
Hosts: *Abramis bjoerkna* (14, 16)
Abramis brama (5, 10, 11)
Alburnus alburnus (26)
Carassius auratus (25)
Leuciscus cephalus (26, 27, 29, 30, 31)
Leuciscus idus (25)
Leuciscus leuciscus (9)
Rutilus pigus (15, 19, 21)
Rutilus rutilus (1, 6, 7, 8, 9, 10, 13, 15, 16, 17, 19, 20, 21, 22, 23, 24, 25, 30, 32)
Scardinius erythrophthalmus (2, 3, 4, 5, 8, 10, 11, 12, 15, 16)
Distribution: **CR:** L – River Malše, Římovská water reservoir, ponds near Jindřichův Hradec and Třeboň; D – Rivers Dyje, Morava, Rokytná, Svitava and Svratka, Kníničská and Vranovská water reservoirs, South-Moravian ponds and pools; O – River Osoblaha; **SR:** D – Rivers Dunaj (= Danube), Hron, Laborec, Latorica and Tisa, Zemplínská Šírava water reservoir, localities Čierna Voda and Senné.
Records: 1. Ergens 1956b; 2. Kašťák 1956d; 3. Kašťák 1957b; 4. Kašťák 1957d; 5. Lucký 1957b; 6. Lucký 1959a; 7. Vojtková 1959b; 8. Dyk 1961a; 9. Ergens 1962a; 10. Dyk and Lucký 1963; 11. Lucký and Dyk 1964; 12. Žitňan 1964f; 13. Ergens 1965a; 14. Žitňan 1965a; 15. Žitňan 1965b; 16. Žitňan 1966a; 17. Lucký and Jílek 1967; 18. Žitňan 1968c; 19. Ergens and Lom 1970; 20. Žitňan 1971c; 21. Ergens et al. 1975; 22. Žitňan 1977; 23. Moravec 1986c; 24. Gelnar et al. 1990; 25. Gelnar et al. 1994; 26. Moravec 1995a; 27. Šebelová et al. 1995; 28. Šindlerová et al. 1996; 29. Gelnar et al. 1997; 30. Kadlec et al. 1998; 31. Dušek et al. 1998; 32. Šimková et al. 2000.

Dactylogyrus cryptomeres Bychowsky, 1934
Site: gills.
Hosts: *Gobio albipinnatus* (9, 11, 12, 13, 17)
Gobio gobio (1, 2, 3, 4, 5, 6, 7, 8, 10, 12, 13, 15, 16, 18)
Distribution: **CR:** D – Rivers Jihlava, Morava and Rokytná, periodical pools of Dyje R.; O – Rivers Moravice and Osoblaha, ponds near Karlovec; **SR:** D – Rivers Dunaj (= Danube), Hron, Laborec, Latorica and Tisa, Dobšínská Šírava water reservoir.
Records: 1. Dyk and Lucký 1956b; 2. Dyk and Lucký 1956c; 3. Dyk and Lucký 1957; 4. Lucký 1957b; 5. Lucký 1957d; 6. Lucký 1959a; 7. Dyk 1964a; 8. Lucký and Dyk 1964; 9. Žitňan 1964f; 10. Ergens 1965a; 11. Žitňan 1965a; 12. Žitňan 1965b; 13. Žitňan 1966a; 14. Žitňan 1968c; 15. Žitňan 1971c; 16. Žitňan 1973a; 17. Ergens et al. 1975; 18. Gelnar et al. 1994.

Dactylogyrus ctenopharyngodonis Akhmerov, 1952
Site: gills.
Host: *Ctenopharyngodon idella* (1, 2, 3, 4)
Distribution: **SR:** D – River Dunaj (= Danube).

Records: 1. Žitňan 1972; 2. Žitňan 1974a; 3. Žitňan 1974b; 4. Žitňan 1981a; 5. Gelnar and Špakulová 1997.

Dactylogyrus difformis Wagener, 1857
Site: gills.
Hosts: *Abramis bjoerkna* (8, 9)
Leuciscus cephalus (6)
**Rutilus rutilus* (16)
Scardinius erythrophthalmus (1, 2, 3, 4, 5, 6, 7, 8, 9, 10, 12, 13, 14, 15)
Distribution: **CR:** L – River Malše, Římovská water reservoir, ponds near Třeboň; D – periodical pools of River Dyje; **SR:** D – Rivers Dunaj (= Danube), Hron, Laborec, Latorica and Tisa, localities Čierna Voda and Senné, Zemplínská Šírava water reservoir.
Records: 1. Kašťák 1956d; 2. Kašťák 1957b; 3. Kašťák 1957d; 4. Lucký 1959b; 5. Vojtek 1959b; 6. Ergens 1962a; 7. Žitňan 1964f; 8. Žitňan 1965a; 9. Žitňan 1965b; 10. Žitňan 1966a; 11. Žitňan 1968c; 12. Ergens and Lom 1970; 13. Ergens et al. 1975; 14. Žitňan 1977; 15. Gelnar et al. 1990; 16. Moravec 1995a.

Dactylogyrus difformoides Gläser et Gussev, 1971
Site: gills.
Host: *Scardinius erythrophthalmus* (1, 2)
Distribution: **SR:** D – River Latorica.
Records: 1. Ergens and Lom 1970; 2. Ergens et al. 1975; 3. Gelnar and Špakulová 1997.

Dactylogyrus dirigerus Gussev, 1966
Site: gills.
Host: *Chondrostoma nasus* (1, 2)
Distribution: **CR:** D – River Morava.
Records: 1. Gelnar et al. 1994; 2. Baruš et al. 1997b.

Dactylogyrus distinguendus Nybelin, 1937
Site: gills.
Hosts: *Abramis bjoerkna* (1, 2, 3)
Vimba vimba (1)
Distribution: **CR:** L – localities not given; D – River Morava; **SR:** D – River Latorica.
Records: 1. Ergens and Lom 1970; 2. Ergens et al. 1975; 3. Gelnar et al. 1994.

Dactylogyrus dulkeiti Bychowsky, 1936
Site: gills.
Host: *Carassius auratus gibelio* (1)
Distribution: **CR:** D – pools of Dyje River.
Record: 1. Ondračková and Jurajda 2000.

Dactylogyrus dyki Ergens et Lucký, 1959
Site: gills.
Hosts: *Barbus barbus* (1, 2, 3, 4, 5, 7, 9, 10, 11)
Barbus petenyi (4, 7, 8, 10, 11)
Distribution: **CR:** L – localities not specified; D – Rivers Rokytná and Želetavka, Dyje R. and Svratka R. drainage systems; O – localities not specified; **SR:** D – Rivers Hornád, Hron, Latorica and Tisa.
Records: 1. Ergens and Lucký 1959; 2. Lucký 1959a; 3. Lucký and Dyk 1964; 4. Žitňan 1965b; 5. Žitňan 1966a; 6. Žitňan 1968c; 7. Ergens and Lom 1970; 8. Žitňan 1970–1971; 9. Ergens et al. 1970; 10. Moravec 1990a; 11. Baruš et al. 1997b.

Dactylogyrus ergensi Molnár, 1964
Syn. *Dactylogyrus nybelini* Markewitsch, 1933 of Ergens (1959c) and Lucký (1959a, b).
Site: gills.
Host: *Chondrostoma nasus* (1, 2, 3, 4, 5, 6, 7)
Distribution: **CR:** L – River Želivka; D – Rivers Rokytná and Svratka, fish farm on Jihlava R.; O – localities not specified; **SR:** D – River Latorica.
Records: 1. Ergens 1959c; 2. Lucký 1959a; 3. Lucký 1959b; 4. Ergens and Lom 1970; 5. Ergens et al. 1975; 6. Navrátil 1991; 7. Baruš et al. 1997b.

Dactylogyrus extensus Mueller et Van Cleave, 1932
Syn. *Dactylogyrus solidus* Achmerow, 1948; *D. hovorkai* Kašťák, 1955
Site: gills.
Hosts: *Cyprinus carpio* (1, 2, 3, 4, 6, 7, 8, 9, 10, 11, 12, 13, 14, 15, 16, 17, 18, 19, 20, 21, 22, 23, 24, 25, 26, 28, 29, 30, 31, 32, 33, 35, 36, 37, 38, 40, 41, 42, 43, 44, 45, 46)
Misgurnus fossilis (4, 10, 19, 20)

Distribution: **CR:** L – ponds near Jindřichův Hradec and Třeboň, Máchovo jezero pond system; D – Rivers Dyje, Morava and Svratka, Kníničská and Vranovská water reservoirs, South- and North-Moravian ponds; O – localities not specified; **SR:** D – Rivers Dunaj (= Danube), Bodva, Hornád, Hron, Ipeľ, Laborec, Latorica, Ondava, Nitra, Tisa, Torysa, Topľa, Uh, Váh and Žitava, Domaša, Oravská and Zemplínská Šírava water reservoirs, ponds near Košice, Levice and Trebišov, localities Čierna Voda and Senné; V – River Dunajec.
Records: 1. Vojtek et al. 1954; 2. Dyk 1955a; 3. Kašťák 1955e; 4. Lucký 1955b; 5. Ergens 1956b; 6. Kašťák 1956b; 7. Kašťák 1956c; 8. Kašťák 1957b; 9. Kašťák 1957d; 10. Lucký 1957b; 11. Lucký 1957d; 12. Vojtek 1959b; 13. Vojtková 1959b; 14. Dyk 1961a; 15. Dyk 1961b; 16. Lucký 1961; 17. Ergens 1962; 18. Dyk and Lucký 1963; 19. Lucký and Dyk 1964; 20. Lucký and Jílek 1964; 21. Žitňan 1964a; 22. Žitňan 1964f; 23. Žitňan 1965a; 24. Žitňan 1965b; 25. Žitňan 1966a; 26. Žitňan 1966c; 27. Žitňan 1968c; 28. Žitňan 1969a; 29. Ergens and Lom 1970; 30. Žitňan 1971c; 31. Žitňan 1972b; 32. Žitňan 1973b; 33. Žitňan 1974c; 34. Žitňan 1977; 35. Žitňan 1978; 36. Žitňan 1979b; 37. Žitňan 1979c; 38. Žitňan 1979d; 39. Žitňan 1981a; 40. Žitňan and Hanzelová 1981; 41. Žitňan et al. 1981; 42. Žitňan 1982a; 43. Hanzelová and Žitňan 1983; 44. Moravec 1984e; 45. Žitňan 1985; 46. Gelnar et al. 1994.

Dactylogyrus falcatus (Wedl, 1857)
Site: gills.
Host: *Abramis bjoerkna* (9)
Abramis brama (1, 2, 3, 4, 6, 7, 8)
Distribution: **CR:** D – River Morava; **SR:** D – Rivers Dunaj (= Danube), Latorica and Tisa, periodical pools of Váh River.
Records: 1. Vojtek 1959b; 2. Dyk 1961a; 3. Žitňan 1966a; 4. Žitňan 1966c; 5. Žitňan 1968c; 6. Ergens and Lom 1970; 7. Ergens et al. 1975; 8. Gelnar et al. 1994; 9. Moravec 1995a.

Dactylogyrus fallax Wagener, 1857
Site: gills.
Hosts: *Alburnus alburnus* (6)
Leuciscus cephalus (5, 6, 7, 8, 9, 10, 11)
Leuciscus leuciscus (6)
Rutilus rutilus (1, 2, 3, 4, 12)
Distribution: **CR:** L – River Malše, Římovská water reservoir, surroundings of Karlovy Vary; D – Rivers Dyje and Morava, South-Moravian ponds and pools.
Records: 1. Kohn 1929; 2. Lucký 1957b; 3. Dyk 1961a; 4. Lucký and Dyk 1964; 5. Ergens and Lom 1970; 6. Gelnar et al. 1990; 7. Gelnar et al. 1994; 8. Koníček and Gelnar 1994; 9. Šebelová et al. 1995; 10. Gelnar et al. 1997; 11. Dušek et al. 1998; 12. Kadlec et al. 1998.

Dactylogyrus finitimus Gussev, 1966
Site: gills.
Host: *Gobio albipinnatus* (1, 2, 3)
Distribution: **SR:** D – River Latorica.
Records: 1. Ergens and Lom 1970; 2. Ergens et al. 1975; 3. Baruš et al. 1997.

Dactylogyrus folkmanovae Ergens, 1956
Site: gills.
Hosts: *Leuciscus cephalus* (1, 2, 3, 4, 5, 7, 8, 9, 10, 11, 13, 14, 15, 16, 17, 18, 19, 20, 21, 22, 23)
Rutilus rutilus (1, 4, 6)
Distribution: **CR:** L – River Malše, Římovská water reservoir; D – Rivers Dřevnice, Dyje, Morava, Rokytná, Svitava, Svratka and Želetavka, Vranovská water reservoir, fish farm on Jihlava R.; O – River Osoblaha; **SR:** D – Rivers Hron, Latorica and Tisa.
Records: 1. Ergens 1956b; 2. Lucký 1957b; 3. Lucký 1959a; 4. Vojtková 1959b; 5. Ergens 1960b; 6. Dyk 1961a; 7. Ergens 1962a; 8. Lucký and Dyk 1964; 9. Ergens 1965a; 10. Žitňan 1965b; 11. Žitňan 1966c; 12. Žitňan 1968c; 13. Ergens and Lom 1970; 14. Ergens et al. 1975; 15. Gelnar et al. 1990; 16. Moravec 1990a; 17. Navrátil 1991; 18. Gelnar et al. 1994; 19. Koníček and Gelnar 1994; 20. Šebelová et al. 1995;

21. Gelnar et al. 1997; 22. Divílek et al. 1998; 23. Dušek et al. 1998.

Dactylogyrus formosus Kulwiec, 1927
Site: gills.
Host: *Carassius carassius* (1, 2, 4, 5)
Distribution: **CR:** L – localities not specified; D – River Morava; O – localities not specified; **SR:** D – River Latorica.
Records: 1. Žitňan 1966c; 2. Žitňan 1967d; 3. Žitňan 1968c; 4. Gelnar et al. 1994; 5. Baruš et al. 1997b.

Dactylogyrus fraternus Wegener, 1910
Site: gills.
Hosts: *Alburnus alburnus* (1, 2, 3, 4, 5, 6, 7, 8, 10, 11, 12, 14, 15)
Leucaspius delineatus (8, 13)
Distribution: **CR:** L – River Malše, Římovská water reservoir, ponds near Třeboň; D – Rivers Dyje, Morava and Svratka; O – River Osoblaha; **SR:** D – Rivers Dunaj (= Danube), Hron, Latorica and Tisa.
Records: 1. Lucký 1957b; 2. Vojtková 1959b; 3. Dyk 1961a; 4. Ergens 1962a; 5. Lucký and Dyk 1964; 6. Ergens 1965a; 7. Žitňan 1965b; 8. Žitňan 1966c; 9. Žitňan 1968c; 10. Ergens and Lom 1970; 11. Žitňan 1971c; 12. Ergens et al. 1975; 13. Žitňan 1986; 14. Gelnar et al. 1990; 15. Gelnar et al. 1994; 16. Šindlerová et al. 1996.

Dactylogyrus haplogonoides Gussev, 1966
Site: gills.
Host: *Vimba vimba* (1, 2)
Distribution: **CR:** L – localities not specified; O – localities not specified; **CR** and **SR:** D – localities not specified.
Records: 1. Ergens and Lom 1970; 2. Baruš et al. 1997b.

Dactylogyrus haplogonus Bychowsky, 1933
Site: gills.
Host: *Gymnocephalus cernuus* (1, 2)
Distribution: **CR:** D – River Dyje and pool near Protivín.
Record: 1. Dyk 1955a; 2. Dyk 1961a.

Dactylogyrus hemiamphibothrium Ergens, 1956
Site: gills.
Host: *Gymnocephalus cernuus* (1, 2, 3, 4, 5, 6, 7, 9, 10, 11)
Distribution: **CR:** L – River Lužnice, ponds near Třeboň and Jindřichův Hradec; D – River Želetavka, Jihlava R. basin, canal near Pohořelice, Kníničská and Vranovská water reservoirs; O – localities not specified; **SR:** D – Rivers Latorica and Tisa.
Records: 1. Ergens 1956b; 2. Lucký 1959b; 3. Ergens 1960b; 4. Ergens 1962a; 5. Lucký and Dyk 1964; 6. Žitňan 1966a; 7. Přibyslavský and Lucký 1967; 8. Žitňan 1968; 9. Ergens and Lom 1970; 10. Ergens et al. 1975; 11. Moravec 1990a.

Dactylogyrus hypophthalmichthys Achmerow, 1952
Site: gills.
Host: not given.
Distribution: **CR** and **SR:** localities not given.
Record: 1. Gelnar and Špakulová 1997.

Dactylogyrus inexpectatus Izjumova in Gussev, 1955
Site: gills.
Host: *Carassius auratus gibelio* (1)
Distribution: **CR:** D – pools of Dyje River.
Record: 1. Ondračková and Jurajda 2000.

Dactylogyrus intermedius Wegener, 1910
Site: gills.
Hosts: *Carassius carassius* (1, 2, 3, 4, 5, 6, 8)
Carassius auratus gibelio (4)
Distribution: **CR:** L – localities not specified; D – River Dyje and its pools; O – localities not specified; **SR:** D – Rivers Hron, Latorica and Tisa.
Records: 1. Lucký 1957b; 2. Lucký and Dyk 1964; 3. Žitňan 1965b; 4. Žitňan 1966a; 5. Žitňan 1968c; 6. Ergens and Lom 1970; 7. Žitňan 1974b; 8. Ergens et al. 1975.

Dactylogyrus izjumovae Gussev, 1966
Site: gills.
Host: *Scardinius erythrophthalmus* (1, 2, 3)

Distribution: **CR:** O – localities not specified; **SR:** D – River Latorica, Zemplínská Šírava water reservoir.
Records: 1. Ergens and Lom 1970; 2. Ergens et al. 1975; 3. Žitňan 1977.

Dactylogyrus lamellatus Achmerow, 1952
Site: gills.
Host: *Ctenopharyngodon idella* (1, 2, 3, 4, 5, 6)
Distribution: **CR:** D – ponds near Pohořelice; **SR:** D – River Dunaj (= Danube), ponds near Nové Zámky and Hroboňov, irrigation canal near Kamenná Moľva (Latorica R. drainage system), Tisa R. basin.
Records: 1. Žitňan 1970b; 2. Žitňan 1972; 3. Žitňan 1973b; 4. Žitňan 1974a; 5. Žitňan 1974b; 6. Lucký 1981b; 7. Žitňan 1981a; 8. Gelnar and Špakulová 1997.

Dactylogyrus macracanthus Wegener, 1910
Syn. *Dactylogyrus monocornis* Ergens, 1956
Site: gills.
Hosts: **Rutilus rutilus* (9)
Tinca tinca (1)
Distribution: **CR:** L – Máchovo jezero pond system, ponds near Třeboň; D – Rivers Dyje and Morava; **SR:** Slovak waters (localities not given).
Records: 1. Ergens 1956b; 2. Ergens 1960b; 3. Ergens 1962a; 4. Lucký and Dyk 1964; 5. Ergens and Lom 1970; 6. Žitňan 1973b; 7. Moravec 1985a; 8. Gelnar et al. 1994; 9. Moravec 1995a.

Dactylogyrus mallaeus Linstow, 1877
Site: gills.
Hosts: *Barbus barbus* (1, 2, 3, 4, 5, 6, 7, 8, 10, 11, 12, 13, 14, 15)
Barbus petenyi (6)
Distribution: **CR:** L – Rivers Labe (= Elbe) and Malše, Římovská water reservoir; D – Rivers Jihlava and Morava, canal near Pohořelice; O – localities not specified; **SR:** D – Rivers Hron, Latorica and Tisa, southern Slovakia, localities Čierna Voda and Senné (Laborec R. and Uh R. basins).
Records: 1. Šrámek 1903; 2. Kašťák 1956d; 3. Kašťák 1957b; 4. Kašťák 1957d; 5. Lucký 1959b; 6. Dyk 1961a; 7. Lucký and Dyk 1964; 8. Žitňan 1965b; 9. Žitňan 1968c; 10. Ergens and Lom 1970; 11. Ergens et al. 1975; 12. Gelnar et al. 1990; 13. Navrátil 1991; 14. Gelnar et al. 1994; 15. Baruš et al. 1997b.

Dactylogyrus micracanthus Nybelin, 1937
Site: gills.
Hosts: *Leuciscus cephalus* (3, 5, 6, 7)
Rutilus rutilus (1, 2, 3)
Scardinius erythrophthalmus (3, 4)
Distribution: **CR:** L – River Malše, Římovská water reservoir, Máchovo jezero pond system, ponds near Jindřichův Hradec; D – Rivers Morava and Svratka.
Records: 1. Vojtková 1959b; 2. Ergens 1962a; 3. Ergens and Lom 1970; 4. Moravec 1986c; 5. Gelnar et al. 1990; 6. Gelnar et al. 1994; 7. Koníček and Gelnar 1994.

Dactylogyrus minor Wagener, 1857
Site: gills.
Host: *Alburnus alburnus* (1, 2, 3, 4, 5, 6, 7, 8, 9, 10, 11, 12)
Distribution: **CR:** L – River Malše, Římovská water reservoir, ponds near Třeboň; D – Rivers Dyje, Morava, Rokytná and Svratka; O – localities not specified; **SR:** D – Rivers Hornád and Latorica.
Records: 1. Ergens 1959b; 2. Lucký 1959a; 3. Lucký 1959b; 4. Vojtková 1959b; 5. Žitňan 1960c; 6. Dyk 1961a; 7. Ergens 1962a; 8. Lucký and Dyk 1964; 9. Ergens and Lom 1970; 10. Ergens et al. 1975; 11. Gelnar et al. 1990; 12. Gelnar et al. 1994.

Dactylogyrus minutus Kulwiec, 1927
Site: gills.
Host: *Cyprinus carpio* (1, 2, 3, 4, 5, 6, 7, 8, 9, 10, 11, 12, 13, 14, 15, 16, 17, 18, 19, 20)
Distribution: **CR:** L – South-Bohemian ponds near Jindřichův Hradec and Třeboň; D – River Svratka, Kníničská water reservoir, South-Moravian ponds; O – localities not specified; **SR:** D – Rivers Dunaj (= Danube), Hron and Tisa, Slovak ponds.
Records: 1. Kašťák 1955d; 2. Ergens 1956b; 3. Kašťák 1956d; 4. Kašťák 1957d; 5. Lucký

1959b; 6. Dyk 1961a; 7. Dyk 1961b; 8. Lucký 1961; 9. Ergens 1962a; 10. Dyk and Lucký 1963; 11. Lucký and Dyk 1964; 12. Žitňan 1964a; 13. Žitňan 1965b; 14. Lucký and Jílek 1967; 15. Ergens and Lom 1970; 16. Žitňan 1979b; 17. Žitňan 1979c; 18. Žitňan 1979d; 19. Žitňan 1982a; 20. Žitňan 1985.

Dactylogyrus nanoides Gussev, 1966
Site: gills.
Host: *Leuciscus cephalus* (1, 2, 3, 4, 6, 7, 8)
Distribution: **CR:** L – River Malše, Římovská water reservoir; D – Rivers Dyje and Morava.
Records: 1. Gelnar et al. 1990; 2. Gelnar et al. 1994; 3. Koníček and Gelnar 1994; 4. Šebelová et al. 1995; 5. Šindlerová et al. 1996; 6. Gelnar et al. 1997; 7. Divílek et al. 1998; 8. Dušek et al. 1998.

Dactylogyrus nanus Dogiel et Bychowsky, 1934
Site: gills.
Hosts: *Abramis bjoerkna* (9)
Alburnus alburnus (1, 3)
Leuciscus cephalus (10)
Rutilus rutilus (1, 2, 3, 4, 6, 7, 8, 9, 12, 13)
Distribution: **CR:** L – River Malše, Římovská water reservoir, South-Bohemian ponds near Jindřichův Hradec and Třeboň; D – Rivers Dyje and Morava; **SR:** D – Rivers Dunaj (= Danube), Latorica and Tisa.
Records: 1. Ergens 1959a; 2. Dyk 1961a; 3. Ergens 1962a; 4. Žitňan 1966a; 5. Žitňan 1968c; 6. Ergens and Lom 1970; 7. Žitňan 1985; 8. Gelnar et al. 1990; 9. Gelnar et al. 1994; 10. Koníček and Gelnar 1994; 11. Šindlerová et al. 1996; 12. Kadlec et al. 1998; 13. Šimková et al. 2000.

Dactylogyrus navicularis Gussev, 1955
Site: gills.
Host: *Leuciscus cephalus* (1, 2)
Distribution: **CR:** L – River Malše, Římovská water reservoir; D – Vranovská water reservoir.
Records: 1. Ergens 1956b; 2. Gelnar et al. 1990.

Dactylogyrus naviculoides Ergens, 1956
Site: gills.
Host: *Leuciscus cephalus* (1, 2, 3, 4, 5, 6, 7)
Distribution: **CR:** D – Rivers Morava and Želetavka, Vranovská water reservoir.
Records: 1. Ergens 1956b; 2. Ergens and Lom 1970; 3. Moravec 1990a; 4. Koníček and Gelnar 1994; 5. Šebelová et al. 1995; 6. Gelnar et al. 1997; 7. Dušek et al. 1998.
Remark: According to Gussev (1985), the validity of this species is questionable.

Dactylogyrus nobilis Long et Yu, 1958
Site: gills.
Host: *Aristichthys nobilis* (1, 2, 3, 4)
Distribution: **CR:** D – ponds near Pohořelice; **SR:** D – ponds near Hroboňovo.
Records: 1. Žitňan 1971; 2. Žitňan 1974a; 3. Žitňan 1974b; 4. Lucký 1981; 5. Gelnar and Špakulová 1997.

Dactylogyrus nybelini Markewitsch, 1933
Site: gills.
Host: *Chondrostoma nasus* (1)
Distribution: **CR** and **SR** (localities not given).
Record: 1. Moravec 1995a.

Dactylogyrus parvus Wegener, 1910
Site: gills.
Host: *Alburnus alburnus* (1, 2, 3, 4, 5, 7, 8, 9, 10, 11, 12)
Distribution: **CR:** L – River Malše, Římovská water reservoir, ponds near Třeboň; D – Rivers Dyje, Morava and Svratka; O – localities not specified; **SR:** D – Rivers Dunaj (= Danube), Latorica and Tisa; V – River Dunajec.
Records: 1. Lucký 1957b; 2. Vojtková 1959b; 3. Ergens 1962a; 4. Lucký and Dyk 1964; 5. Žitňan 1966a; 6. Žitňan 1968c; 7. Ergens and Lom 1970; 8. Žitňan 1971c; 9. Žitňan 1974c; 10. Ergens et al. 1975; 11. Gelnar et al. 1990; 12. Gelnar et al. 1994.

Dactylogyrus petenyi Kašťák, 1957
Site: gills.
Host: *Barbus petenyi* (1, 2, 3, 4, 6)
Distribution: **SR:** D – Rivers Hornád and Tisa, brook Myslavský potok.

Records: 1. Kašťák 1956d; 2. Kašťák 1957a; 3. Kašťák 1957d; 4. Žitňan 1966a; 5. Žitňan 1968c; 6. Žitňan 1970–1971.

Dactylogyrus phoxini Malewitzkaja, 1949
Site: gills.
Host: *Cyprinus carpio* (1, 2, 3)
Distribution: **SR:** D – ponds near Košice.
Records: 1. Kašťák 1955d; 2. Kašťák 1956d; 3. Kašťák 1957d.
Remark: Since in Europe *D. phoxini* is known as a specific parasite of *Phoxinus phoxinus*, its occurrence on carp should be verified.

Dactylogyrus propinquus Bychowsky, 1931
Site: gills.
Hosts: *Abramis brama* (5)
Abramis sapa (1, 2, 3, 5, 6)
Distribution: **CR:** D – River Morava; **SR:** D – Rivers Dunaj (= Danube), Latorica and Tisa.
Records: 1. Žitňan 1966a; 2. Žitňan 1967d; 3. Žitňan 1968c; 4. Ergens and Lom 1970; 5. Žitňan 1986; 6. Gelnar et al. 1994.

Dactylogyrus prostae Molnár, 1964
Site: gills.
Host: *Leuciscus çephalus* (1, 2, 3, 4, 5, 6, 7, 8, 9, 10, 11, 12, 13, 14)
Distribution: **CR:** L – River Malše, Římovská water reservoir; D – Rivers Dřevnice, Dyje and Morava; O – River Osoblaha; **SR:** D – Rivers Hornád and Latorica; V – River Dunajec.
Records: 1. Ergens 1965a; 2. Ergens and Gussev 1965; 3. Ergens and Lom 1970; 4. Žitňan 1970–1971; 5. Žitňan 1974c; 6. Ergens et al. 1975; 7. Gelnar et al. 1990; 8. Gelnar et al. 1994; 9. Koníček and Gelnar 1994; 10. Šebelová et al. 1995; 11. Šindlerová et al. 1996; 12. Gelnar et al. 1997; 13. Divílek et al. 1998; 14. Dušek et al. 1998.

Dactylogyrus rarissimus Gussev, 1966
Site: gills.
Host: *Rutilus rutilus* (1, 2, 3)
Distribution: **CR:** L – River Malše, Římovská water reservoir; D – Morava River basin.
Records: 1. Ergens and Lom 1970; 2. Gelnar et al. 1990; 3. Šimková et al. 2000.

Dactylogyrus rutili Gläser, 1965
Site: gills.
Host: *Leuciscus cephalus* (3)
Rutilus rutilus (1, 2, 4, 5)
Distribution: **CR:** L – localities not specified; D – River Morava, Morava R. basin.
Records: 1. Ergens and Lom 1970; 2. Gelnar et al. 1994; 3. Šebelová et al. 1995; 4. Kadlec et al. 1998; 5. Šimková et al. 2000.

Dactylogyrus similis Wegener, 1910
Site: gills.
Hosts: *Abramis bjoerkna* (4, 6)
Alburnus alburnus (11)
Aspius aspius (4, 6, 9)
Chondrostoma nasus (1, 7, 11)
Leuciscus cephalus (2, 3, 4, 5, 6, 7, 8, 11)
Leuciscus idus (5)
Leuciscus leuciscus (11)
Rutilus rutilus (10, 12, 13, 14)
Vimba vimba (4, 5, 6)
Distribution: **CR:** L – Máchovo jezero pond system, South-Bohemian ponds near Jindřichův Hradec; D – Rivers Dyje, Rokytná and Svratka, Morava River basin, Vranovská water reservoir; **SR:** D – River Hornád, Torysa R. and tributaries, Oravská water reservoir, ponds near Jasov, localities Čierna Voda and Senné (Laborec R. and Uh R. basins).
Records: 1. Dyk 1955a; 2. Ergens 1956a; 3. Ergens 1956b; 4. Kašťák 1956d; 5. Kašťák 1957b; 6. Kašťák 1957d; 7. Lucký 1957b; 8. Vojtková 1959b; 9. Dyk 1961a; 10. Ergens 1962a; 11. Hradil 1964; 12. Ergens and Lom 1970; 13. Moravec 1986c; 14. Kadlec et al. 1998.

Dactylogyrus simplicimallaeata Bychowsky, 1931
Site: gills.
Host: *Pelecus cultratus* (1, 2, 3, 4, 5, 6, 7, 8, 9, 10)
Distribution: **CR:** D – River Morava; **SR:** D – Rivers Dunaj (= Danube) and Tisa, pools of Váh R.

Records: 1. Vojtková 1959a; 2. Dyk 1961a; 3. Žitňan 1966a; 4. Žitňan 1967d; 5. Žitňan 1968; 6. Ergens and Lom 1970; 7. Žitňan 1985; 8. Jurajda et al. 1992; 9. Gelnar et al. 1994; 10. Baruš et al. 1997b.

Dactylogyrus sphyrna Linstow, 1878
Site: gills.
Hosts: *Abramis bjoerkna* (3, 5, 6, 7, 9, 12, 13, 15, 17, 18, 19, 22, 24, 26, 28)
Abramis brama (2, 6, 7, 12, 13, 15, 18, 25)
Chondrostoma nasus (1, 14)
Leuciscus cephalus (1, 6, 11, 13, 14, 15, 18, 19, 23, 27)
Leuciscus idus (28)
Leuciscus leuciscus (18)
Rutilus rutilus (1, 2, 6, 8, 10, 12, 13, 15, 16, 18, 19, 20, 22, 28, 29)
Scardinius erythrophthalmus (3, 4, 5, 18)
Vimba vimba (14, 18, 19, 26)
Distribution: **CR:** L – River Malše, Římovská water reservoir, South-Bohemian ponds near Jindřichův Hradec and Třeboň; D – Rivers Dyje, Morava, Rokytná, Svitava and Svratka, Kníničská and Vranovská water reservoirs, fish farm on Jihlava R., South-Moravian ponds and pools; O – localities not specified; **SR:** D – Rivers Dunaj (= Danube), Hornád, Hron, Laborec, Latorica and Tisa, Torysa R. and tributaries, Oravská and Zemplínská Šírava water reservoirs, localities Čierna Voda and Senné (Laborec R. and Už R. basins).
Records: 1. Dyk 1954c; 2. Ergens 1956b; 3. Kašťák 1956d; 4. Kašťák 1957b; 5. Kašťák 1957d; 6. Lucký 1957b; 7. Lucký 1957d; 8. Lucký 1959a; 9. Vojtek 1959b; 10. Vojtková 1959b; 11. Žitňan 1960c; 12. Ergens 1962a; 13. Dyk and Lucký 1963; 14. Hradil 1964; 15. Lucký and Dyk 1964; 16. Žitňan 1964f; 17. Žitňan 1965a; 18. Žitňan 1965b; 19. Žitňan 1966a; 20. Lucký and Jílek 1967; 21. Žitňan 1968c; 22. Ergens and Lom 1970; 23. Žitňan 1971c; 24. Ergens et al. 1975; 25. Žitňan 1977; 26. Gelnar et al. 1990; 27. Navrátil 1991; 28. Gelnar et al. 1994; 29. Kadlec et al. 1998.

Dactylogyrus suecicus Nybelin, 1937
Site: gills.
Hosts: *Abramis brama* (8)
Rutilus rutilus (1, 2, 3, 4, 5, 6, 7, 8, 9, 10)
Distribution: **CR:** L – River Malše, Římovská water reservoir, South-Bohemian ponds near Jindřichův Hradec and Třeboň; D – River Morava and its drainage system, River Svratka; **SR:** D – River Latorica.
Records: 1. Ergens 1959a; 2. Vojtková 1959b; 3. Dyk 1961a; 4. Ergens 1962a; 5. Ergens and Lom 1970; 6. Ergens et al. 1975; 7. Gelnar et al. 1990; 8. Gelnar et al. 1994; 9. Kadlec et al. 1998; 10. Šimková et al. 2000.

Dactylogyrus tincae Gussev, 1965
Site: gills.
Host: *Tinca tinca* (1, 2, 3)
Distribution: **CR:** L – River Malše, Římovská water reservoir; **SR:** D – localities not specified.
Records: 1. Ergens and Lom 1970; 2. Žitňan 1973b; 3. Gelnar et al. 1990.

Dactylogyrus tissensis Zachvatkin, 1951
Site: gills.
Hosts: *Alburnoides bipunctatus* (1, 2, 4, 5)
Alburnus alburnus (1, 2)
Distribution: **SR:** D – Rivers Dunaj (= Danube), Hornád and Tisa.
Records: 1. Žitňan 1966a; 2. Žitňan 1967d; 3. Žitňan 1968c; 4. Žitňan 1970–1971; 5. Žitňan 1986.

Dactylogyrus tuba Linstow, 1878
Site: gills.
Hosts: *Aspius aspius* (5, 9, 10, 12, 13, 14, 16)
Leuciscus cephalus (1, 2, 3, 6, 15)
Leuciscus idus (4, 8, 9, 10, 13, 14, 16)
Leuciscus leuciscus (7, 16)
Vimba vimba (3, 5, 6)
Distribution: **CR:** L – River Malše, Římovská water reservoir, ponds near Třeboň; D – Rivers Dyje and Morava;

O – localities not specified; **SR:** D – Rivers Dunaj (= Danube), Hornád, Hron, Latorica, Tisa and Torysa, localities Čierna Voda and Senné.
Records: 1. Kašťák 1956d; 2. Kašťák 1957b; 3. Kašťák 1957d; 4. Lucký 1957b; 5. Žitňan 1960c; 6. Dyk 1961a; 7. Ergens 1962a; 8. Lucký and Dyk 1964; 9. Žitňan 1965b; 10. Žitňan 1966a; 11. Žitňan 1968; 12. Ergens and Lom 1970; 13. Žitňan 1971; 14. Ergens et al. 1975; 15. Gelnar et al. 1990; 16. Gelnar et al. 1994.

Dactylogyrus vastator Wagener, 1857
Site: gills.
Hosts: **Barbus barbus* (41)
Carassius auratus gibelio (29, 42)
Carassius carassius (21, 22, 27)
Cyprinus carpio (1, 2, 3, 4, 5, 6, 7, 9, 10, 11, 12, 13, 14, 15, 16, 17, 18, 19, 20, 21, 22, 23, 24, 26, 27, 28, 30, 31, 32, 33, 34, 35, 37, 38, 39, 40)
**Vimba vimba* (8, 43)
Distribution: **CR:** L – South-Bohemian ponds near Jindřichův Hradec; D – River Morava, Kníničská water reservoir, fish farm on Jihlava R., fish hatchery Třebíč, South-Moravian ponds and pools; O – ponds of Odra (= Oder) R. basin; **SR:** D – Rivers Dunaj (= Danube), Hron, Laborec, Latorica, and Tisa, Domaša, Oravská, Ružín, Sľňava and Zemplínská Šírava water reservoirs, Slovak ponds, locality Čierna Voda, rice field near Vojčice.
Records: 1. Dyk 1939; 2. Dyk 1944; 3. Berner 1948; 4. Lucký 1953a; 5. Volf 1953; 6. Dyk 1954c; 7. Lucký 1955b; 8. Kašťák 1956d; 9. Kašťák 1957d; 10. Lucký 1957b; 11. Lucký 1957d; 12. Dyk 1961a; 13. Dyk 1961b; 14. Lucký 1961; 15. Ergens 1962a; 16. Dyk and Lucký 1963; 17. Dyk 1964a; 18. Lucký and Dyk 1964; 19. Žitňan 1964a; 20. Žitňan 1964f; 21. Žitňan 1965b; 22. Žitňan 1966a; 23. Žitňan 1966c; 24. Lucký and Jílek 1967; 25. Žitňan 1968; 26. Žitňan 1969; 27. Ergens and Lom 1970; 28. Žitňan 1971; 29. Žitňan 1974b; 30. Žitňan 1974d; 31. Žitňan 1977; 32. Žitňan 1978; 33. Žitňan 1979a; 34. Žitňan 1979b; 35. Žitňan 1979c; 36. Žitňan 1981; 37. Žitňan et al. 1981; 38. Žitňan 1982; 39. Hanzelová and Žitňan 1983; 40. Žitňan 1985; 41. Navrátil 1991; 42. Gelnar et al. 1994; 43. Moravec 1995a.

Dactylogyrus vistulae Prost, 1957
Site: gills.
Hosts: *Abramis bjoerkna* (29)
Alburnoides bipunctatus (1, 5, 9, 14, 24)
Alburnus alburnus (1, 5, 6, 9, 10)
Chondrostoma nasus (3, 5, 7, 8, 9, 10, 13, 14, 15, 22)
Ctenopharyngodon idella (16, 19, 20)
Leuciscus cephalus (1, 2, 3, 4, 5, 6, 7, 8, 9, 10, 13, 14, 15, 17, 21, 22, 25, 26, 27, 28, 30, 31, 32, 33, 34)
Leuciscus leuciscus (3, 7, 9, 10, 21, 25)
Rutilus pigus (9, 10)
Rutilus rutilus (5, 6, 10, 11, 23)
Scardinius erythrophthalmus (29)
Vimba vimba (13)
Distribution: **CR:** L – River Malše, Římovská water reservoir, ponds near Třeboň; D – Rivers Dřevnice, Dyje, Morava and Rokytná, fish farm on Jihlava R., Kníničská water reservoir, South Moravia; O – River Osoblaha; **SR:** D – Rivers Dunaj (= Danube), Hornád, Hron, Laborec, Latorica and Tisa, Dobšínská water reservoir, ponds near Hroboňovo; V – River Dunajec, lake Vrbické pleso.
Records: 1. Lucký 1959a; 2. Kašťák and Žitňan 1960a; 3. Žitňan 1960c; 4. Ergens 1962a; 5. Lucký and Dyk 1964; 6. Žitňan 1964f; 7. Ergens 1965a; 8. Žitňan 1965a; 9. Žitňan 1965b; 10. Žitňan 1966a; 11. Lucký and Jílek 1967; 12. Žitňan 1968c; 13. Ergens and Lom 1970; 14. Žitňan 1970–1971; 15. Žitňan 1971c; 16. Žitňan 1971b; 17. Žitňan 1973a; 18. Žitňan 1973b; 19. Žitňan 1974a; 20. Žitňan 1974b; 21. Žitňan 1974c; 22. Ergens et al. 1975; 23. Žitňan 1985; 24. Žitňan 1986; 25. Gelnar et al. 1990; 26. Navrátil 1991; 27. Gelnar et al. 1994; 28. Koníček and Gelnar 1994;

29. Moravec 1995a; 30. Šebelová et al. 1995; 31. Šindlerová et al. 1996; 32. Gelnar et al. 1997; 33. Divílek et al. 1998; 34. Dušek et al. 1998.

Dactylogyrus vranoviensis Ergens, 1956
Site: gills.
Host: *Leuciscus cephalus* (1, 2, 3, 4, 5, 6, 7, 8, 9, 10, 11, 12, 13, 14, 15)
Distribution: **CR:** L – River Malše, Římovská water reservoir; D – Rivers Dřevnice, Dyje and Morava, Vranovská water reservoir; O – River Osoblaha; **SR:** D – River Hron.
Records: 1. Ergens 1956b; 2. Lucký 1959b; 3. Ergens 1960b; 4. Lucký and Dyk 1964; 5. Ergens 1965a; 6. Žitňan 1965b; 7. Ergens and Lom 1970; 8. Gelnar et al. 1990; 9. Moravec 1990a; 10. Gelnar et al. 1994; 11. Koníček and Gelnar 1994; 12. Šebelová et al. 1995; 13. Gelnar et al. 1997; 14. Divílek et al. 1998; 15. Dušek et al. 1998.

Dactylogyrus wegeneri Kulwiec, 1927
Site: gills.
Host: *Carassius carassius* (1, 2, 3, 4, 6, 7, 9, 10, 11)
Distribution: **CR:** L – localities not specified; D – River Dyje and its periodical pools; O – localities not specified; **SR:** D – Rivers Latorica and Tisa.
Records: 1. Dyk 1955a; 2. Lucký 1955b; 3. Lucký 1957b; 4. Lucký 1957d; 5. Dyk 1961a; 6. Lucký and Dyk 1964; 7. Žitňan 1966a; 8. Žitňan 1968c; 9. Ergens and Lom 1970; 10. Ergens et al. 1975; 11. Baruš et al. 1997b.

Dactylogyrus wunderi Bychowsky, 1931
Site: gills.
Hosts: *Abramis bjoerkna* (9, 12)
Abramis brama (1, 2, 3, 4, 5, 6, 7, 8, 10, 11, 12, 13, 14, 16, 17, 19, 20, 21, 22)
Abramis sapa (11)
Distribution: **CR:** L – River Malše, Římovská water reservoir; D – Rivers Dyje, Morava and Svratka, Kníničská and Vranovská water reservoirs; O – localities not specified; **SR:** D – Rivers Dunaj (= Danube), Hron, Laborec, Latorica and Tisa, Zemplínská Šírava water reservoir.
Records: 1. Ergens 1956a; 2. Ergens 1956b; 3. Lucký 1957b; 4. Vojtek 1959b; 5. Vojtková 1959b; 6. Dyk 1961a; 7. Dyk and Lucký 1963; 8. Lucký and Dyk 1964; 9. Žitňan 1964f; 10. Žitňan 1965a; 11. Žitňan 1965b; 12. Žitňan 1966a; 13. Žitňan 1966c; 14. Lucký and Jílek 1967; 15. Žitňan 1968c; 16. Ergens and Lom 1970; 17. Žitňan 1971c; 18. Žitňan 1973b; 19. Ergens et al. 1975; 20. Žitňan 1977; 21. Gelnar et al. 1990; 22. Gelnar et al. 1994; 23. Šindlerová et al. 1996.

Dactylogyrus yinwenyingae Gussev, 1962
Syn. *Dactylogyrus nasalis* Strelkov et Ha Ky, 1964
Site: nasal cavities.
Host: *Scardinius erythrophthalmus* (1, 2)
Distribution: **SR:** D – River Latorica.
Records: 1. Ergens 1966d; 2. Ergens et al. 1975.

Dactylogyrus zandti Bychowsky, 1933
Site: gills.
Hosts: *Abramis bjoerkna* (6)
Abramis brama (1, 2, 3, 4, 5, 6, 7, 10, 11, 13, 14, 15, 16, 17)
Rutilus rutilus (8)
Distribution: **CR:** L – River Malše, Římovská water reservoir, Máchovo jezero pond system, South-Bohemian ponds near Jindřichův Hradec; D – Rivers Morava and Svratka, Kníničská and Vranovská water reservoirs; O – localities not specified; **SR:** D – Rivers Dunaj (= Danube), Latorica and Tisa, Zemplínská Šírava water reservoir.
Records: 1. Ergens 1956b; 2. Lucký 1959b; 3. Vojtek 1959b; 4. Vojtková 1959b; 5. Ergens 1962a; 6. Žitňan 1966a; 7. Žitňan 1966c; 8. Lucký and Jílek 1967; 9. Žitňan 1968; 10. Ergens and Lom 1970; 11. Žitňan 1971; 12. Žitňan 1973; 13. Ergens et al. 1975; 14. Žitňan 1977; 15. Moravec 1986; 16. Gelnar et al. 1990; 17. Gelnar et al. 1994.

Dactylogyrus spp.
Site: gills.
Hosts: *Abramis bjoerkna* (1, 8)
Abramis brama (8)
Alburnoides bipunctatus (10)
Carassius carassius (1, 6)
Chondrostoma nasus (8, 9)
Ctenopharyngodon idella (11, 12)
Cyprinus carpio (2, 3, 4)
Hypophthalmichthys molitrix (13)
Leuciscus cephalus (15)
Leuciscus idus (1)
Rutilus rutilus (3, 4, 5, 6, 7, 14)
Scardinius erythrophthalmus (1, 3, 4, 8)
Distribution (findings):
CR: L – South-Bohemian ponds near Jindřichův Hradec and Třeboň, Bohemia; D – Rivers Dřevnice, Dyje, Jihlava, Morava, Rokytná and Svratka, South-Moravian ponds; O – River Osoblaha; **SR:** D – River Hron.
Records: 1. Frič 1908; 2. Schäferna 1927; 3. Dyk and Lucký 1954a; 4. Dyk and Lucký 1954c; 5. Lucký et al. 1955; 6. Dyk 1956a; 7. Ergens 1962a; 8. Lucký and Dyk 1964; 9. Ergens 1965a; 10. Žitňan 1965b; 11. Willomitzer 1980a; 12. Willomitzer 1980b; 13. Lucký 1981b; 14. Gelnar et al. 1994; 15. Divílek et al. 1998.

Pellucihaptor merus (Zaika, 1961)
Syn. *Dactylogyrus merus* Zaika, 1961
Site: nasal cavities and mouth cavity, less often skin and gills.
Host: *Phoxinus phoxinus* (1, 2, 3, 4, 5, 6)
Distribution: **CR:** L – localities not given; D – localities not given; O – Hořina Brook; **SR:** D – Dobšínská water reservoir.
Records: 1. Žitňan 1967d; 2. Řehulka 1968; 3. Ergens and Lom 1970; 4. Žitňan 1973a; 5. Žitňan 1986; 6. Baruš et al. 1997b.

Pseudacolpenteron pavlovskii Bychowsky et Gussev, 1955
Site: gills.
Host: *Cyprinus carpio* (1, 2, 3, 4)
Distribution: **CR:** D – pond near Brno; **SR:** D – River Latorica.
Records: 1. Ergens 1958; 2. Dyk 1961a; 3. Žitňan 1966a; 4. Ergens and Lom 1970.

Fam. Ancyrocephalidae Bychowsky, 1937

Ancyrocephalus cruciatus (Wedl, 1857)
Site: gills.
Hosts: **Ictalurus nebulosus* (2)
Misgurnus fossilis (1, 3, 4, 5, 6, 7, 8, 9, 10, 11, 13, 14, 15)
**Perca fluviatilis* (2)
**Stizostedion lucioperca* (2)
Distribution: **CR:** L – localities not specified; D – River Dyje, South-Moravian pools; O – localities not specified; **SR:** D – Rivers Dunaj (= Danube), Hron and Tisa, Váh R. and its periodical pools, irrigation canal near Váh R., Oravská water reservoir, Slovak ponds, rice field near Vojčice.
Records: 1. Lucký 1955b; 2. Kašťák 1956d; 3. Lucký 1957b; 4. Lucký 1957d; 5. Vojtek 1959b; 6. Žitňan 1960c; 7. Dyk 1961a; 8. Lucký and Dyk 1964; 9. Žitňan 1965b; 10. Ergens 1966b; 11. Žitňan 1966a; 12. Žitňan 1968c; 13. Ergens and Lom 1970; 14. Žitňan 1986; 15. Baruš et al. 1997b.

Ancyrocephalus gussevi Dontsov, 1972
Site: gills.
Host: *Stizostedion volgense* (1)
Distribution: **SR:** D – Rivers Dunaj (= Danube) and Tisa.
Record: 1. Žitňan 1986.

Ancyrocephalus paradoxus Creplin, 1839
Site: gills.
Hosts: **Cyprinus carpio* (4)
**Esox lucius* (13)
Perca fluviatilis (1, 5, 6, 7, 9, 10, 11, 12, 16, 17, 19)
Stizostedion lucioperca (1, 2, 3, 5, 6, 8, 10, 11, 12, 14, 15, 16, 17, 18, 19, 20, 22, 24, 25)
Stizostedion volgense (17, 19, 22)
Distribution: **CR:** L – South-Bohemian ponds; D – Rivers Dyje, Morava and Svratka, Vranovská water reservoir, South-Moravian ponds; O – localities not specified;

SR: D – Rivers Hornád, Hron, Laborec, Latorica, Tisa, Torysa and Váh, Oravská and Zemplínská Šírava water reservoirs.
Records: 1. Vojtek et al. 1954; 2. Lucký et al. 1955; 3. Dyk and Štědronský 1956; 4. Kašťák 1956d; 5. Kašťák 1957d; 6. Lucký 1957b; 7. Lucký 1957d; 8. Vojtková 1959b; 9. Žitňan 1960c; 10. Dyk 1961a; 11. Ergens 1962a; 12. Dyk and Lucký 1963; 13. Hradil 1964; 14. Lucký and Dyk 1964; 15. Žitňan 1964f; 16. Žitňan 1965a; 17. Žitňan 1965b; 18. Ergens 1966b; 19. Žitňan 1966a; 20. Žitňan 1966c; 21. Žitňan 1968c; 22. Ergens and Lom 1970; 23. Žitňan 1973b; 24. Ergens et al. 1975; 25. Žitňan 1977; 26. Žitňan 1981a; 27. Gelnar et al. 1994.

Ancyrocephalus percae Ergens, 1966
Site: gills.
Hosts: *Perca fluviatilis* (1, 2, 3, 4)
Stizostedion lucioperca (5)
Distribution: **CR:** L – River Teplá near Karlovy Vary; D – River Morava; O – localities not given; **SR:** D – River Dunaj (= Danube).
Records: 1. Ergens 1966b; 2. Ergens and Lom 1970; 3. Žitňan 1971c; 4. Moravec 1990a; 5. Gelnar et al. 1994.

Cleidodiscus pricei Mueller, 1936
Syn. *Ancyrocephalus* sp. of Kašťák (1957d)
Site: gills.
Host: *Ictalurus nebulosus* (1, 2, 3, 4, 5, 7, 8, 9)
Distribution: **CR:** L – localities not specified; **SR:** D – Rivers Dunaj (= Danube), Hron, Laborec, Latorica and Tisa, rice field near Vojčice.
Records: 1. Kašťák 1957d; 2. Žitňan 1964f; 3. Žitňan 1965a; 4. Žitňan 1965b; 5. Žitňan 1966a; 6. Žitňan 1968c; 7. Ergens and Lom 1970; 8. Žitňan 1974b; 9. Ergens et al. 1975; 10. Žitňan 1981a; 11. Gelnar and Špakulová 1997.

Pseudodactylogyrus anguillae (Yin et Sproston, 1948)
Site: gills.
Host: *Anguilla anguilla* (1, 2)
Distribution: **CR:** L – Rivers Lužnice, Metuje, Ohře, Otava, Sázava, Úhlavka and Želivka, Orlická and Želivská water reservoirs, Máchovo jezero pond system, small warm-water reservoir at Tisová in North Bohemia; D – small water reservoir Fryšták.
Records: 1. Škoríková et al. 1996a; 2. Škoríková et al. 1996b; 3. Gelnar and Špakulová 1997.

Pseudodactylogyrus bini (Kikuchi, 1929)
Site: gills.
Host: *Anguilla anguilla* (1)
Distribution: **CR:** L – Rivers Lužnice, Ohře, Otava, Ploučnice and Úhlavka, Orlická water reservoir, Máchovo jezero pond system, small warm-water reservoir at Tisová in North Bohemia.
Records: 1. Škoríková et al. 1996a; 2. Škoríková et al. 1996b; 3. Gelnar and Špakulová 1997.

Silurodiscoides siluri (Zandt, 1924)
Syn. *Ancylodiscoides siluri* (Zandt, 1924)
Site: gills.
Host: *Silurus glanis* (1, 2, 3, 4, 5, 6, 7, 9)
Distribution: **CR:** D – Vranovská water reservoir, South-Moravian ponds and pools; **SR:** D – Rivers Dunaj (= Danube) and Tisa, Tisa R. lowlands.
Records: 1. Ergens 1956a; 2. Lucký 1957b; 3. Dyk 1961a; 4. Dyk and Lucký 1963; 5. Lucký and Dyk 1964; 6. Žitňan 1966a; 7. Žitňan 1966c; 8. Žitňan 1968c; 9. Ergens and Lom 1970; 10. Žitňan 1973b; 11. Žitňan 1981a.

Silurodiscoides vistulensis (Siwak, 1932)
Syn. *Ancylodiscoides vistulensis* (Siwak, 1932)
Site: gills.
Host: *Silurus glanis* (1, 2, 3, 4, 6, 8, 10)
Distribution: **CR:** L – localities not specified; D – River Morava; **SR:** D – Rivers Dunaj (= Danube), Hron, Latorica and Tisa.
Records: 1. Žitňan 1965a; 2. Žitňan 1965b; 3. Žitňan 1966a; 4. Žitňan 1966c; 5. Žitňan 1968c; 6. Ergens and Lom 1970; 7. Žitňan

1973b; 8. Ergens et al. 1975; 9. Žitňan 1981a; 10. Gelnar 1994.

Urocleidus dispar (Mueller, 1936)
Site: gills.
Host: *Lepomis gibbosus* (1, 2, 3, 4, 5, 6, 7)
Distribution: **SR:** D – Rivers Dunaj (= Danube) and Hron, periodical pools and drainage canal of Váh R. near Komárno.
Records: 1. Vojtek 1958b; 2. Vojtek 1959b; 3. Dyk 1961a; 4. Žitňan 1965b; 5. Ergens and Lom 1970; 6. Žitňan 1974b; 7. Žitňan 1981a; 8. Gelnar and Špakulová 1997.

Urocleidus similis (Mueller, 1936)
Site: gills.
Host: *Lepomis gibbosus* (1, 2, 3, 4, 5, 7, 8)
Distribution: **SR:** D – Rivers Dunaj (= Danube), Hron and Tisa, periodical pools and drainage canals of Váh R. near Komárno.
Records: 1. Vojtek 1958b; 2. Vojtek 1959b; 3. Dyk 1961a; 4. Žitňan 1965b; 5. Žitňan 1966a; 6. Žitňan 1968c; 7. Ergens and Lom 1970; 8. Žitňan 1974b; 9. Žitňan 1981a; 10. Gelnar and Špakulová 1997.

Fam. Tetraonchidae Bychowsky, 1937

Tetraonchus borealis (Olsson, 1893)
Site: gills.
Hosts: *Thymallus baicalensis* (8, 13)
Thymallus thymallus (1, 2, 3, 4, 5, 6, 7, 9, 10, 11, 14, 15)
Distribution: **CR:** L – River Malše, Římovská water reservoir; D – localities not specified; O – localities not specified; **SR:** D – Rivers Hnilec, Hornád, Hron, Muráňka and Váh, brook and pond near Demänová, Dobšínská water reservoir; V – Rivers Dunajec and Poprad.
Records: 1. Pacák 1957; 2. Pacák 1958; 3. Pacák 1960; 4. Žitňan 1960c; 5. Dyk 1961a; 6. Žitňan 1965b; 7. Dyk 1965b; 8. Žitňan 1969b; 9. Ergens and Lom 1970; 10. Žitňan 1970–1971; 11. Žitňan 1973a; 12. Žitňan 1973b; 13. Žitňan 1974b; 14. Žitňan 1980b; 15. Gelnar et al. 1990.

Tetraonchus monenteron (Wagener, 1857)
Syn. *Ancyrocephalus monenteron* (Wagener, 1857)
Site: gills.
Hosts: *Chondrostoma nasus* (38)
Cyprinus carpio (21)
Esox lucius (1, 2, 3, 4, 5, 6, 7, 8, 9, 10, 11, 12, 13, 14, 15, 16, 17, 18, 19, 20, 21, 22, 23, 24, 25, 26, 27, 28, 29, 31, 32, 34, 35, 36, 39, 40)
Gobio gobio (21)
Leuciscus cephalus (7, 11, 21)
Leuciscus leuciscus (41)
Rutilus rutilus (21)
Scardinius erythrophthalmus (7, 11, 21)
Distribution: **CR:** L – River Malše, Římovská water reservoir, Máchovo jezero pond system, pond near Lnáře, South-Bohemian ponds near Jindřichův Hradec and Třeboň; D – Rivers Dyje, Morava, Rokytná and Želetavka, Vranovská water reservoir, South-Moravian ponds and pools; O – River Moravice; **SR:** D – Rivers Dunaj (= Danube), Bodrog, Bodva, Hornád, Hron, Laborec, Latorica, Ondava, Tisa and Torysa, periodical pools and irrigation canal near Váh R., brook near Demänová, Oravská and Zemplínská Šírava water reservoirs, locality Senné.
Records: 1. Jírovec 1940; 2. Lucký 1954; 3. Vojtek et al. 1954; 4. Dyk 1955a; 5. Dyk 1955b; 6. Dyk and Lucký 1956b; 7. Kašťák 1956d; 8. Dyk and Lucký 1957; 9. Kašťák 1957b; 10. Kašťák 1957c; 11. Kašťák 1957d; 12. Lucký 1957b; 13. Lucký 1957d; 14. Pacák 1958; 15. Lucký 1959a; 16. Vojtek 1959b; 17. Žitňan 1960c; 18. Dyk 1961a; 19. Ergens 1962a; 20. Dyk and Lucký 1963; 21. Dyk 1964a; 22. Hradil 1964; 23. Lucký and Dyk 1964; 24. Žitňan 1964f; 25. Žitňan 1965a; 26. Žitňan 1965b; 27. Ergens 1966a; 28. Žitňan 1966a; 29. Žitňan 1966c; 30. Žitňan 1968c; 31. Ergens and Lom 1970; 32. Žitňan 1971c; 33. Žitňan 1973b; 34. Ergens et al. 1975; 35. Žitňan 1977; 36. Moravec 1979a; 37. Žitňan 1981a; 38. Kirka et al. 1982; 39. Gelnar et al. 1990; 40. Gelnar et al. 1994; 41. Moravec 1995a.

Tetraonchus sp.
Site: gills.
Host: *Salmo trutta* m. *fario* (1)
Distribution: **CR** and **SR** (localities not given).
Record: 1. Moravec 1995a.

Fam. Gyrodactylidae Van Beneden et Hesse, 1863

Gyrodactylus aphyae Malmberg, 1957
Syn. *Gyrodactylus wageneri aphyae* Malmberg, 1957
Site: fins, skin, less often gills and nasal cavities.
Host: *Phoxinus phoxinus* (1, 2, 3, 4, 5, 6, 7, 8, 9, 10, 11, 12)
Distribution: **CR:** L – brook near Prague; D – localities not given; O – River Osoblaha; **SR:** D – Rivers Dunaj (= Danube), Hornád and Hron, Dobšínská water reservoir, lake Morské oko in Vihorlat; V – River Dunajec.
Records: 1. Ergens 1959a; 2. Žitňan 1960c; 3. Ergens 1965a; 4. Žitňan 1965b; 5. Ergens and Lom 1970; 6. Žitňan 1970–1971; 7. Žitňan 1973a; 8. Žitňan 1974c; 9. Ergens 1975a; 10. Ergens 1976a; 11. Žitňan 1986; 12. Baruš et al. 1997b.

Gyrodactylus arcuatus Bychowsky, 1933
Site: gills, skin and fins.
Host: *Proterorhinus marmoratus* (1)
Distribution: **SR:** D – Dunajská Streda.
Record: 1. Ergens 1962b.

Gyrodactylus barbatuli Achmerow, 1952
Site: fins and gills.
Host: *Barbatula barbatula* (1, 2)
Distribution: **CR:** D – River Dyje.
Records: 1. Ergens 1962–1963a; 2. Ergens 1969; 3. Ergens and Lom 1970.
Remark: Ergens (1969) reports this species from former Czechoslovakia as *G. barbatuli* form D.

Gyrodactylus barbi Ergens, 1976
Site: fins.
Host: *Barbus barbus* (1, 2, 3)
Distribution: **CR:** L – River Želivka; O – localities not specified; **CR** and **SR** (former Czechoslovakia): D – localities not given.
Records: 1. Ergens 1976; 2. Moravec 1990a; 3. Baruš et al. 1997b; 4. Gelnar and Špakulová 1997.

Gyrodactylus bliccensis Gläser, 1974
Site: fins and gills.
Hosts: not given.
Distribution: **CR** and **SR**.
Record: 1. Gelnar and Špakulová 1997.

Gyrodactylus bohemicus Ergens, 1992
Site: fins.
Hosts: *Oncorhynchus mykiss* (1)
Salvelinus fontinalis (1)
Distribution: **CR:** L – fish farm in Kaplice.
Records: 1. Ergens 1992; 2. Gelnar and Špakulová 1997.

Gyrodactylus carassii Malmberg, 1957
Syn. *Gyrodactylus decorus* Malmberg, 1957
Site: fins, skin, gills and nasal cavities.
Hosts: *Abramis bjoerkna* (12)
Alburnus alburnus (1, 4, 6, 8, 9, 10, 12, 16)
Carassius auratus gibelio (18)
Carassius carassius (18)
Leucaspius delineatus (7, 12, 19)
Leuciscus cephalus (6, 16, 18, 19, 20, 21)
Leuciscus leuciscus (4, 16)
**Phoxinus phoxinus* (12, 15, 19)
Rutilus rutilus (1, 4, 7, 16, 19)
Scardinius erythrophthalmus (1, 2, 3, 4, 5, 9, 10, 12, 13, 14, 17, 18, 19)
Distribution: **CR:** L – River Malše, Římovská water reservoir, Máchovo jezero pond system, South-Bohemian ponds near Jindřichův Hradec and Třeboň; D – Rivers Jihlava and Morava, South-Moravian ponds; O – River Osoblaha; **SR:** D – Rivers Dunaj (= Danube), Hron, Laborec, Latorica and Tisa, Zemplínská Šírava water reservoir.
Records: 1. Ergens 1959a; 2. Lucký 1959b; 3. Dyk 1961a; 4. Ergens 1962a; 5. Lucký and Dyk 1964; 6. Ergens 1965a; 7. Ergens 1965b;

8. Žitňan 1965a; 9. Žitňan 1965d; 10. Žitňan 1966a; 11. Žitňan 1968c; 12. Ergens and Lom 1970; 13. Žitňan 1977; 14. Moravec 1986c; 15. Žitňan 1986; 16. Ergens 1988; 17. Gelnar et al. 1990; 18. Gelnar et al. 1994; 19. Moravec 1995a; 20. Gelnar et al. 1997; 21. Dušek et al. 1998.

Gyrodactylus cernuae Malmberg, 1957
Syn. *Gyrodactylus wageneri cernuae* Malmberg, 1957
Site: fins, skin and gills.
Hosts: *Gymnocephalus cernuus* (1, 2, 3, 4, 5, 6, 7)
Perca fluviatilis (8)
Distribution: **CR:** L – South-Bohemian ponds near Třeboň; D – Rivers Jihlava, Morava and Rokytná, Kníničská water reservoir; O – localities not specified; **SR:** D – River Latorica.
Records: 1. Ergens 1959a; 2. Lucký 1959b; 3. Ergens 1962a; 4. Lucký and Dyk 1964; 5. Přibyslavský and Lucký 1967; 6. Ergens and Lom 1970; 7. Ergens et al. 1975; 8. Gelnar et al. 1994.

Gyrodactylus cobitis Bychowsky, 1933
Site: gills, fins and skin.
Hosts: *Cobitis taenia* (5, 6)
Misgurnus fossilis (1, 2, 3, 4, 5)
Distribution: **CR:** L – localities not specified; D – River Dyje and its periodical pools; O – localities not specified; **SR:** River Dunaj (= Danube).
Records: 1. Lucký 1957b; 2. Lucký 1957d; 3. Dyk 1961a; 4. Lucký and Dyk 1964; 5. Ergens and Lom 1970; 6. Žitňan 1985.

Gyrodactylus cotti Roman, 1956
Site: gills.
Hosts: *Cottus gobio* (1, 2, 3)
Cottus poecilopus (1, 2)
Distribution: **CR:** L – localities not specified; D – localities not specified; O – localities not specified; **SR:** D – River Dunaj (= Danube).
Records: 1. Ergens 1962b; 2. Ergens and Lom 1970; 3. Žitňan 1986.

Gyrodactylus cyprini Diarova, 1964
Site: skin, fins and gills.
Hosts: *Abramis brama* (7)
Cyprinus carpio (1, 2, 3, 4, 5, 6, 8)
Distribution: **CR:** L – Máchovo jezero pond system; D – River Morava; **SR:** D – River Dunaj (= Danube) and Tisa, Slovak ponds.
Records: 1. Ergens and Lom 1970; 2. Žitňan 1979c; 3. Žitňan 1979d; 4. Žitňan 1981a; 5. Žitňan 1982a; 6. Žitňan 1985; 7. Moravec 1986c; 8. Gelnar et al. 1994; 9. Gelnar and Špakulová 1997.

Gyrodactylus dykovae Ergens, 1991
Site: fins.
Host: *Gobio gobio* (1)
Distribution: **CR:** D – River Rokytná.
Record: 1. Ergens 1991b.

Gyrodactylus elegans Nordmann, 1832
Syn. *Gyrodactylus parvicopula* Bychowsky, 1933; *G. bychowskyi* Sproston, 1946 of Lucký (1957b, d), Dyk (1961a) and Dyk and Lucký (1963), partly (only records from *A. brama*), and that of Žitňan (1965a, b)
Site: gills.
Hosts: *Abramis ballerus* (33, 34, 36)
Abramis brama (1, 12, 16, 22, 23, 24, 25, 26, 29, 31, 32)
Abramis sapa (24, 25, 29, 31, 35, 37)
**Carassius carassius* (2, 10, 11, 13, 17, 23)
**Cyprinus carpio* (2, 7, 9, 10, 11, 13, 14, 15, 17, 19, 20, 21, 22, 23, 25, 26, 27)
**Gobio gobio* (5, 6, 8, 15, 18, 23)
**Misgurnus fossilis* (2, 3, 4, 13)
Rutilus rutilus (12, 25)
**Tinca tinca* (23, 25)
Vimba vimba (23, 25)
Distribution: **CR:** L – Rivers Labe (= Elbe) and Malše, South-Bohemian ponds near Jindřichův Hradec; D – Rivers Dyje and Morava, Kníničská water reservoir, South-Moravian ponds and pools; O – River Moravice, pond near Karlovec; **SR:** Rivers Dunaj (= Danube), Hron, Laborec, Latorica and Tisa, Zemplínská Šírava water reservoir, Slovak ponds, rice field near Vojčice.

Records: 1. Šrámek 1901, 1903; 2. Dyk 1955a; 3. Lucký 1955b; 4. Dyk 1956a; 5. Dyk and Lucký 1956b; 6. Dyk and Lucký 1956c; 7. Kašťák 1956d; 8. Dyk and Lucký 1957; 9. Kašťák 1957d; 10. Lucký 1957b; 11. Lucký 1957d; 12. Ergens 1959a; 13. Dyk 1961a; 14. Lucký 1961; 15. Ergens 1962a; 16. Pacák 1962; 17. Dyk and Lucký 1963; 18. Dyk 1964a; 19. Lucký and Dyk 1964; 20. Žitňan 1964a; 21. Žitňan 1964f; 22. Žitňan 1965a; 23. Žitňan 1965b; 24. Ergens 1966c; 25. Žitňan 1966a; 26. Žitňan 1966c; 27. Lucký and Jílek 1967; 28. Žitňan 1968; 29. Ergens and Lom 1970; 30. Žitňan 1973b; 31. Ergens et al. 1975; 32. Žitňan 1977; 33. Žitňan 1981a; 34. Ergens 1986; 35. Žitňan 1986; 36. Gelnar et al. 1990; 37. Gelnar et al. 1994.
Remark: According to Gusev (1985), *G. elegans* occurs only on *Abramis* spp., whereas previous records (mainly untill 1964) of *G. elegans* from other fishes concerned, in fact, other congeneric species; those from *Cyprinus carpio* concerned *G. katharineri*.

Gyrodactylus fairporti Van Cleave, 1921
Site: gills and nasal cavities.
Host: not given (probably *Ictalurus nebulosus*) (1)
Distribution: **SR:** D – River Dunaj (= Danube) near Komárno.
Record: 1. Žitňan 1981a; Gelnar and Špakulová 1997.

Gyrodactylus fossilis Lupu et Roman, 1956
Site: skin and gills.
Host: *Misgurnus fossilis* (1, 2, 3)
Distribution: **CR:** L – localities not specified; D – localities not specified; O – localities not specified; **SR:** River Dunaj (= Danube).
Records: 1. Ergens 1973a; 2. Žitňan 1986; 3. Baruš et al. 1997b.

Gyrodactylus gasterostei Gläser, 1974
Site: gills, fins, skin.
Host: *Leuciscus cephalus* (1, 2, 4)
Distribution: **CR:** D – River Morava, fish farm on Jihlava R.
Records: 1. Navrátil 1991; 2. Gelnar et al. 1997; 3. Gelnar and Špakulová 1997; 4. Dušek et al. 1998.

Gyrodactylus glaeseri Ergens, 1980
Site: fins.
Host: *Alburnus alburnus* (1, 2)
Distribution: **CR:** L – Rivers Malše and Želivka, Římovská water reservoir.
Records: 1. Ergens 1980b; 2. Moravec 1990a; 3. Gelnar et al. 1990; 4. Gelnar and Špakulová 1997.

Gyrodactylus gobiensis Gläser, 1974
Site: fins, skin, less often gills.
Hosts: *Gobio gobio* (2, 3, 4, 5)
Gobio kessleri (1)
Distribution: **CR:** L – River Malše, Římovská water reservoir; D – River Morava; **SR:** D – River Tisa.
Records: 1. Žitňan 1985; 2. Gelnar 1987b; 3. Gelnar et al. 1990; 4. Gelnar 1991, 5. Gelnar et al. 1994; 6. Gelnar and Špakulová 1997.

Gyrodactylus gobii Schulman, 1953
Syn: *Gyrodactylus parvicopula* Bychowsky, 1933 of Ergens and Lom (1970) and Ergens et al. (1975)
Site: fins.
Hosts: *Gobio albipinnatus* (1, 2, 3)
Gobio gobio (1, 3)
Distribution: **CR:** L – localities not specified; D – River Morava; O – localities not specified; **SR:** D – River Latorica.
Records: 1. Ergens and Lom 1970; 2. Ergens et al. 1975; 3. Gelnar et al. 1994.

Gyrodactylus gobioninum Gussev, 1955
Site: fins.
Hosts: *Gobio albipinnatus* (3, 5, 6)
Gobio gobio (1, 2, 3, 5)
Gobio kessleri (6)
Distribution: **CR:** L – South-Bohemian ponds near Jindřichův Hradec; D – River Želetavka; O – River Osoblaha; **SR:** D – Rivers Hron, Laborec and Tisa.

Records: 1. Ergens 1959a; 2. Ergens 1962a; 3. Žitňan 1964f; 4. Ergens 1965a; 5. Žitňan 1965b; 6. Žitňan 1966a; 7. Žitňan 1968c.

Gyrodactylus gracilihamatus Malmberg, 1964
Site: fins and skin.
Hosts: *Alburnoides bipunctatus* (2)
Alburnus alburnus (1)
Leuciscus cephalus (1, 3, 4, 5, 6)
Rutilus rutilus (1, 7)
Distribution: **CR:** D – River Morava and its basin; O – localities not specified; **SR:** D – River Tisa.
Records: 1. Ergens and Lom 1970; 2. Žitňan 1986; 3. Gelnar et al. 1994; 4. Koníček and Gelnar 1994; 5. Gelnar et al. 1997; 6. Dušek et al. 1998; 7. Kadlec et al. 1998.

Gyrodactylus gurlei Price, 1937
Site: fins.
Host: *Carassius auratus gibelio* (1)
Distribution: **CR:** D – River Morava.
Record: 1. Gelnar et al. 1994; 2. Gelnar and Špakulová 1997.

Gyrodactylus hemibarbi Ergens, 1980
Site: fins and gills.
Hosts: *Barbus barbus* (1, 2)
Leuciscus cephalus (1, 4, 5)
Distribution: **CR:** D – Rivers Bečva and Morava; O – River Osoblaha.
Records: 1. Ergens 1980b; 2. Moravec 1990a; 3. Gelnar and Špakulová 1997; 4. Gelnar et al. 1997; 5. Dušek et al. 1998.

Gyrodactylus hrabei Ergens, 1957
Site: fins.
Hosts: *Cottus gobio* (1, 3, 5, 6)
Cottus poecilopus (3, 4, 5)
Distribution: **CR:** L – localities not specified; D – localities not specified; O – localities not specified; **SR:** D – Rivers Dunaj (= Danube), Hnilec and Hron, brook Vojsovský potok, Dobšinská water reservoir.
Records: 1. Ergens 1957; 2. Dyk 1961a; 3. Ergens and Lom 1970; 4. Žitňan 1973a; 5. Žitňan 1986; 6. Moravec 1990a.

Gyrodactylus hronosus Žitňan, 1964
Site: fins.
Hosts: *Alburnoides bipunctatus* (1, 2, 3, 4)
Alburnus alburnus (1, 2, 3)
Distribution: **SR:** D – Rivers Dunaj (= Danube) and Hron.
Records: 1. Žitňan 1964e; 2. Žitňan 1965b; 3. Ergens and Lom 1970; 4. Žitňan 1986.
Remark: According to Gusev (1985), *G. hronosus* should be considered a *species inquirenda*.

Gyrodactylus jiroveci Ergens et Bykhovsky, 1967
Syn. *Gyrodactylus papernai* Ergens et Bykhovsky, 1967
Site: gills, fins and skin.
Host: *Barbatula barbatula* (1, 2, 3, 4, 5, 6, 7)
Distribution: **CR:** L – pond near Písek; D – Rivers Dyje and Morava; O – localities not specified; **SR:** D – Rivers Dunaj (= Danube) and Tisa, Dobšínská water reservoir.
Records: 1. Ergens and Bykhovsky 1967; 2. Ergens and Lom 1970; 3. Žitňan 1973a; 4. Žitňan 1986; 5. Moravec 1990a, 6. Gelnar et al. 1994; 7. Moravec 1995a.

Gyrodactylus katharineri Malmberg, 1964
Site: gills, skin and fins.
Hosts: *Barbus barbus* (18)
Cyprinus carpio (1, 2, 4, 5, 6, 7, 9, 10, 11, 13, 14, 15, 16, 17, 19, 20)
Gobio gobio (1)
Distribution: **CR:** L – River Malše, Římovská water reservoir, Máchovo jezero pond system; D – River Morava; O – localities not specified; **SR:** D – Rivers Dunaj (= Danube), Hron and Tisa, Domaša, Liptovská Mara, Oravská, Ružín, Sľňava and Zemplínská Šírava water reservoirs, Slovak ponds.
Records: 1. Ergens and Lom 1970; 2. Žitňan 1971c; 3. Žitňan 1973b; 4. Žitňan 1977; 5. Žitňan 1978; 6. Žitňan 1979b; 7. Žitňan 1979d; 8. Žitňan 1981a; 9. Žitňan et al. 1981; 10. Hanzelová and Žitňan 1982; 11. Žitňan 1982b; 12. Ergens 1983b; 13. Hanzelová and Žitňan 1983; 14. Moravec 1984e; 15. Žitňan

1985; 16. Gelnar 1987a; 17. Gelnar 1987c; 18. Gelnar et al. 1990; 19. Navrátil 1991; 20. Gelnar et al. 1994.

Gyrodactylus kearni Ergens, 1990
Site: fins.
Host: *Leuciscus cephalus* (1, 2, 3, 5, 6)
Distribution: **CR:** D – Rivers Dřevnice, Morava and Rokytná, brook Štěpánovický potok.
Records: 1. Ergens 1990; 2. Moravec 1990a; 3. Gelnar et al. 1997; 4. Gelnar and Špakulová 1997; 5. Divílek et al. 1998; 6. Gelnar et al. 1998.

Gyrodactylus kherulensis Ergens, 1974
Site: fins.
Host: *Cyprinius carpio* (1)
Distribution: **CR:** L – River Vltava.
Records: 1. Gelnar and Lux 1991; 2. Gelnar and Špakulová 1997.

Gyrodactylus kobayashii Hukuda, 1940
Site: skin and fins.
Host: not given *(? Carassius auratus gibelio)*.
Distribution: **CR** and **SR**.
Record: 1. Gelnar and Špakulová 1997.

Gyrodactylus laevis Malmberg, 1957
Syn. *Gyrodactylus paralaevis* Ergens, 1966
Site: gills.
Hosts: *Abramis ballerus* (3)
Abramis bjoerkna (3, 4, 7)
Abramis brama (4)
Alburnus alburnus (3)
Leucaspius delineatus (3, 4)
Leuciscus cephalus (3, 9, 10, 12, 14, 15, 16)
Phoxinus phoxinus (1, 2, 3, 4, 5, 6, 8, 11, 13)
Rhodeus sericeus (3, 4)
Rutilus rutilus (3, 9)
Scardinius erythrophthalmus (9)
Distribution: **CR:** L – localities not specified; D – Rivers Dřevnice and Morava, Morava R. basin; O – localities not specified; **SR:** D – River Hron and Latorica, Dobšínská water reservoir, lake Morské oko (Vihorlat); V – River Dunajec.
Records: 1. Ergens 1959b; 2. Žitňan 1965b; 3. Ergens 1966c; 4. Ergens and Lom 1970; 5. Žitňan 1973a; 6. Žitňan 1974c; 7. Ergens et al. 1975; 8. Žitňan 1986; 9. Gelnar et al. 1994; 10. Koníček and Gelnar 1994; 11. Moravec 1995a; 12. Gelnar et al. 1997; 13. Baruš et al. 1997b; 14. Divílek et al. 1998; 15. Dušek et al. 1998; 16. Kadlec et al. 1998.

Gyrodactylus lamberti Ergens, 1990
Site: fins.
Host: *Leuciscus cephalus* (1, 2, 3, 5, 6)
Distribution: **CR:** D – Rivers Dřevnice, Morava and Rokytná, brook Štěpánovický potok (Dyje R. basin).
Records: 1. Ergens 1990; 2. Moravec 1990a; 3. Gelnar et al. 1997; 4. Gelnar and Špakulová 1997; 5. Divílek et al. 1998; 6. Dušek et al. 1998.

Gyrodactylus latus Bychowsky, 1933
Site: skin, fins and gills.
Host: *Cobitis taenia* (1, 2, 3, 4, 5)
Distribution: **CR:** L – River Lužnice, ponds near Třeboň; D – River Morava; O – localities not specified.
Records: 1. Ergens 1962a; 2. Ergens and Lom 1970; 3. Ergens 1978a; 4. Gelnar et al. 1994; 5. Baruš et al. 1997b.

Gyrodactylus leucisci Žitňan, 1964
Site: fins.
Host: *Leuciscus cephalus* (1, 2, 3, 5, 6, 7, 8, 9)
Leuciscus leuciscus (4)
Distribution: **CR:** L – Rivers Malše and Želivka, Římovská water reservoir; D – Rivers Bečva, Morava, Olešnice and Rokytná, brook Štěpánovický potok; **SR:** D – River Hron.
Records: 1. Žitňan, 1964e; 2. Žitňan 1965b; 3. Ergens and Lom 1970; 4. Gelnar et al. 1990; 5. Ergens 1991a; 6. Gelnar et al. 1994; 7. Koníček and Gelnar 1994; 8. Gelnar et al. 1997; 9. Dušek et al. 1998.

Gyrodactylus limneus Malmberg, 1964
Syn. *Gyrodactylus phoxini* Malmberg, 1957 of Ergens (1962), Ergens and Lom (1970), Žitňan (1965, 1974, 1986) and Baruš et al. (1997b)

Site: fins, skin, less often gills and nasal cavities.
Host: *Phoxinus phoxinus* (1, 2, 3, 4, 5, 6, 7, 8)
Distribution: **CR:** L – Běchovický Brook near Prague; D – localities not specified; O – localities not specified; **SR:** D – River Hron, Dobšínská water reservoir; V – River Dunajec.
Records: 1. Ergens 1962b; 2. Žitňan 1965b; 3. Ergens and Lom 1970; 4. Žitňan 1973a; 5. Žitňan 1974c; 6. Žitňan 1986; 7. Moravec 1995a; 8. Baruš et al. 1997b.
Remark: According to Gusev (1985), it may well be a synonym of *G. phoxini* Malmberg, 1957 with which it was frequently confused; the validity of both species should be verified by experiments.

Gyrodactylus lomi Ergens et Gelnar, 1988
Site: fins.
Host: *Leuciscus cephalus* (1, 2, 3, 4, 5, 7, 8)
Distribution: **CR:** L – River Malše; D – Rivers Dřevnice and Morava.
Records: 1. Ergens and Gelnar 1988; 2. Moravec 1990a; 3. Gelnar et al. 1994; 4. Koníček and Gelnar 1994; 5. Gelnar et al. 1997; 6. Gelnar and Špakulová 1997; 7. Divílek et al. 1998; 8. Dušek et al. 1998.

Gyrodactylus longiradix Malmberg, 1957
Site: fins and skin.
Hosts: *Gymnocephalus cernuus* (1, 2, 3, 4, 5, 6, 7, 8)
Perca fluviatilis (1, 2, 4, 5)
Stizostedion lucioperca (1, 4)
Distribution: **CR:** L – South-Bohemian ponds near Jindřichův Hradec and Třeboň; D – River Jihlava; O – localities not specified; **SR:** D – Rivers Hron and Latorica.
Records: 1. Ergens 1959a; 2. Lucký 1959b; 3. Dyk 1961a; 4. Ergens 1962a; 5. Lucký and Dyk 1964; 6. Žitňan 1965b; 7. Ergens and Lom 1970; 8. Ergens et al. 1975.

Gyrodactylus longoacuminatus Žitňan, 1964
Syn. *Gyrodactylus mutabilitas* Bychowsky, 1957 of Ergens (1962–1963a, 1965a)
Site: fins and skin.
Host: *Carassius auratus gibelio* (8)
Carassius carassius (1, 2, 3, 4, 5, 6, 7)
Distribution: **CR:** L – Máchovo jezero pond system; D – pools of Dyje River; O – River Osoblaha; **SR:** D – River Hron.
Records: 1. Ergens 1962–1963a; 2. Žitňan 1964e; 3. Ergens 1965a; 4. Žitňan 1965b; 5. Ergens and Lom 1970; 6. Ergens 1980; 7. Baruš et al. 1997b; 8. Ondračková and Jurajda 2000.

Gyrodactylus lotae Gussev, 1953
Site: gills, skin and fins.
Host: *Lota lota* (1, 2, 3, 4)
Distribution: **CR:** L – River Vltava; **SR:** D – River Dunaj (= Danube).
Records: 1. Ergens 1962–1963a; 2. Ergens and Lom 1970; 3. Žitňan 1986; 4. Baruš et al. 1997b.

Gyrodactylus lucii Kulakowskaja, 1951
Syn. *Gyrodactylus wageneri lucii* Kulakowskaja, 1951
Site: skin and fins.
Hosts: **Anguilla anguilla* (16)
Esox lucius (2, 3, 4, 5, 6, 7, 9, 10, 11, 12, 13, 15, 16, 18, 19, 22, 23)
Perca fluviatilis (4, 8, 10, 12, 13)
Stizostedion lucioperca (1)
Distribution: **CR:** L – Lipno water reservoir, Máchovo jezero pond system, South-Bohemian ponds near Jindřichův Hradec and Třeboň; D – Rivers Morava and Rokytná, Vranovská water reservoir; O – River Osoblaha; **SR:** D – Rivers Dunaj (= Danube), Hron, Laborec, Latorica and Tisa, Zemplínská Šírava water reservoir.
Records: 1. Ergens 1959a; 2. Lucký 1959a; 3. Lucký 1959b; 4. Ergens 1962a; 5. Dyk and Lucký 1963; 6. Lucký and Dyk 1964; 7. Žitňan 1964; 8. Ergens 1965a; 9. Žitňan 1965a; 10. Žitňan 1965b; 11. Ergens 1966a; 12. Žitňan 1966a; 13. Žitňan 1966c; 14. Žitňan 1968c; 15. Ergens and Lom 1970; 16. Žitňan 1971c; 17. Žitňan 1973b; 18. Ergens et al. 1975; 19. Žitňan 1977; 20. Žitňan 1981a; 21. Moravec 1985b; 22. Gelnar et al. 1994; 23. Moravec 1995a.

Gyrodactylus luciopercae Gussev, 1962
Site: fins, skin and gills.
Hosts: *Perca fluviatilis* (1, 7, 8)
Stizostedion lucioperca (1, 3, 4, 7, 8)
Stizostedion volgense (3)
Zingel streber (5, 6)
Distribution: **CR:** L – River Malše, Římovská water reservoir; D – River Morava; **SR:** D – River Dunaj (= Danube) and Latorica, Zemplínská Šírava water reservoir.
Records: 1. Ergens and Lom 1970; 2. Žitňan 1973b; 3. Ergens et al. 1975; 4. Žitňan 1977; 5. Žitňan 1981a; 6. Žitňan 1985; 7. Gelnar et al. 1990; 8. Gelnar et al. 1980.

Gyrodactylus macrocornis Ergens, 1963
Syn. *Gyrodactylus chondrostomatis* Žitňan, 1964
Site: fins.
Host: *Chondrostoma nasus* (1, 2, 3, 4, 5, 6, 7)
Distribution: **CR:** O – River Osoblaha; **SR:** D – Rivers Hron and Topľa.
Records: 1. Ergens 1962–1963b; 2. Žitňan 1964e; 3. Ergens 1965a; 4. Žitňan 1965b; 5. Ergens and Lom 1970; 6. Baruš et al. 1997b.

Gyrodactylus macronychus Malmberg, 1957
Site: fins, skin, gills and nasal cavities.
Host: *Phoxinus phoxinus* (1, 2, 3, 4, 5, 6, 7, 8, 9, 10)
Distribution: **CR:** L – brook near Prague; D – localities not given; O – River Osoblaha; **SR:** D – Rivers Dunaj (= Danube), Hornád and Hron, lake Morské oko, Dobšínská water reservoir; V – River Dunajec.
Records: 1. Ergens 1959a; 2. Ergens 1965a; 3. Žitňan 1965b; 4. Ergens and Lom 1970; 5. Žitňan 1971–1972; 6. Žitňan 1973a; 7. Žitňan 1974c; 8. Ergens 1976a; 9. Žitňan 1986; 10. Baruš et al. 1997b.

Gyrodactylus magnificus Malmberg, 1957
Site: fins, skin and gills.
Host: *Rutilus rutilus* (1, 2)
Distribution: **CR:** D – Kníničská water reservoir; O – River Osoblaha.
Records: 1. Ergens 1965a; 2. Lucký and Jílek 1967.

Gyrodactylus malmbergi Ergens, 1961
Site: skin, fins.
Hosts: *Barbus barbus* (3, 6, 8, 9)
Barbus petenyi (2, 3, 4, 5, 6, 7, 8, 9)
Distribution: **SR:** D – Rivers Hron, Tisa and Topľa; V – River Dunajec.
Records: 1. Dyk 1961a; 2. Ergens 1961a; 3. Žitňan 1965b; 4. Žitňan 1966b; 5. Žitňan 1968c; 6. Ergens and Lom 1970; 7. Žitňan 1970–1971; 8. Žitňan 1974; 9. Baruš et al. 1997b.

Gyrodactylus markakulensis Gvosdev, 1950
Site: gills, skin, fins and less often nasal cavities.
Hosts: *Gobio albipinnatus* (3)
Gobio gobio (1, 2, 3, 4)
Gobio uranoscopus (5, 6)
Distribution: **CR:** L – South-Bohemian ponds near Jindřichův Hradec and Třeboň; D – localities not specified; O – River Osoblaha; **SR:** D – Rivers Dunaj (= Danube) and Hron.
Records: 1. Ergens 1962a; 2. Ergens 1965a; 3. Žitňan 1965b; 4. Ergens and Lom 1970; 5. Žitňan 1981a; 6. Žitňan 1985.

Gyrodactylus markewitschi Kulakowskaja, 1951
Syn. *Gyrodactylus albaniensis* Ergens, 1960
Site: gills.
Hosts: *Barbus barbus* (2, 4, 6, 7)
Barbus petenyi (1, 3, 4, 5)
Distribution: **CR:** D – Rivers Morava and Rokytná; **SR:** D – River Hron, locality near Bardejov.
Records: 1. Ergens 1962b; 2. Lucký and Dyk 1964; 3. Žitňan 1964c; 4. Ergens 1965c; 5. Žitňan 1965b; 6. Gelnar et al. 1994; 7. Baruš et al. 1997b.

Gyrodactylus medius Kathariner, 1893
Site: gills, less often nasal cavities.
Hosts: **Abramis brama* (3)
**Carassius auratus gibelio* (6)
**Carassius carassius* (6)

Cyprinus carpio (1, 2, 3, 4, 5, 6, 7, 8, 10, 11, 13, 14, 15, 16, 17, 18, 19)
**Misgurnus fossilis* (3)
Distribution: **CR:** L – South-Bohemian ponds near Jindřichův Hradec; D – River Dyje, Knínička water reservoir, South-Moravian ponds; O – localities not specified; **SR:** D – Rivers Dunaj (= Danube), Laborec, Latorica and Tisa, Slovak pondws; V – River Dunajec.
Records: 1. Lucký 1961; 2. Ergens 1962a; 3. Lucký and Dyk 1964; 4. Žitňan 1964a; 5. Žitňan 1964f; 6. Žitňan 1966a; 7. Žitňan 1966c; 8. Lucký and Jílek 1967; 9. Žitňan 1968c; 10. Ergens and Lom 1970; 11. Žitňan 1971c; 12. Žitňan 1973b; 13. Žitňan 1974b; 14. Žitňan 1974c; 15. Žitňan 1979b; 16. Žitňan 1979c; 17. Žitňan 1979d; 18. Žitňan 1982a; 19. Žitňan 1985.
Remark: According to Gusev (1985), this species is a specific parasite of *C. carpio* and, therefore, its records from other hosts are either misidentifications or accidental findings.

Gyrodactylus menschikowi Gvosdev, 1950
Site: gills.
Host: *Barbatula barbatula* (1, 2, 3, 4, 5)
Distribution: **CR:** L – South-Bohemian ponds near Jindřichův Hradec; D – localities not specified; O – localities not specified; **SR:** D – Rivers Dunaj (= Danube), Hornád and Tisa, Dobšínská water reservoir.
Records: 1. Ergens 1962a; 2. Ergens and Lom 1970; 3. Žitňan 1970–1971; 4. Žitňan 1973a; 5. Žitňan 1985.

Gyrodactylus minimus Malmberg, 1957
Site: gills.
Host: *Phoxinus phoxinus* (1, 2, 3, 4, 5)
Distribution: **CR:** L – localities not specified; D – localities not specified; O – localities not specifid; **SR:** River Dunaj (= Danube), Dobšínská water reservoir.
Records: 1. Ergens 1966c; 2. Ergens and Lom 1970; 3. Žitňan 1973a; 4. Žitňan 1986; 5. Baruš et al. 1997b.

Gyrodactylus misgurni Ling, 1962
Syn. *Gyrodactylus bychowskyi* Sproston, 1946 of Lucký (1957b), Dyk (1961a) and Dyk and Lucký (1963), partly (only records from *M. fossilis*).
Site: gills.
Host: *Misgurnus fossilis* (1, 2, 3, 4, 5, 6, 7)
Distribution: **CR:** D – River Dyje, South-Moravian ponds; **SR:** D – River Dunaj (= Danube).
Records: 1. Lucký 1957b; 2. Dyk 1961a; 3. Dyk and Lucký 1963; 4. Žitňan 1981a; 5. Ergens 1986; 6. Žitňan 1986; 7. Baruš et al. 1997b.

Gyrodactylus osoblahensis Ergens, 1963
Site: fins.
Host: *Leuciscus cephalus* (5)
Leuciscus leuciscus (1, 2, 3, 4)
Distribution: **CR:** D – River Dřevnice; O – River Osoblaha.
Records: 1. Ergens 1962–1963b; 2. Ergens 1965a; 3. Ergens and Lom 1970; 4. Moravec 1990a; 5. Divílek et al. 1998.

Gyrodactylus pannonicus Molnár, 1968
Site: fins, skin, less often gills.
Hosts: **Barbus barbus* (2)
Phoxinus phoxinus (1)
Distribution: **SR:** D – River Dunaj (= Danube).
Records: 1. Žitňan 1986; 2. Baruš et al. 1997b.

Gyrodactylus paraminimus Ergens, 1966
Site: gills.
Host: *Chondrostoma nasus* (1, 2, 3, 4)
Distribution: **SR:** D – River Latorica.
Records: 1. Ergens 1966c; 2. Ergens and Lom 1970; 3. Moravec 1990a; 4. Baruš et al. 1997b.

Gyrodactylus pavlovskyi Ergens et Bykhovsky, 1967
Site: fins.
Host: *Barbatula barbatula* (1, 2, 3, 4)
Distribution: **CR:** L – ponds near Mladá Boleslav; **SR:** D – River Dunaj (= Danube).
Records: 1. Ergens and Bykhovsky 1967;

2. Ergens and Lom 1970; 3. Žitňan 1986; 4. Moravec 1990a.

Gyrodactylus prostae Ergens, 1963
Site: fins, skin, less often gills.
Hosts: *Abramis ballerus* (8)
Abramis bjoerkna (1, 2, 3)
Abramis brama (1)
Leucaspius delineatus (1)
Leuciscus cephalus (1, 2, 4, 6, 7)
Leuciscus idus (2, 3, 6)
Leuciscus leuciscus (1, 2, 3, 6)
Rutilus rutilus (1, 2, 3, 5, 6, 9)
Vimba vimba (4)
Distribution: **CR:** L – River Malše, Římovská water reservoir, ponds near Jindřichův Hradec; D – River Morava and its basin; O – localities not specified; **SR:** D – River Latorica.
Records: 1. Ergens 1962–1963b; 2. Ergens 1966c; 3. Ergens and Lom 1970; 4. Gelnar et al. 1990; 5. Moravec 1990a; 6. Gelnar et al. 1994; 7. Koníček and Gelnar 1994; 8. Moravec 1995a; 9. Kadlec et al. 1998.

Gyrodactylus proterorhini Ergens, 1967
Site: gills, fins, skin, nasal cavities.
Host: *Proterorhinus marmoratus* (1, 2, 3, 4)
Distribution: **SR:** D – River Dunaj (= Danube) and its periodical pools.
Records: 1. Ergens 1967; 2. Ergens and Lom 1970; 3. Žitňan 1986; 4. Moravec 1990a.

Gyrodactylus rhodei Žitňan, 1964
Site: fins and skin.
Host: *Rhodeus sericeus* (1, 2, 3, 5, 6, 7, 8, 9)
Distribution: **CR:** L – periodical pools of Labe (= Elbe) River; D – River Morava; O – localities not specified; **SR:** D – River Dunaj (= Danube), Hron and Tisa.
Records: 1. Žitňan 1964d; 2. Žitňan 1965b; 3. Žitňan 1966a; 4. Žitňan 1968c; 5. Ergens and Lom 1970; 6. Ergens 1975b; 7. Žitňan 1986; 8. Gelnar et al. 1994; 9. Baruš et al. 1997b.

Gyrodactylus rutilensis Gläser, 1974
Site: fins.
Host: *Rutilus rutilus* (1, 2)
Distribution: **CR:** L – River Malše, Římovská water reservoir; D – River Morava.
Records: 1. Gelnar et al. 1990; 2. Gelnar 1994; 3. Gelnar and Špakulová 1997.

Gyrodactylus scardiniensis Gläser, 1974
Site: fins.
Host: *Leuciscus cephalus* (1, 3, 4)
Distribution: **CR:** D – Rivers Dřevnice and Morava.
Records: 1. Gelnar et al. 1997; 2. Gelnar and Špakulová 1997; 3. Divílek et al. 1998; 4. Dušek et al. 1998.

Gyrodactylus sedelnikowi Gvosdev, 1950
Site: gills.
Host: *Barbatula barbatula* (1, 2, 3, 4, 5, 6, 7)
Distribution: **CR:** L – localities not specified; D – River Morava, ponds near Brno; O – River Osoblaha; **SR:** D – River Hornád, Dobšinská water reservoir, lake Morské oko in Vihorlat.
Records: 1. Ergens 1962b; 2. Ergens 1965a; 3. Ergens and Lom 1970; 4. Žitňan 1970–1971; 5. Žitňan 1973a; 6. Žitňan 1986; 7. Gelnar et al. 1994.

Gyrodactylus shulmani Ling, 1962
Site: gills.
Hosts: *Carassius auratus gibelio* (3)
Carassius carassius (1)
Cyprinus carpio (1, 2, 4, 5, 6, 7, 8, 9, 11, 12, 13, 14)
Distribution: **CR:** L – localities not specified; **SR:** D – Rivers Dunaj (= Danube), Hron and Tisa, Domaša, Oravská and Zemplínská Šírava water reservoirs, ponds in western, central and eastern Slovakia.
Records: 1. Ergens 1970; 2. Ergens and Lom 1970; 3. Žitňan 1974b; 4. Žitňan 1977; 5. Žitňan 1978; 6. Žitňan 1979a; 7. Žitňan 1979b; 8. Žitňan 1979c; 9. Žitňan 1979d; 10. Žitňan 1981a; 11. Žitňan 1982a; 12. Žitňan and Hanzelová 1982b; 13. Hanzelová and Žitňan 1983; 14. Žitňan 1985; 15. Gelnar and Špakulová 1997.

Gyrodactylus slovacicus Ergens, 1963
Site: gills, fins and skin.
Host: *Umbra krameri* (1, 2, 4, 5, 6)
Distribution: **SR:** D – Rivers Dunaj (= Danube), Latorica and Tisa.
Records: 1. Ergens 1962–1963b; 2. Žitňan 1966a; 3. Žitňan 1968; 4. Ergens and Lom 1970; 5. Žitňan 1985; 6. Moravec 1990a.

Gyrodactylus sprostonae Ling, 1962
Site: gills.
Hosts: *Carassius auratus gibelio* (6)
Cyprinus carpio (1, 2, 4, 5)
Distribution: **CR:** D – River Morava; **SR:** D – Rivers Dunaj (= Danube) and Tisa, Slovak ponds.
Records: 1. Žitňan 1979c; 2. Žitňan 1979d; 3. Žitňan 1981a; 4. Žitňan 1982a; 5. Žitňan 1985; 6. Gelnar et al. 1994.

Gyrodactylus stankovici Ergens, 1970
Site: fins, gills, skin and nasal cavities.
Host: *Cyprinus carpio* (1, 2, 3, 4, 6, 7, 8)
Distribution: **SR:** D – River Dunaj (= Danube), Liptovská Mara and Ružín water reservoirs, Slovak ponds.
Records: 1. Ergens 1971a; 2. Žitňan 1978; 3. Žitňan 1979c; 4. Žitňan 1979d; 5. Žitňan 1981a; 6. Žitňan et al. 1981; 7. Žitňan 1982a; 8. Žitňan 1985; 9. Gelnar and Špakulová 1997.

Gyrodactylus thymalli Žitňan, 1960
Site: fins.
Host: *Thymallus thymallus* (1, 2, 3, 4, 5, 6, 8, 9)
Distribution: **SR:** D – Rivers Hnilec, Hornád, Hron, Dobšínská water reservoir; V – River Poprad.
Records: 1. Žitňan 1960a; 2. Žitňan 1965b; 3. Žitňan 1965c; 4. Ergens and Lom 1970; 5. Žitňan 1970–1971; 6. Žitňan 1973a; 7. Žitňan 1973b; 8. Žitňan 1980b; 9. Ergens 1983a.

Gyrodactylus tincae Malmberg, 1957
Syn. *Gyrodactylus wageneri tincae* Malmberg, 1957
Site: fins, fins, skin.
Host: *Tinca tinca* (1, 2, 3, 4, 5, 6, 8, 11, 12)
Distribution: **CR:** L – Máchovo jezero pond system, South-Bohemian ponds near Jindřichův Hradec and Třeboň; D – River Morava, South-Moravian ponds; **SR:** D – River Dunaj (= Danube), Hron and Tisa.
Records: 1. Lucký 1957b; 2. Ergens 1962a; 3. Lucký and Dyk 1964; 4. Žitňan 1965b; 5. Žitňan 1966a; 6. Žitňan 1966c; 7. Žitňan 1968c; 8. Ergens and Lom 1970; 9. Žitňan 1973b; 10. Žitňan 1981a; 11. Moravec 1985a; 12. Gelnar et al. 1994.

Gyrodactylus truttae Gläser, 1974
Syn. *Gyrodactylus salaris* Malmberg, 1957 of Ergens (1961a, 1965a), Lucký (1963) and Žitňan (1965a, b, 1970–1971, 1973a, b, 1974b)
Site: fins and skin.
Hosts: *Oncorhynchus mykiss* (2, 6, 7, 9, 13, 14)
Salmo trutta m. *fario* (1, 3, 4, 5, 6, 7, 10, 11, 13, 14, 15)
Salmo trutta m. *lacustris* (7)
Salvelinus fontinalis (6, 13, 14)
Thymallus thymallus (15)
Distribution: **CR:** L – River Malše, Římovská water reservoir, brook in South Bohemia,pond near Český Krumlov; D – localities not specified; O – River Osoblaha; **SR:** D – Rivers Dunaj (= Danube), Hnilec, Hornád, Hron, Topľa and Váh, Dobšínská water reservoir; V – localities not specified.
Records: 1. Ergens 1961a; 2. Lucký 1963; 3. Ergens 1965a; 4. Žitňan 1965b; 5. Žitňan 1970–1971; 6. Řehulka 1973; 7. Žitňan 1973a; 8. Žitňan 1973b; 9. Žitňan 1974b; 10. Žitňan 1980a; 11. Ergens 1981; 12. Žitňan 1981a; 13. Ergens 1983a; 14. Ergens 1992a; 15. Gelnar et al. 1990; 16. Gelnar and Špakulová 1997.

Gyrodactylus vimbi Schulman, 1953
Syn. *Gyrodactylus wageneri scardinii* Malmberg, 1957; *G. scardinii* Malmberg, 1957
Site: fins, skin, nasal cavities, less often gills.
Hosts: *Abramis bjoerkna* (10)
Abramis brama (16)
Abramis sapa (18)
Alburnus alburnus (9,10,17)

Barbus petenyi (10)
Carassius carassius (9)
Cyprinus carpio (4, 7, 10)
Gobio gobio (4)
Leuciscus cephalus (4, 9, 10, 20, 21, 22)
Leuciscus leuciscus (4, 9, 10)
Rutilus pigus (10)
Rutilus rutilus (1, 4, 5, 9, 10, 11, 14, 15, 23)
Scardinius erythrophthalmus (1, 4, 5, 8, 10, 13, 15)
Vimba vimba (1, 2, 3, 6, 10, 12, 17)

Distribution: **CR:** L – River Malše, Římovská water reservoir, Máchovo jezero pond system, South-Bohemian ponds near Jindřichův Hradec and Třeboň; D – Rivers Dřevnice, Morava and Rokytná, Morava R. basin, South-Moravian ponds and canals; O – River Osoblaha; **SR:** D – Rivers Dunaj (= Danube), Hron, Laborec, Latorica and Tisa, Slovak ponds.

Records: 1. Lucký 1959a; 2. Lucký 1959b; 3. Dyk 1961a; 4. Ergens 1962a; 5. Dyk and Lucký 1963; 6. Lucký and Dyk 1964; 7. Žitňan 1964a; 8. Žitňan 1964f; 9. Ergens 1965a; 10. Žitňan 1965b; 11. Žitňan 1966a; 12. Žitňan 1968c; 13. Ergens and Lom 1970; 14. Žitňan 1971c; 15. Ergens et al. 1975; 16. Moravec 1986c; 17. Gelnar et al. 1990; 18. Gelnar et al. 1994; 19. Moravec 1995a; 20. Gelnar et al. 1997; 21. Divílek et al. 1998; 22. Dušek et al. 1998; 23. Kadlec et al. 1998.

Gyrodactylus spp.
Site: fins, skin, gills.
Hosts: *Abramis brama* (1)
Aspius aspius (22)
Barbatula barbatula (10)
Barbus barbus (5, 8)
Carassius carassius (9, 15)
Chondrostoma nasus (6)
Cobitis taenia (20)
Ctenopharyngodon idella (12, 16, 17, 18)
Cyprinus carpio (2, 3, 9, 15)
Gobio gobio (4, 5, 7, 8, 14)
Hypophthalmichthys molitrix (19)
Ictalurus nebulosus (9, 15, 18)
Leucaspius delineatus (9, 15, 21)
Leuciscus cephalus (4, 5, 6, 7, 13)
Pelecus cultratus (22)
Rutilus rutilus (20, 22)
Vimba vimba (5, 7)

Distribution (findings): **CR:** D – Rivers Dřevnice, Jihlava, Morava and Rokytná, South-Moravian ponds and pools; O – River Osoblaha; **SR:** D – Rivers Dunaj (= Danube), Hornád and Tisa, Dobšinská water reservoir, Slovak ponds.

Records: 1. Frič 1908; 2. Schäferna 1927; 3. Dyk 1939; 4. Lucký 1957b; 5. Lucký 1959a; 6. Žitňan 1960c; 7. Lucký and Dyk 1964; 8. Ergens 1965a; 9. Žitňan 1966a; 10. Žitňan 1967a; 11. Žitňan 1968c; 12. Žitňan 1970b; 13. Žitňan 1970–1971; 14. Žitňan 1971c; 15. Žitňan 1971d; 16. Žitňan 1972a; 17. Žitňan 1974a; 18. Žitňan 1974b; 19. Lucký 1981b; 20. Žitňan 1985; 21. Žitňan 1986; 22. Gelnar et al. 1994; 23. Šebelová et al. 1995; 24. Divílek et al. 1998; 25. Dušek et al. 1998,

Misidentified and questionable species of *Gyrodactylus*:

Gyrodactylus gracilis Kathariner, 1894 (*nomen dubium*)
Site: fins.
Host: *Abramis brama* (1)
Distribution: **SR:** D – River Dunaj (= Danube).
Record: 1. Pacák 1962.

Gyrodactylus nemachili Bychowsky, 1936
Site: gills.
Hosts: *Barbatula barbatula* (1, 2, 3, 4, 5, 6, 7, 8)
Eudontomyzon danfordi (8)
Phoxinus phoxinus (8)

Distribution: **CR:** L – South-Bohemian ponds near Třeboň; D – River Rokytná; O – River Osoblaha; **SR:** D – River Hron, lake Morské oko.

Records: 1. Ergens 1959a; 2. Lucký 1959a; 3. Lucký 1959b; 4. Dyk 1961a; 5. Ergens 1962a; 6. Lucký and Dyk 1964; 7. Ergens 1965a; 8. Žitňan 1965.

Remark: Monogeneans reported as *G. nemachili* from *B. barbatula* might belong to *G. barbatuli*, *G. jiroveci*, *G. menschikowi* or *G. sedelnikowi*. *G. nemachili* occurs only in Central Asia and does not occur in Europe (Gusev 1985).

Gyrodactylus rarus Wegener, 1909
Site: fins, skin.
Hosts: *Perca fluviatilis* (1, 3)
cyprinids (2)
Distribution: **CR:** D – River Dyje.
Records: 1. Lucký 1957b; 2. Dyk 1961a; 3. Lucký and Dyk 1964.
Remak: *G. rarus* is a specific parasite of sticklebacks (Gusev 1985) and, therefore, its records from percids and cyprinids in Moravia are, undoubtedly, misidentifications.

Fam. Mazocraeidae Price, 1936

Mazocraes alosae Hermann, 1782
Syn. *Octobothrium lanceolatum* Leuckart, 1827
Site: gills.
Host: *Alosa alosa* (1, 2)
Distribution: **CR:** L – Bohemia.
Records: 1. John 1877; 2. Šrámek 1900.

Fam. Diplozoidae Palombi, 1949

Diplozoon paradoxum Nordmann, 1832
Site: gills.
Hosts: *Abramis bjoerkna* (11, 13, 16, 17, 19, 20, 23, 28, 30, 33)
Abramis brama (1, 3, 4, 7, 8, 9, 13, 20, 23, 24, 29, 30, 33, 36, 37, 38, 39, 40, 41, 43, 44, 45, 46, 47)
Alburnoides bipunctatus (12, 14, 24, 32)
Alburnus alburnus (12, 14, 25, 28, 30, 32)
Aspius aspius (17, 18, 19, 20, 25, 30, 32, 33)
Barbus barbus (24)
Carassius carassius (8)
Ctenopharyngodon idella (48)
Cyprinus carpio (7, 8, 9, 10, 20, 30, 32, 34, 35)
Gobio gobio (9, 12, 13, 14, 28, 35)
Gymnocephalus cernuus (23)
Leuciscus cephalus (12, 14, 17, 19, 25, 28, 31, 32, 35)
Leuciscus idus (6, 27)
Leuciscus leuciscus (17, 19, 28, 35)
Phoxinus phoxinus (1)
Rhodeus sericeus (13)
Rutilus rutilus (5, 9, 10, 11, 12, 13, 15, 16, 20, 22, 23, 24, 28, 30, 33, 35, 39)
Scardinius erythrophthalmus (9, 10, 20, 30, 33, 35)
Vimba vimba (1)
Distribution: **CR:** L – Rivers Labe (= Elbe) and Vltava (Prague), South-Bohemian ponds near Jindřichův Hradec and Třeboň, surroundings of Karlovy Vary; D – Rivers Dyje, Morava, Rokytná, Svitava and Svratka, Kníničská and Vranovská water reservoirs, South-Moravian ponds and pools; O – River Opava; **SR:** D – Rivers Dunaj (= Danube), Bodva, Hornád, Hron, Ipeľ, Latorica, Morava, Tisa and Torysa, Zemplínská Šírava water reservoir, Slovak ponds, irrigation canal near Váh R., localities Čierna Voda and Senné (Laborec R. and Uh R. basins).
Records: 1. John 1877; 2. Šrámek 1900; 3. Šrámek 1901, 1903; 4. Frič 1908; 5. Kohn 1929; 6. Dyk 1952a; 7. Dyk 1953a; 8. Lucký 1953a; 9. Dyk 1954c; 10. Dyk and Lucký 1954a; 11. Dyk and Lucký 1954c; 12. Gotthardt 1954; 13. Vojtek et al. 1954; 14. Dyk 1955b; 15. Lucký et al. 1955; 16. Dyk 1956a; 17. Kašťák 1956d; 18. Kašťák 1957b; 19. Kašťák 1957d; 20. Lucký 1957b; 21. Lucký 1957d; 22. Lucký 1959a; 23. Vojtek 1959b; 24. Vojtková 1959b; 25. Žitňan 1960c; 26. Dyk 1961a; 27. Pacák 1961; 28. Ergens 1962a; 29. Pacák 1962; 30. Dyk and Lucký 1963; 31. Dyk 1964a; 32. Hradil 1964; 33. Lucký and Dyk 1964; 34. Žitňan 1964a; 35. Rodák et al. 1965; 36. Žitňan 1965a; 37. Žitňan 1966a; 38. Žitňan 1966c; 39. Lucký and Jílek 1967; 40. Žitňan

1947; 41. Pejčoch 1968; 42. Žitňan 1968; 43. Ergens and Lom 1970; 44. Žitňan 1971; 45. Ergens et al. 1975; 46. Žitňan 1977; 47. Gelnar et al. 1994; 48. Moravec 1995a; 49. Koubková et al. 1997.
Remark: Older records probably include also other species of Diplozoidae.

Diplozoon sp.
Site: gills.
Hosts: *Abramis ballerus* (5)
Abramis bjoerkna (4, 5, 7)
Abramis sapa (5)
Alburnus alburnus (3)
Barbus barbus (3)
Cyprinus carpio (2, 5)
Gobio albipinnatus (5)
Gobio kessleri (8)
Gobio uranoscopus (8)
Leuciscus cephalus (4, 5)
Pelecus cultratus (5, 8)
Phoxinus phoxinus (9)
Rhodeus sericeus (5, 9)
Vimba vimba (1)
Distribution (findings): **CR:** L – Bohemia; O – River Osoblaha; **SR:** D – Rivers Dunaj (= Danube), Hron, Laborec, Latorica and Tisa.
Records: 1. Frič 1908; 2. Žitňan 1964f; 3. Ergens 1965a; 4. Žitňan 1965a; 5. Žitňan 1966a; 6. Žitňan 1968c; 7. Žitňan 1971c; 8. Žitňan 1985; 9. Žitňan 1986.

Eudiplozoon nipponicum (Goto, 1891)
Site: gills.
Host: *Cyprinus carpio* (1, 2, 3)
Distribution: **CR:** L – River Blanečnice and ponds near Vodňany; D – River Morava.
Records: 1. Gelnar et al. 1989; 2. Gelnar et al.1994; 3. Šimková et al. 1997; 4. Gelnar and Špakulová 1997.

Paradiplozoon alburni Khotenovsky, 1982
Site: gills.
Host: *Alburnus alburnus* (1)
Distribution: **CR:** D – River Morava.
Record: 1. Gelnar et al. 1994; 2. Gelnar and Špakulová 1997.

Paradiplozoon bliccae (Reichenbach–Klinke, 1961)
Syn. *Diplozoon gussevi* Gläser, 1964; *D. markewitschi* Bychowsky, Gyntovt et Koval, 1964
Site: gills.
Hosts: *Abramis bjoerkna* (1, 2, 3, 5, 6)
Abramis brama (6)
Leuciscus cephalus (6)
Scardinius erythrophthalmus (1, 2)
Distribution: **CR:** L – localities not specified; D – River Morava, pond Zámecký near Lednice, South Moravia; **SR:** D – Rivers Dunaj (= Danube) and Latorica.
Records: 1. Pejčoch 1968; 2. Ergens and Lom 1970; 3. Ergens et al. 1975; 4. Žitňan 1981a; 5. Gelnar et al. 1994; 6. Moravec 1995a.

Paradiplozoon ergensi (Pejčoch, 1968)
Syn. *Diplozoon ergensi* Pejčoch, 1968
Site: gills.
Host: *Leuciscus cephalus* (1, 2, 3, 4, 5, 6, 8, 9, 10)
Distribution: **CR:** L – localities not specified; D – River Morava; O – localities not specified; **SR:** D – River Latorica.
Records: 1. Pejčoch 1968; 2. Ergens and Lom 1970; 3. Gelnar et al. 1994; 4. Koníček and Gelnar 1994; 5. Šebelová et al. 1995; 6. Gelnar et al. 1997; 7. Gelnar and Špakulová 1997; 8. Koubková et al. 1997; 9. Dušek et al. 1998; 10. Šebelová et al. 2000.

Paradiplozoon homoion (Bychowsky et Nagibina, 1959)
Syn. *Diplozoon homoion* Bychowsky et Nagibina, 1959
Site: gills.
Hosts: *Abramis bjoerkna* (12)
Abramis brama (11, 12)
Alburnoides bipunctatus (14)
Alburnus alburnus (4, 12)
Barbus barbus (12)
Carassius carassius (4, 6, 12)
Gobio albipinnatus (12)
Gobio gobio (5, 12)
Hypophthalmichthys molitrix (8, 9)

Leucaspius delineatus (14)
Leuciscus cephalus (4, 6, 12, 13, 15, 16, 17, 18, 20)
Leuciscus leuciscus (6, 12)
Rhodeus sericeus (12)
Rutilus pigus (6)
Rutilus rutilus (1, 2, 3, 4, 6, 7, 10, 11, 12, 19)
Scardinius erythrophthalmus (4)
Distribution: **CR:** L – Rivers Malše, Lužnice and Teplá, Římovská water reservoir; D – River Morava and its basin, South-Moravian ponds and pools; O – localities not specified; **SR:** D – Rivers Dunaj (= Danube), Laborec, Latorica and Tisa.
Records: 1. Žitňan 1964f; 2. Žitňan 1965a; 3. Žitňan 1966a; 4. Pejčoch 1968; 5. Žitňan 1968c; 6. Ergens and Lom 1970; 7. Ergens et al. 1975; 8. Lucký 1981a; 9. Lucký 1981b; 10. Žitňan 1985; 11. Gelnar et al. 1990; 12. Gelnar et al. 1994; 13. Koníček and Gelnar 1994; 14. Moravec 1995a; 15. Šebelová et al. 1995; 16. Gelnar et al. 1997; 17. Koubková et al. 1997; 18. Dušek et al. 1998; 19. Kadlec et al. 1998; 20. Šebelová et al. 2000.

Paradiplozoon megan (Bychowsky et Nagibina, 1959)
Syn. *Diplozoon megan* Bychowsky et Nagibina, 1959
Site: gills.
Host: *Leuciscus cephalus* (4, 5, 6, 7, 8)
Leuciscus idus (1, 3)
Distribution: **CR:** D – River Morava and its basin; **SR:** D – Rivers Dunaj (= Danube), Latorica and Tisa.
Records: 1. Žitňan 1966a; 2. Žitňan 1968c; 3. Žitňan 1971c; 4. Šebelová et al. 1995; 5. Gelnar et al. 1997; 6. Koubková et al. 1997; 7. Dušek et al. 1998; 8. Šebelová et al. 2000.

Paradiplozoon nagibinae (Gläser, 1965)
Syn. *Diplozoon nagibinae* Gläser, 1965
Site: gills.
Host: *Abramis ballerus* (1, 2, 3)
Distribution: **CR:** D – Nové Mlýny water reservoir; **SR:** D – River Latorica, irrigation canal near Komárno.
Records: 1. Pejčoch 1968; 2. Ergens and Lom 1970; 3. Koubková et al. 1997.

Paradiplozoon pavlovskii (Bychowsky et Nagibina, 1959)
Syn. *Diplozoon pavlovskii* Bychowsky et Nagibina, 1959
Site: gills.
Host: *Aspius aspius* (1, 2, 4)
Distribution: **SR:** D – Rivers Latorica and Tisa.
Records: 1. Žitňan 1965a; 2. Žitňan 1966a; 3. Žitňan 1968; 4. Ergens et al. 1975.

Paradiplozoon rutili (Gläser, 1967)
Site: gills.
Host: *Leuciscus cephalus* (1, 2)
Distribution: **CR:** D – River Morava.
Records: 1. Gelnar et al. 1997; 2. Gelnar et al. 1998.

Paradiplozoon sapae (Reichenbach–Klinke, 1961)
Syn. *Diplozoon bergi* Gavrilova, 1964
Site: gills.
Host: *Abramis sapa* (2, 3)
Distribution: **CR:** D – River Morava; **SR:** D – Rivers Dunaj (= Danube) and Tisa.
Record: 1. Žitňan 1981a; 2. Žitňan 1986; 3. Gelnar et al. 1994.

Paradiplozoon tisae Khotenovsky, 1982
Site: gills.
Host: *Barbus barbus* (1)
Distribution: **CR:** L – localities not specified; O – localities not specified; **CR** and **SR** (former Czechoslovakia): D – localities not specified.
Records: 1. Baruš et al. 1997; 2. Gelnar and Špakulová 1997.

Paradiplozoon vojteki (Pejčoch, 1968)
Site: gills.
Hosts: *Aspius aspius* (2)
Pelecus cultratus (1, 2, 3)
Distribution: **CR:** D – River Morava; **CR** and **SR** (former Czechoslovakia): D – localities not given.
Records: 1. Jurajda et al. 1992; 2. Gelnar et

al. 1994; 3. Baruš et al. 1997b; 4. Gelnar and Špakulová 1997.

Paradiplozoon sp.
Site: gills.
Hosts: *Alburnus alburnus* (1)
Carassius auratus gibelio (3)
Chondrostoma nasus (3)
Leuciscus cephalus (3, 4, 5, 6)
Distribution (findings): **CR:** L – River Malše, Římovská water reservoir; D – River Morava, fish hatchery Třebíč.
Records: 1. Gelnar et al. 1990; 2. Navrátil 1991; 3. Gelnar et al. 1994; 4. Koníček and Gelnar 1994; 5. Šebelová et al. 1995; 6. Dušek et al. 1998.

Fam. Discocotylidae Diesing, 1850

Octomacrum europaeum Roman et Bychowsky, 1956
Syn. *Discocotyle sagittata* (Leuckart, 1842) of Lucký (1957b)
Site: gills.
Host: *Alburnoides bipunctatus* (1, 2, 3, 4, 5, 7, 8)
Distribution: **CR:** D – Rivers Dyje and Rokytná; **SR:** D – Rivers Dunaj (= Danube) and Tisa.
Records: 1. Lucký 1957b; 2. Lucký 1959a; 3. Dyk 1961a; 4. Lucký and Dyk 1964; 5. Žitňan 1966a; 6. Žitňan 1968c; 7. Žitňan 1986; 8. Baruš et al. 1997b.

CESTODA

Fam. Amphilinidae Claus, 1879

Amphilina foliacea (Rudolphi, 1819)
Site: abdominal cavity.
Host: *Acipenser ruthenus* (1, 2, 3, 4, 5, 6)
Distribution: **SR:** D – River Dunaj (= Danube).
Records: 1. Rašín 1930c; 2. Rašín 1931; 3. Žitňan 1979e; 4. Žitňan 1985; 5. Moravec et al. 1989; 6. Scholz 1989a; 7. Macko et al. 1993.

Fam. Acrobothriidae Olsson, 1872

Cyathocephalus truncatus (Pallas, 1781)
Site: intestine.
Hosts: *Acipenser ruthenus* (5, 18, 20)
**Barbus barbus* (2)
Cottus gobio (14, 19, 20)
Cottus poecilopus (2, 3, 10, 19, 20)
Hucho hucho (13, 20)
Oncorhynchus mykiss (1, 9, 10, 12, 20)
Salmo trutta m. *fario* (1, 2, 3, 4, 9, 10, 14, 20, 15)
Salmo trutta m. *lacustris* (10, 20)
Salvelinus fontinalis (12, 20)
Thymallus baicalensis (7, 10, 12, 20)
Thymallus thymallus (8, 10, 16, 20)
Distribution: **SR:** D – Rivers Dunaj (= Danube), Hnilec, Hornád, Hron, Orava, Revúca, Váh and Tisa, brooks Demänovský potok and Mošnica, Dobšinská water reservoir, trout farm in Horní Orešany; V – River Poprad.
Records: 1. Pacák 1957; 2. Pacák 1958; 3. Pacák 1960; 4. Dyk 1961a; 5. Žitňan 1966b; 6. Žitňan 1968; 7. Žitňan 1969; 8. Žitňan 1970–1971; 9. Žitňan 1971; 10. Žitňan 1973a; 11. Žitňan 1973b; 12. Žitňan 1974; 13. Žitňan 1976; 14. Žitňan 1979; 15. Žitňan 1980a; 16. Žitňan 1980b; 17. Žitňan 1981; 18. Žitňan 1985; 19. Žitňan 1986; 20. Macko et al. 1993.

Fam. Lytocestidae Hunter, 1927

Caryophyllaeides fennica (Schneider, 1902)
Site: intestine.
Hosts: *Abramis bjoerkna* (1, 2, 4, 9, 11, 12, 15, 17, 18, 19, 21, 26, 30)
Abramis brama (1, 8, 15, 17, 18, 20, 30)
Abramis sapa (15, 21, 25, 30, 32)
Alburnus alburnus (21, 30)
Aspius aspius (12, 30)
Barbus barbus (17)

Chondrostoma nasus (12, 18, 30)
Ctenopharyngodon idella (26)
Cyprinus carpio (3, 4)
Gobio albipinnatus (28)
Gobio gobio (10, 26, 28, 30, 33)
Leuciscus cephalus (1, 2, 3, 4, 7, 9, 10, 12, 15, 16, 17, 21, 26,27, 29, 30, 31, 32, 34, 35, 36)
Leuciscus leuciscus (10, 12, 26, 30)
**Perca fluviatilis* (2, 3, 4)
Rutilus pigus (12, 21, 30)
Rutilus rutilus (1, 5, 7, 13, 15, 17, 19, 20, 21, 22, 23, 24, 26, 30, 32, 36, 37)
Scardinius erythrophthalmus (9, 11, 12, 19, 20, 22, 24, 26)
Vimba vimba (12, 21, 26, 30)

Distribution: **CR:** L – Rivers Labe, Lužnice, Malše, Nežárka and Vltava, Bystřička Brook near Hradec Králové, Lipno water reservoir, Římovská reservoir, Máchovo jezero pond system, ponds near Třeboň, Jindřichův Hradec and České Budějovice; D – Rivers Bečva, Dyje, Morava, Olešnice, Rokytná, Svitava and Svratka, Morava R. basin, Kníničská and Vranovská water reservoirs, fish farm on Jihlava R., South-Moravian pools; O – River Osoblaha; **SR:** D – Rivers Dunaj (= Danube), Hornád, Hron, Laborec, Latorica, Nitra and Tisa, Oravská and Zemplínská Šírava water reservoirs, localities Čierna Voda and Senné, eastern Slovakia.

Records: 1. Vojtek et al. 1954; 2. Kašťák 1956d; 3. Kašťák 1957b; 4. Kašťák 1957d; 5. Vojtková 1959b; 6. Dyk 1961a; 7. Ergens 1961b; 8. Pacák 1962; 9. Žitňan 1964f; 10. Ergens 1965a; 11. Žitňan 1965a; 12. Žitňan 1966b; 13. Lucký and Jílek 1967; 14. Žitňan 1968c; 15. Ergens and Lom 1970; 16. Žitňan 1970–1971; 17. Žitňan 1971b; 18. Žitňan 1971c; 19. Ergens et al. 1975; 20. Žitňan 1977; 21. Žitňan 1979e; 22. Moravec 1978d; 23. Žitňan 1985; 24. Moravec 1986c; 25. Žitňan 1986; 26. Scholz 1989a; 27. Gelnar et al. 1990; 28. Moravec and Scholz 1991; 29. Navrátil 1991; 30. Macko et al. 1993; 31. Koníček and Gelnar 1994; 32. Gelnar et al. 1994; 33. Moravec 1995a; 34. Šebelová et al. 1995; 35. Gelnar et al. 1997; 36. Moravec et al. 1997; 37. Kadlec et al. 1998.

Khawia baltica Szidat, 1942
Site: intestine.
Host: *Tinca tinca* (1, 2, 3, 4)
Distribution: **CR:** L – Máchovo jezero pond system, South-Bohemian ponds near Kardašova Řečice, Jindřichův Hradec and Třeboň.
Records: 1. Moravec 1984b; 2. Moravec 1985a; 3. Scholz 1989a; 4. Scholz 1993a.

Khawia rossittensis (Szidat, 1937)
Site: intestine.
Host: *Carassius carassius* (1, 3, 4)
Distribution: **SR:** D – Rivers Dunaj (= Danube) and Tisa.
Records: 1. Žitňan 1966b; 2. Žitňan 1968c; 3. Macko et al. 1993; 4. Baruš et al. 1997b.

Khawia sinensis Hsü, 1935
Syn. *Caryophyllaeus fimbriceps* Annenkova-Khlopina, 1919 of Moravec (1984e) – see Scholz (1989a)
Site: intestine.
Host: *Cyprinus carpio* (1, 2, 3, 4, 5, 6, 7, 8, 9, 10, 11, 12, 13, 14)
Distribution: **CR:** L – River Vltava, Hostivařská water reservoir, Máchovo jezero pond system, South-Bohemian ponds; D – River Morava, Kníničská water reservoir, pond Nesyt, North-Moravian ponds; O – pond systems in Ostrava and Karviná region.
Records: 1. Přibyslavský et al. 1965; 2. Lucký and Jílek 1967; 3. Tesarčík and Prouza 1974; 4. Prouza and Tesarčík 1977; 5. Prouza 1982a; 6. Prouza 1982c; 7. Moravec 1984e, 8. Scholz 1989a; 9. Scholz et al. 1990; 10. Scholz 1991b; 11. Scholz 1991c; 12. Gelnar et al. 1994; 13. Moravec et al. 1997; 14. Kodedová et al. 2000.

Fam. Caryophyllaeidae Leuckart, 1878

Archigetes sieboldi Leuckart, 1878
Syn.: *Archigetes appendiculatus* Mrázek, 1897; *Biacetabulum appendiculatum* Janiszewska, 1950

Site: intestine.
Hosts: *Abramis bjoerkna* (3)
Abramis brama (3)
Cyprinus carpio (2, 3)
Tinca tinca (1, 2, 5)
Distribution: **CR:** L – pond Rožmberk near Třeboň, small pond in Vodňany, Hostivař water reservoir; **SR:** D – Rivers Hron and Latorica.
References: 1. Žitňan 1971b; 2. Moravec 1986a; 3. Scholz 1989a; 4. Macko et al. 1993; 5. Moravec 1995a.

Caryophyllaeus brachycollis Janiszewska, 1951
Syn. *Caryophyllaeus laticeps* (Pallas, 1781) of Lucký (1959a) – see Scholz (1989a)
Site: intestine.
Hosts: *Barbus barbus* (1, 2, 3, 4, 8, 11, 15)
Barbus petenyi (8, 15)
Cyprinus carpio (15)
Leuciscus cephalus (6, 7, 8, 9, 10, 13, 14, 15, 17, 18, 19, 20)
Leuciscus idus (4, 15, 16)
Rutilus pigus (15)
Rutilus rutilus (11, 12)
Distribution: **CR:** L – River Vltava; D – Rivers Bečva, Dřevnice, Morava and Rokytná; O – localities not specified; **SR:** D – Rivers Dunaj (= Danube), Hornád, Hron, Laborec, Latorica, Ondava, Tisa and Topľa; V – River Dunajec.
Records: 1. Lucký 1959a; 2. Žitňan 1964f; 3. Žitňan 1965a; 4. Žitňan 1966b; 5. Žitňan 1968c; 6. Ergens and Lom 1970; 7. Žitňan 1970–1971; 8. Žitňan 1971b; 9. Žitňan 1971c; 10. Žitňan 1974c; 11. Žitňan 1979e; 12. Žitňan 1985; 13. Scholz 1989a; 14. Moravec and Scholz 1991; 15. Macko et al. 1993; 16. Gelnar et al. 1994; 17. Šebelová et al. 1995; 18. Gelnar et al. 1997; 19. Moravec et al. 1997; 20. Divílek et al. 1998.

Caryophyllaeus fimbriceps Annenkova-Khlopina, 1919
Site: intestine.
Hosts: **Abramis brama* (2, 4)
Alburnus alburnus (23)
Cyprinus carpio (1, 6, 7, 8, 11, 12, 13, 14, 15, 16, 17, 18,19, 20, 21)
**Leuciscus leuciscus* (3, 5)
Rutilus rutilus (22)
Tinca tinca (1, 6, 10, 20)
Distribution: **CR:** L – Rivers Labe (= Elbe), Kocába and Vltava, Mácha Lake pond system, South-Bohemian ponds; D – River Dyje; O – River Opava; **SR:** D – Rivers Dunaj (= Danube) and Hron, Ružínská water reservoir, Slovak ponds.
Records: 1. Ergens 1961b; 2. Dyk and Lucký 1963; 3. Dyk 1964a; 4. Lucký and Dyk 1964; 5. Rodák et al. 1965; 6. Ergens and Lom 1979; 7. Žitňan 1971b; 8. Tesarčík 1973; 9. Žitňan 1973b; 10. Moravec 1978d; 11. Žitňan 1978; 12. Žitňan 1979c; 13. Žitňan 1979d; 14. Žitňan 1979e; 15. Žitňan 1981a; 16. Žitňan 1982a; 17. Žitňan 1982b; 18. Mitterpák et al. 1984; 19. Žitňan 1985; 20. Scholz 1989a; 21. Macko et al. 1993; 22. Moravec 1995a; 23. Moravec et al. 1997.

Caryophyllaeus laticeps (Pallas, 1781)
Syn. *Caryophyllaeus mutabilis* Rudolphi, 1802; *Caryophyllaeus fimbriceps* Annenkova-Khlopina, 1919 of Vojtek et al. (1954), Pacák (1962) and Lucký and Jílek (1967) – see Scholz (1989)
Site: intestine.
Hosts: *Abramis bjoerkna* (3, 21, 22, 24, 29, 30, 35, 41, 42, 43)
Abramis bjoerkna x *Rutilus rutilus* (4)
Abramis brama (2, 3, 6, 7, 8, 9, 15, 20, 21, 22, 23, 24, 25, 26, 29, 30, 32, 33, 34, 35, 36, 39, 41, 42, 43, 44, 46)
Abramis brama x *Rutilus rutilus* (3)
Abramis sapa (24, 31, 32, 35, 40, 43, 44)
Abramis spp. (28)
Alburnoides bipunctatus (43)
Alburnus alburnus (29, 43)
Aspius aspius (29, 43)
Barbus barbus (17, 21, 22, 29, 43)
Barbus petenyi (29, 37, 43)
Carassius auratus gibelio (44)
Carassius carassius (29)

Chondrostoma nasus (14, 29, 43)
**Cyprinus carpio* (6, 8, 18, 21, 22)
Gobio gobio (44)
**Gymnocephalus schraetser* (10, 12)
Leuciscus cephalus (3, 9, 16, 17, 29, 43, 44, 45)
Leuciscus idus (3, 19, 29, 35, 43)
Leuciscus leuciscus (3, 13, 14, 29, 43, 44)
**Perca fluviatilis* (44)
Rutilus pigus (29, 43)
Rutilus rutilus (5, 21, 22, 26, 31, 32, 38, 43)
Scardinius erythrophthalmus (3, 44)
**Tinca tinca* (2, 3, 11, 21, 22)
Vimba vimba (8, 29, 35, 41, 43)

Distribution: **CR:** L – Rivers Labe (= Elbe), Malše and Vltava, Lipno water reservoir, Římovská reservoir, Máchovo jezero pond system, South-Bohemian ponds Velký Tisý and Rožmberk; D – Rivers Dyje, Morava, Svitava, Svratka, Vranovská water reservoir, Mostiště reservoir, South-Moravian ponds and pools (Pohořelice, Lednice); O – localities not given; **SR:** D – Rivers Dunaj (= Danube), Cirocha, Hornád, Hron, Latorica, Morava, Ondava, Orava, Tisa, Topľa and Váh.

Records: 1. John 1877; 2. Šrámek 1900; 3. Šrámek 1901, 1903; 4. Frič 1908; 5. Kohn 1929; 6. Dyk 1953b; 7. Lucký 1953a; 8. Dyk 1954c; 9. Vojtek et al. 1954; 10. Dyk 1955b; 11. Lucký et al. 1955; 12. Vejnar 1956; 13. Pacák 1957; 14. Pacák 1958; 15. Vojtek 1959b; 16. Vojtková 1959b; 17. Žitňan 1960c; 18. Dyk 1961a; 19. Pacák 1961; 20. Pacák 1962; 21. Dyk and Lucký 1963; 22. Lucký and Dyk 1964; 23. Žitňan 1965a; 24. Žitňan 1966b; 25. Žitňan 1966c; 26. Lucký and Jílek 1967; 27. Žitňan 1968c; 28. Ergens and Lom 1970; 29. Žitňan 1971a; 30. Žitňan 1971c; 31. Žitňan 1973b; 32. Ergens et al. 1975; 33. Žitňan 1977; 34. Moravec 1978d; 35. Žitňan 1979e; 36. Žitňan 1981a; 37. Kirka et al. 1982; 38. Žitňan 1985; 39. Moravec 1986; 40. Žitňan 1986; 41. Scholz 1989a; 42. Gelnar et al. 1990; 43. Macko et al. 1993; 44. Gelnar et al. 1994; 45. Koníček and Gelnar 1994; 46. Moravec et al. 1997 .

Caryophyllaeus sp.
Site: intestine.
Hosts: *Abramis bjoerkna* (5)
Abramis brama (2)
Alburnus alburnus (3)
Cyprinus carpio (2)
Gobio gobio (5)
Leuciscus cephalus (2)
Leuciscus idus (2)
Leuciscus leuciscus (2, 4)
Rutilus rutilus (2)
Scardinius erythrophthalmus (2)
Tinca tinca (2)

Distribution: **CR:** L – Prague (?); D – Rivers Dyje and Morava; **SR:** D – River Ondava.
Records: 1. Mrázek 1901b; 2. Frič 1908; 3. Lucký and Dyk 1964; 4. Kirka et al. 1982; 5. Gelnar et al. 1994.

Monobothrium wageneri Nybelin, 1922
Site: intestine.
Host: *Tinca tinca* (1, 2, 3)
Distribution: **CR:** L – Máchovo jezero pond system; D – River Morava.
Records: 1. Scholz 1987a; 2. Scholz 1989a; 3. Gelnar et al. 1994.

Paracaryophyllaeus gotoi (Motomura, 1927)
Syn. *Paracaryophyllaeus dubininae* Kulakovskaya, 1961
Site: intestine.
Hosts: *Cobitis taenia* (2)
Phoxinus phoxinus (1)
Distribution: **CR:** O – Hořina Brook; **SR:** D – locality not given.
Records: 1. Řehulka 1968; 2. Baruš et al. 1997b.

Paraglaridacris limnodrili (Yamaguti, 1934)
Syn. *Brachyurus gobii* (Szidat, 1938) (see Mackiewicz in Khalil et al. 1994)
Site: intestine.
Host: *Gobio gobio* (1)
Distribution: **SR:** D – River Dunaj (= Danube).
Record: 1. Žitňan 1979e.

Caryophyllaeidae gen. sp.
Site: intestine.
Hosts: *Abramis sapa* (1)
Alburnus alburnus (5)
Gobio albipinnatus (4)
Leuciscus cephalus (3)
Misgurnus fossilis (7)
Scardinius erythrophthalmus (4)
Tinca tinca (6)
Distribution: **SR:** D – Rivers Dunaj (= Danube), Hornád, Hron and Tisa.
Records: 1. Žitňan 1966b; 2. Žitňan 1968c; 3. Žitňan 1970–1971; 4. Žitňan 1971b; 5. Žitňan 1971c; 6. Žitňan 1979e; 7. Žitňan 1986.

Fam. Bothriocephalidae Blanchard, 1849

Bothriocephalus acheilognathi Yamaguti, 1934
Syn. *Bothriocephalus gowkongensis* Yeh, 1955; *Schyzocotyle fluviatilis* Akhmerov, 1960
Site: intestine.
Hosts: *Carassius auratus gibelio* (26, 32)
Carassius carassius (22, 26)
Ctenopharyngodon idella (2, 4, 5, 6, 15, 16, 25, 26, 32, 33)
Cyprinus carpio (7, 8, 9, 10, 11, 12, 13, 14, 15, 16, 18, 20, 21, 22, 23, 24, 26, 27, 28, 29, 30, 32)
**Esox lucius* (26)
Gobio albipinnatus (16, 32)
Gobio gobio (1)
Leuciscus cephalus (31)
Silurus glanis (30)
Distribution: **CR:** L – River Malše, Římovská water reservoir, South-Bohemian ponds near Vodňany and Třeboň; D – South-Moravian pond Nesyt; **SR:** D – Rivers Dunaj (= Danube) and Tisa, Mankovce, Velký Ďúr and Zemplínská Šírava water reservoirs, Perín pond system near Košice, southern and eastern Slovakia.
Records: 1. Ergens and Lom 1970; 2. Žitňan 1972a; 3. Žitňan 1973b; 4. Žitňan 1973c; 5. Žitňan 1974a; 6. Žitňan 1974b; 7. Pár 1975; 8. Pár and Párová 1976; 9. Žitňan 1977; 10. Pár 1978; 11. Svobodová 1978; 12. Žitňan 1978; 13. Žitňan 1979b; 14. Žitňan 1979c; 15. Žitňan 1979d; 16. Žitňan 1979e; 17. Pár 1980; 18. Žitňan et al. 1980; 19. Žitňan 1981a; 20. Žitňan et al. 1981; 21. Mitterpák 1982; 22. Prouza 1982a; 23. Prouza 1982b; 24. Žitňan 1982a; 25. Žitňan 1982b; 26. Žitňan 1983; 27. Žitňan 1985; 28. Hanzelová and Žitňan 1986; 29. Hanzelová and Žitňan 1987; 30. Scholz 1989a; 31. Gelnar et al. 1990; 32. Macko et al. 1993; 33. Moravec 1995a.

Bothriocephalus claviceps (Goeze, 1782)
Syn. *Eubothrium crassum* (Bloch, 1779) of Lucký (1953b)
Site: intestine.
Host: *Anguilla anguilla* (1, 2, 3, 4, 5, 6, 7, 8, 9, 10, 11, 12, 13, 14, 15, 16, 17, 18, 19, 20)
Distribution: **CR:** L – Rivers Labe (= Elbe), Lužnice, Malše, Sázava and Vltava, Orlická and Římovská water reservoirs, Mrázov water reservoir near Mariánské Lázně, Máchovo jezero pond system, South-Bohemian ponds near České Budějovice, Jindřichův Hradec and Třeboň; D – River Svratka; O – localities not given, **SR:** D – River Dunaj (= Danube).
Records: 1. Šrámek 1901, 1903; 2. Frič 1908; 3. Lucký 1953b; 4. Volf and Smíšek 1955; 5. Vojtková 1959b; 6. Ergens 1961a; 7. Dyk and Lucký 1963; 8. Ergens and Lom 1970; 9. Moravec 1978d; 10. Žitňan 1979e; 11. Pár 1980; 12. Moravec 1985b; 13. Scholz 1989a; 14. Gelnar et al. 1990; 15. Scholz and Moravec 1990; 16. Macko 1993; 17. Moravec et al. 1997; 18. Scholz 1997; 19. Žďárská and Nebesářová 1997; 20. Kodedová et al. 2000.

Fam. Triaenophoridae Lönnberg, 1889

Bathybothrium rectangulum (Bloch, 1782)
Syn. *Bothriocephalus rectangulus* (Bloch, 1782); *Abothrium rectangulum* (Bloch, 1782)
Site: intestine.

Hosts: *Barbus barbus* (2, 3, 4, 5, 7, 9, 10, 11, 12, 13, 14, 15, 16, 18, 20, 21, 23, 24, 25, 26, 27, 28, 29, 30)
Barbus petenyi (6, 8, 12, 16, 18, 19, 20, 21, 22, 27)
**Cyprinus carpio* (1)
**Leuciscus cephalus* (20, 27)

Distribution: **CR:** L – Rivers Labe (= Elbe), Blanice and Malše, Římovská water reservoir; D – Rivers Dyje, Jihlava and Rokytná, fish farm on Jihlava R., Vranovská water reservoir; O – Rivers Moravice and Osoblaha; **SR:** D – Rivers Dunaj (= Danube), Hornád, Hron, Okna, Orava and Tisa, brook Myslava; V – River Dunajec.

Records: 1. John 1877; 2. Mrázek 1901a; 3. Šrámek, 1901, 1903; 4. Dyk, 1954c; 5. Dyk and Lucký 1956b; 6. Kašťák 1956d; 7. Dyk and Lucký 1957; 8. Kašťák 1957d; 9. Lucký 1958; 10. Pacák 1958; 11. Lucký 1959a; 12. Dyk 1961a; 13. Dyk 1964a; 14. Lucký and Dyk 1964; 15. Ergens 1965a; 16. Žitňan 1966b; 17. Žitňan 1968c; 18. Ergens and Lom 1970; 19. Žitňan 1970–1971; 20. Žitňan 1971b; 21. Žitňan 1974c; 22. Žitňan 1979e; 23. Scholz 1989a; 24. Gelnar et al. 1990; 25. Moravec and Scholz 1991; 26. Navrátil 1991; 27. Macko et al. 1993; 28. Schoz and Moravec 1996; 29. Moravec et al. 1997; 30. Špakulová and Scholz 1999.

Eubothrium crassum (Bloch, 1779)
Syn. *Bothriocephalus proboscideus* (Batsch, 1786); *Bothriocephalus infundibuliformis* Bellingham, 1844
Site: intestine.
Host: *Salmo salar* (1, 2, 3, 4)
Distribution: **CR:** L – Rivers Labe (= Elbe) and Vltava (Prague).
Records: 1. John 1877; 2. Frič 1893; 3. Šrámek 1900; 4. Frič 1908.

Triaenophorus crassus Forel, 1868
Site: intestine.
Hosts: *Acipenser ruthenus* (1, 2)
Aspius aspius (1, 2)
Esox lucius (1, 2, 3, 4, 5, 6, 7, 8, 9, 10, 11, 12, 13, 14, 16, 18)
Perca fluviatilis (2)
Silurus glanis (1)

Distribution: **CR:** L – localities not given; D – River Dyje; O – localities not given; **SR:** D – Rivers Dunaj (= Danube), Bodva, Hron, Váh and Tisa, drainage canals near Komárno, Tisa R. basin.

Records: 1. Michalovič 1954; 2. Dyk 1955b; 3. Čákay 1958; 4. Lucký 1958; 5. Vojtek 1959b; 6. Žitňan 1960c; 7. Dyk and Lucký 1963; 8. Lucký and Dyk 1964; 9. Žitňan 1966b; 10. Žitňan 1966c; 11. Žitňan 1968c; 12. Ergens and Lom 1970; 13. Žitňan 1971b; 14. Žitňan 1971c; 15. Žitňan 1973b; 16. Žitňan 1979e; 17. Žitňan 1981a; 18. Macko et al. 1993.

Triaenophorus nodulosus (Pallas, 1781)
Site: intestine.
Hosts: *Anguilla anguilla* (28, 36, 43)
**Barbus petenyi* (8, 11)
Esox lucius (1, 2, 3, 4, 5, 6, 7, 8, 9, 10, 11, 12, 13, 14, 15, 16, 17, 18, 19, 20, 21, 22, 23, 24, 25, 26, 28, 29, 30, 32, 33, 34, 35, 38, 41, 42, 43)
Hucho hucho (39)
Lota lota (36, 40)
Perca fluviatilis (13, 26, 43)
Stizostedion lucioperca (24, 25, 43)

Distribution: **CR:** L – Rivers Labe (= Elbe), Malše and Vltava, Lipno water reservoir, Římovská reservoir, Máchovo jezero pond system, ponds near Lnáře, Třeboň and Jindřichův Hradec; D – River Dyje and its periodical pools, Svratka River, Kníničská, Mostiště and Vranovská water reservoirs, South-Moravian pond; O – localities not specified; **SR:** D – Rivers Dunaj (= Danube), Bodrog, Hron, Ipeľ, Laborec, Latorica, Tisa and Torysa, brook Myslava, locality Senné, Domaša, Oravská and Zemplínská Šírava water reservoirs; V – River Poprad.

Records: 1. Šrámek 1900; 2. Šrámek 1901, 1903; 3. Frič 1908; 4. Jírovec 1940; 5. Lucký 1953a; 6. Dyk 1954c; 7. Vojtek et al. 1954; 8. Kašťák 1956d; 9. Kašťák 1957b; 10. Kašťák

1957c; 11. Kašťák 1957d; 12. Čákay 1958; 13. Lucký 1958; 14. Vojtek 1959b; 15. Žitňan 1960c; 16. Dyk 1961a; 17. Ergens 1961b; 18. Dyk and Lucký 1963; 19. Hradil 1964; 20. Lucký and Dyk 1964; 21. Žitňan 1964f; 22. Žitňan 1965a; 23. Ergens 1966a; 24. Žitňan 1966b; 25. Žitňan 1966c; 26. Přibyslavský and Lucký 1967; 27. Žitňan 1968c; 28. Ergens and Lom 1970; 29. Žitňan 1971b; 30. Žitňan 1971c; 31. Žitňan 1973b; 32. Ergens et al. 1975; 33. Žitňan 1977; 34. Moravec 1978d; 35. Moravec 1979a; 36. Žitňan 1979e; 37. Žitňan 1981a; 38. Žitňan 1982b; 39. Žitňan 1985; 40. Žitňan 1986; 41. Scholz 1989a; 42. Gelnar et al. 1990, 43. Macko et al. 1993.

Triaenophorus sp.
Site: intestine.
Host: *Esox lucius* (1)
Distribution: **SR:** trout farms in Slovakia.
Record: 1. Pacák 1960.

Fam. Proteocephalidae La Rue, 1911

Proteocephalus cernuae (Gmelin, 1790)
Syn. *Proteocephalus longicollis* (Zeder, 1800) of Přibyslavský et al. (1965) and Přibyslavský and Lucký (1967) – see Ergens and Lom (1970) and Scholz (1989)
Site: intestine.
Hosts: *Gymnocephalus cernuus* (1, 2, 3, 4, 5, 6, 7, 8, 9, 10, 11, 13, 14, 15, 16)
Gymnocephalus schraetser (9, 11, 12, 14, 16)
Perca fluviatilis (6, 7, 14)
Distribution: **CR:** L – Orlická water reservoir, Máchovo jezero pond system, South-Bohemian ponds (Velký Tisý, Dehtář, pond system of Třeboň); D – Kníničská water reservoir, South-Moravian ponds (Novoveský rybník); **SR:** D – Rivers Dunaj (= Danube), Hron, Laborec, Latorica and Tisa.
Records: 1. Lucký et al. 1955; 2. Lucký 1958; 3. Dyk 1961a; 4. Ergens 1961b; 5. Lucký and Dyk 1964; 6. Přibyslavský et al. 1965; 7. Přibyslavský and Lucký 1967; 8. Ergens and Lom 1970; 9. Žitňan 1971b; 10. Ergens et al. 1975; 11. Žitňan 1979e; 12. Žitňan 1986; 13. Scholz 1989a; 14. Macko et al. 1993; 15. Scholz et al. 1998; 16. Scholz and Hanzelová 1998.

Proteocephalus longicollis (Zeder, 1800)
Syn. *Proteocephalus exiguus* La Rue, 1911; *Proteocephalus neglectus* La Rue, 1911
Site: intestine.
Hosts: *Oncorhynchus mykiss* (1, 2, 3, 5, 6, 7, 8, 9, 11, 12, 13, 14, 15, 16, 17, 18, 19, 20, 21, 22, 23, 24, 25, 26, 27, 28, 29, 30, 32)
**Perca fluviatilis* (23)
Salmo trutta m. *fario* (4, 16, 23, 25)
Salvelinus fontinalis (23, 25, 30)
Distribution: **CR:** L – River Kamenice, Jesenická water reservoir, ponds in western and southern Bohemia; D – ponds in southern Moravia; **SR:** D – Rivers Hnilec, Hornád, Ondava, Slaná, Turiec, Uh, Bukovec and Dobšinská water reservoirs, lake Morské oko, trout farms, ponds, small reservoirs and streams in Slovakia; V – River Poprad.
Records: 1. Král 1977; 2. Prouza 1978; 3. Mitterpák 1982; 4. Moravec 1982a; 5. Prouza 1982a; 6. Žitňan 1982b; 7. Prouza 1983; 8. Žitňan and Hanzelová 1987; 9. Scholz 1989a; 10. Hanzelová et al. 1990, 11. Scholz 1991a; 12. Hanzelová 1992; 13. Hanzelová and Scholz 1992; 14. Hanzelová and Špakulová 1992; 15. Sysoev et al. 1992; 16. Hanzelová and Scholz 1993; 17. Macko et al. 1993; 18. Šnábel et al. 1994; 19. Gustafsson et al. 1995; 20. Hanzelová et al. 1995; 21. Siles-Lucas et al. 1995; 22. Turčeková and Kráľová 1995; 23. Hanzelová et al. 1996; 24. Kráľová 1996; 25. Kráľová and Špakulová 1996; 26. Šnábel et al. 1996; 27. Bruňanská 1997; 28. Kráľová et al. 1997; 29. Bruňanská et al. 1998; 30. Scholz et al. 1998; 31. Scholz and Hanzelová 1998; 32. Bruňanská 1999; 33. Bruňanská et al. 2000; 34. Špakulová 2000.

Proteocephalus macrocephalus (Creplin, 1825)
Site: intestine.
Host: *Anguilla anguilla* (1, 2, 3, 4, 5, 6, 7, 8)

Distribution: **CR:** L – Rivers Lužnice, Ohře and Vltava, Mrázov reservoir near Mariánské Lázně, Orlická water reservoir, Želivská reservoir, canal near Lomnice nad Lužnicí, South-Bohemian ponds.
Records: 1. Scholz and Kepr 1988; 2. Scholz 1989a; 3. Moravec et al. 1997; 4. Scholz et al. 1997; 5. Scholz et al 1998; 6. Scholz and Hanzelová 1998; 7. Žďárská and Nebesářová 1999a; 8. Žďárská and Nebesářová 1999b.

Proteocephalus osculatus (Goeze, 1782)
Syn. *Taenia osculata* Goeze, 1782; *Proteocephalus percae* (Müller, 1780) of Lucký (1958) – see Scholz (1989a)
Site: intestine.
Host: **Gymnocephalus cernuus* (3, 4)
Silurus glanis (1, 2, 5, 6, 7, 8, 9, 10, 12, 13, 14)
Distribution: **CR:** L – Orlická water reservoir, Bohemia; D – River Dyje, Vranovská water reservoir, South-Moravian ponds and pools; **SR:** D – Rivers Dunaj (= Danube) and Latorica, localities Čierna Voda and Senné.
Records: 1. John 1877; 2. Dyk 1955b; 3. Kašťák 1956d; 4. Kašťák 1957b; 5. Lucký 1958; 6. Dyk and Lucký 1963; 7. Lucký and Dyk 1964; 8. Ergens and Lom 1970; 9. Ergens et al. 1975; 10. Žitňan 1979e; 11. Žitňan 1981a; 12. Macko et al. 1993; 13. Scholz et al. 1998; 14. Scholz and Hanzelová 1998.

Proteocephalus percae (Müller, 1780)
Syn. *Taenia ocellata* Rudolphi, 1802; *Proteocephalus dubius* La Rue, 1911
Site: intestine.
Hosts: **Esox lucius* (6, 7)
Gymnocephalus cernuus (3, 4, 10, 11)
Perca fluviatilis (1, 2, 4, 5, 6, 7, 8, 9, 10, 11, 12, 13, 14, 15, 16, 17, 18, 19, 20, 21, 22, 23, 24, 25)
**Tinca tinca* (11)
Distribution: **CR:** L – River Malše, Římovská water reservoir, Lipno reservoir, Máchovo jezero pond system, ponds near Jindřichův Hradec, Bohemia; D – Kníničská water reservoir, small reservoirs of the Morava R. drainage system; O – localities not specified; **SR:** D – Dobšinská and Ružínská water reservoirs.
Records: 1. Šrámek 1900; 2. Ergens 1961b; 3. Přibyslavský et al. 1965; 4. Přibyslavský and Lucký 1967; 5. Ergens and Lom 1970; 6. Moravec 1978d; 7. Moravec 1979a; 8. Lucký and Navrátil 1983; 9. Lucký and Navrátil 1984; 10. Moravec 1984e; 11. Moravec 1985a; 12. Scholz 1986; 13. Scholz 1987b; 14. Scholz 1989a; 15. Gelnar et al. 1990; 16. Špakulová and Hanzelová 1992; 17. Macko et al. 1993; 18. Scholz et al. 1995; 19. Siles-Lucas et al. 1995; 20. Hanzelová et al. 1996; 21. Kráľová 1996; 22. Hanzelová et al. 1998; 23. Scholz et al. 1998; 24. Scholz and Hanzelová 1998; 25. Hanzelová et al. 1999.

Proteocephalus torulosus (Batsch, 1786)
Syn. *Ichthyotaenia torulosa* (Batsch, 1786); *Ichthyotaenia sagitta* (Batsch, 1786); *Proteocephalus sagittus* (Grimm, 1872); *Proteocephalus* sp. of Dyk and Lucký (1956b), Lucký (1959a), Vojtková (1959b) and Žitňan (1967a)
Site: intestine.
Hosts: *Abramis ballerus* (38)
Abramis bjoerkna (10, 11, 17)
Abramis brama (20)
Abramis sapa (48)
Alburnoides bipunctatus (14, 44, 46)
Alburnus alburnus (14, 15, 34, 38, 41, 43, 46, 48, 50, 53, 57)
Aspius aspius (43, 49, 55)
Barbatula barbatula (3, 6, 7, 17, 29, 31, 35, 39, 40, 42, 48, 56)
Barbus barbus (12, 13, 14, 21, 23, 49, 52, 53, 56, 57)
Barbus petenyi (33, 48)
Chondrostoma nasus (7, 9, 22, 23)
Gobio gobio (48)
**Gymnocephalus cernuus* (8, 10, 11, 17)
Leuciscus cephalus (4, 7, 9, 10, 11, 13, 14, 15, 16, 17, 18, 21, 22, 23, 24, 26,

27, 31, 32, 33, 34, 37, 43, 44, 45, 46, 47, 48, 49, 50, 51, 56, 57)
Leuciscus idus (4, 5, 13, 17, 19, 21, 23, 27, 34, 48, 57)
Leuciscus leuciscus (1, 3, 18, 25, 31, 33, 43, 44, 45, 46)
**Perca fluviatilis* (1, 2)
Rutilus rutilus (27, 28, 33, 43, 48, 56)
Scardinius erythrophthalmus (43)
Distribution: **CR:** L – Rivers Labe (= Elbe), Blanice (near Vodňany), Kocába, Malše and Vltava, Bystřice Brook near Hradec Králové, Rokytka Brook, Římovská water reservoir, ponds near Třeboň and other South-Bohemian ponds; D – Rivers Bečva, Bystřice (near Olomouc), Dřevnice, Dyje, Jihlava, Morava, Olešnice, Rokytná, Svitava and Svratka, brooks Víšňovské potoky, fish farm on Jihlava R., Kníničská water reservoir, ponds near Pohořelice; O – Rivers Moravice, Opava and Osoblaha; **SR:** D – Rivers Dunaj (= Danube), Hornád, Hron, Ondava, Laborec, Latorica, Tisa and Torysa, lake Morské oko, Dobšinská water reservoir, localities Čierna Voda and Senné; V – River Dunajec.
Records: 1. Šrámek 1901, 1903; 2. Frič 1908; 3. Anonym (? Štěpán) 1926; 4. Dyk 1954b; 5. Dyk 1954c; 6. Vojtek et al. 1954; 7. Dyk and Lucký 1956b; 8. Kašťák 1956d; 9. Dyk and Lucký 1957; 10. Kašťák 1957b; 11. Kašťák 1957d; 12. Dyk and Štědronský 1958; 13. Lucký 1958; 14. Lucký 1959a; 15. Vojtková 1959b; 16. Žitňan 1960c; 17. Dyk 1961a; 18. Ergens 1961b; 19. Pacák 1961; 20. Pacák 1962; 21. Dyk and Lucký 1963; 22. Dyk 1964a; 23. Lucký and Dyk 1964; 24. Žitňan 1964f; 25. Ergens 1965a; 26. Rodák et al. 1965; 27. Žitňan 1966b; 28. Lucký and Jílek 1967; 29. Žitňan 1967a; 30. Žitňan 1968c; 31. Ergens and Lom 1970; 32. Žitňan 1970–1971; 33. Žitňan 1971b; 34. Žitňan 1971c; 35. Žitňan 1973a; 36. Žitňan 1973b; 37. Žitňan 1974c; 38. Ergens et al. 1975; 39. Moravec 1977c; 40. Žitňan 1979e; 41. Kirka et al. 1982; 42. Žitňan 1986; 43. Scholz 1989a; 44. Scholz 1989b; 45. Gelnar et al. 1990; 46. Moravec and Scholz 1991b; 47. Navrátil 1991; 48. Macko et al. 1993; 49. Scholz and Moravec 1993; 50. Gelnar et al. 1994; 51. Koníček and Gelnar 1994; 52. Scholz and Moravec 1994; 53. Moravec et al. 1997a; 54. Moravec et al. 1997b; 55. Divílek et al. 1998; 56. Scholz et al. 1998; 57. Scholz and Hanzelová 1998; 58. Špakulová 2000.

Proteocephalus sp.
Site: intestine.
Hosts: *Aspius aspius* (10)
Esox lucius (8)
Gymnocephalus cernuus (4, 5, 6, 8)
Leuciscus cephalus (5, 6)
Lota lota (3)
Perca fluviatilis (5, 6, 8, 11)
Proterorhinus marmoratus (12)
Silurus glanis (1)
Stizostedion lucioperca (2, 3, 11)
Distribution (findings): **CR:** L – River Vltava, ponds near Třeboň; D – Rivers Morava, Svitava and Svratka, Mušov and Vranovská water reservoirs, fish hatchery Třebíč; **SR:** D – Rivers Dunaj (= Danube), Hron, Latorica and Tisa.
Records: 1. Dyk 1954; 2. Vejnar 1956; 3. Ergens 1961; 4. Žitňan 1965; 5. Žitňan 1966; 6. Žitňan 1967; 7. Žitňan 1968; 8. Žitňan 1971; 9. Navrátil 1991, 10.Gelnar et al. 1994; 11. Moravec et al. 1997b; 12. Koubková and Baruš 2000.

Silurotaenia siluri (Batsch, 1786)
Syn. *Proteocephalus osculatus* (Goeze, 1782) of Ergens (1956a) and Lucký (1958); *Gangesia parasiluri* Yamaguti, 1934 of Lucký and Dyk (1964) – see Scholz (1989)
Site: intestine.
Host: *Silurus glanis* (1, 2, 3, 4, 5, 6, 7, 8, 9, 10)
Distribution: **CR:** L – Orlická water reservoir; D – River Dyje, Vranovská water reservoir, ponds near Pohořelice, pool near Přítluky; **SR:** D – Rivers Dunaj (= Danube) and Tisa.
Records: 1. Ergens 1956a; 2. Lucký 1958; 3. Lucký and Dyk 1964; 4. Žitňan 1966b; 5. Žitňan 1968c; 6. Žitňan 1979e; 7. Žitňan 1981a; 8. Macko et al. 1993; 9. Scholz et al. 1999; 10. Kodedová et al. 2000.

Fam. Tentaculariidae Poche, 1926

Tentacularia coryphaenae Bosc, 1797
postlarvae
Syn. *Tetrarhynchus macrobothrius* Rudolphi, 1819
Site: surface of internal organs.
Host: *Salmo salar* (1)
Distribution: **CR:** L – River Labe (= Elbe).
Record: 1. Frič 1893.
Remark: According to Scholz (1989a), it cannot be excluded that the larvae recorded by Frič belonged to the genus *Nybelinia* Poche, 1926.

Tetrarhynchus sp. postlarvae
Site: walls of oesophagus and stomach.
Host: *Salmo salar* (1)
Distribution: **CR:** L – River Labe (= Elbe).
Record: 1. Frič 1908.

Fam. Triaenophoridae Lönnberg, 1889

Triaenophorus crassus Forel, 1868
plerocercoids
Site: muscles.
Hosts: *Esox lucius* (1, 2, 3)
Perca fluviatilis (7)
Proterorhinus marmoratus (4, 6)
Silurus glanis (1, 2)
Distribution: **SR:** D – Rivers Dunaj (= Danube) and Váh, drains near Komárno.
Records: 1. Michalovič 1954; 2. Dyk 1955b; 3. Čákay 1958; 4. Vojtek 1959b; 5. Vojtek 1961; 6. Ergens and Lom 1970; 7. Moravec 1995a.

Triaenophorus nodulosus (Pallas, 1781)
plerocercoids
Site: liver, rarely other internal organs.
Hosts: *Abramis sapa* (34)
Anguilla anguilla (16, 33)
Chondrostoma nasus (12, 17)
Cyprinus carpio (29, 33)
Esox lucius (1, 4, 11, 12, 13, 18, 23, 31, 33)
Hucho hucho (24, 33)
Lota lota (22, 23, 33)
Perca fluviatilis (2, 3, 4, 5, 6, 7, 8, 9, 10, 12, 14, 15, 16, 17, 18, 19, 20, 21, 22, 25, 26, 27, 28, 29, 30, 31, 32, 33, 35)
Salmo trutta m. *fario* (22)
Distribution: **CR:** L – Rivers Malše and Vltava, Máchovo jezero pond system, Lipno, Klíčava and Římovská water reservoirs, South-Bohemian ponds near Třeboň and Jindřichův Hradec; D – Rivers Dyje, Morava and Svratka, Kníničská and Vranovská water reservoirs, South-Moravian ponds and pools, small water reservoirs in Morava R. basin; **SR:** D – Rivers Dunaj (= Danube), Hron, Ipeľ, Laborec, Latorica, Tisa and Torysa, drains near Komárno, Domaša, Oravská and Zemplínská Šírava water reservoirs; V – River Poprad.
Records: 1. Šrámek 1900; Štěpán 1918; 2. Tománek 1952; 3. Dyk 1954c; 4. Dyk and Lucký 1954c; 5. Vojtek et al. 1954; 6. Dyk 1955b; 7. Lucký et al. 1955; 8. Dyk 1956a; 9. Vejnar 1956; 10. Čákay 1958; 11. Lucký 1958; 12. Vojtek 1959b; 13. Vojtková 1959b; 14. Dyk 1961a; 15. Ergens 1961b; 16. Vojtek 1961; 17. Dyk and Lucký 1963; 18. Lucký and Dyk 1964; 19. Ergens 1966a; 20. Žitňan 1966b; 21. Přibyslavský and Lucký 1967; 22. Ergens and Lom 1970; 23. Žitňan 1971c; 24. Žitňan 1976; 25. Moravec 1978d; 26. Moravec 1979a; 27. Lucký and Navrátil 1983; 28. Lucký and Navrátil 1984; 29. Moravec 1984e; 30. Scholz 1987b; 31. Scholz 1989a; 32. Gelnar et al. 1990; 33. Macko et al. 1993; 34. Gelnar et al. 1994; 35. Moravec et al. 1997.

Fam. Bothriocephalidae Blanchard, 1849

Bothriocephalus claviceps (Goeze, 1782)
larvae
Site: intestine.
Host: *Perca fluviatilis* (1, 2, 3)
Distribution: **CR:** L – Máchovo jezero pond system.
Records: 1. Moravec 1985b; 2. Scholz 1986; 3. Scholz 1987b.

Fam. Diphyllobothriidae Lühe, 1910

Diphyllobothrium latum (Linnaeus, 1758) plerocercoids
Site: musculature.
Host (presumed): *Perca fluviatilis* (1, 2)
Distribution: **SR:** D – River Dunaj (= Danube).
Records: 1. Čatár et al. 1967; 2. Macko et al. 1993.
Remark: Plerocercoids of *D. latum* have not yet been recorded from fishes in Slovakia; the only known case of adult *D. latum* in man in southern Slovakia is allegedly associated with the consumption of raw flesh of *P. fluviatilis* originating from the Danube.

Ligula colymbi Zeder, 1803 plerocercoids
Site: abdominal cavity.
Host: *Cobitis elongatoides* (2)
Cobitis elongatoides x *Cobitis* sp. (2)
Cobitis taenia (1)
Distribution: **CR:** D – Nová Říše valley reservoir (Dyje River drainage system); **SR:** D – ponds near Malacky.
Records: 1. Lucký and Dyk 1964; 2. Halačka et al. 2000.

Ligula intestinalis (Linnaeus, 1758) plerocercoids
Syn. *Ligula simplicissima* Rudolphi, 1819; *Ligula* of Frič (1908)
Site: abdominal cavity.
Hosts: *Abramis bjoerkna* (9, 15, 25, 28, 59, 60, 61, 65, 70, 73)
Abramis bjoerkna × *Rutilus rutilus* (62)
Abramis bjoerkna × *Scardinius erythrophthalmus* (62)
Abramis brama (19, 20, 24, 25, 28, 32, 33, 34, 47, 48, 51, 53, 59, 60, 65, 69, 70, 71, 73)
Abramis sapa (32, 54, 59)
Alburnus alburnus (10, 11, 13, 32, 41, 49, 56, 59, 60)
**Cobitis taenia* (5, 6, 13)
Ctenopharyngodon idella (36, 40, 43, 45, 46, 49, 51, 59)
Cyprinus carpio (2)
Esox lucius (66)
Gobio albipinnatus (49, 59)
Gobio gobio (18, 22, 42, 56, 59, 63)
Leuciscus cephalus (16, 17, 31)
Leuciscus leuciscus (57)
Rutilus pigus (32, 59)
Rutilus rutilus (3, 7, 9, 11, 12, 13, 14, 15, 22, 25, 26, 27, 28, 29, 30, 32, 47, 51, 52, 55, 56, 59, 60, 63, 65, 66, 70, 73)
Scardinius erythrophthalmus (4, 6, 7, 8, 9, 25, 27, 28, 47, 51, 59)
Stizostedion lucioperca (72)
Distribution: **CR:** L – Rivers Malše and Vltava, Hostivařská, Slapská and Římovská water reservoirs, Máchovo jezero pond system, South-Bohemian ponds; D – Rivers Dyje and Svratka, Dalešická, Kníničská, Mostiště, Mušov, Nové Mlýny, Věstonice and Vranovská water reservoirs, South-Moravian ponds, fish hatchery Třebíč; O – River Osoblaha; **SR:** D – Rivers Dunaj (= Danube), Ipeľ, Hron, Laborec, Latorica, Tisa and Váh, Zemplínská Šírava water reservoir, ponds near Malacky, localities Čierna Voda and Senné, surroundings of Komárno, eastern Slovakia.
Records: 1. John 1877; 2. Frič 1908; 3. Anonym (? Schäferna) 1933; 4. Dyk 1952b; 5. Dyk 1952a; 6. Dyk 1953d; 7. Dyk 1954c; 8. Dyk and Lucký 1954a; 9. Dyk and Lucký 1954c; 10. Gotthardt 1954; 11. Michalovič 1954; 12. Vojtek et al. 1954; 13. Dyk 1955b; 14. Lucký et al. 1955; 15. Dyk 1956a; 16. Kašťák 1956d; 17. Kašťák 1957b; 18. Podubský 1958; 19. Vojtková 1959b; 20. Dyk and Štědronský 1960; 21. Dyk 1961a; 22. Ergens 1961b; 23. Vojtek 1961; 24. Pacák 1962; 25. Dyk and Lucký 1963; 26. Žitňan 1963; 27. Dyk 1964a; 28. Lucký and Dyk 1964; 29. Žitňan 1964b; 30. Žitňan 1964f; 31. Ergens 1965a; 32. Žitňan 1966b; 33. Žitňan 1966c; 34. Lucký and Jílek 1967; 35. Žitňan 1967b; 36. Tesarčík and Ehl 1968; 37. Žitňan 1968a; 38. Žitňan 1968c; 39. Ergens and Lom 1970; 40. Žitňan 1970b; 41. Žitňan 1971b; 42. Žitňan 1971c;

43. Žitňan 1972a; 44. Žitňan 1973b; 45. Žitňan 1974a; 46. Žitňan 1974b; 47. Žitňan 1977; 48. Moravec 1978d; 49. Žitňan 1979e; 50. Žitňan 1981a; 51. Žitňan 1982b; 52. Žitňan 1985; 53. Moravec 1986c; 54. Žitňan 1986; 55. Hanel 1988; 56. Scholz 1989a; 57. Gelnar et al. 1990; 58. Navrátil 1991; 59. Macko et al. 1993; 60. Baruš and Prokeš 1994a; 61. Baruš and Prokeš 1994b; 62. Baruš and Prokeš 1995a; 63. Baruš and Prokeš 1995b; 64. Baruš and Prokeš 1995c; 65. Baruš et al. 1995; 66. Moravec 1995a; 67. Adámek et al. 1996; 68. Baruš and Prokeš 1996b; 69. Baruš et al. 1997a; 70. Baruš et al. 1997c; 71. Tenora et al. 1997a; 72. Baruš et al. 1998; 73. Tenora et al. 2000.

Fam. Phyllobothriidae Braun, 1900

Scolex pleuronectis Müller, 1788 plerocercoids
Syn. *Scolex polymorphus* Rudolphi, 1819
Site: pylloric caeca, small intestine, posterior part of intestine.
Host: *Salmo salar* (1)
Distribution: **CR:** L – River Labe (= Elbe).
Record: 1. Frič 1893.
Remark: *Scolex* Müller, 1787 is a collective name for larvae. Under the name *S. pleuronectis*, larvae belonging to different tetraphyllidean species and genera have been reported.

Fam. Dilepididae Fuhrmann, 1907

Neogryporhynchus cheilancristrotus (Wedl, 1855) plerocerci
Syn. *Gryporhynchus cheilancristrotus* Wedl, 1855; *Gryporhynchus pusillus* Nordmann, 1832, auctores; *Gryporhynchus macropeos* (Wedl, 1855) of Lucký and Dyk (1964); *Gryporhynchus* sp. of Dyk (1952b), Lucký (1953a), Vojtek et al. (1954) and Vojtková (1959b); *Cysticercus dilepidis* of Dyk (1961a)
Site: intestine.
Hosts: *Abramis ballerus* (15)
Alburnus bjoerkna (11, 26)
Alburnus alburnus (24, 25)
Carassius auratus (16)
Carassius carassius (1, 2, 3, 6, 9, 10)
Ctenopharyngodon idella (17)
Cyprinus carpio (9, 18, 19)
Gymnocephalus cernuus (10)
Leucaspius delineatus (1, 3, 6)
Misgurnus fossilis (9, 10, 11, 16, 21, 22, 26)
Perca fluviatilis (1, 3, 6)
Scardinius erythropterus (10)
Tinca tinca (1, 2, 3, 4, 5, 6, 7, 9, 10, 14, 18, 19, 22, 23)
Zingel streber (16, 20)
Distribution: **CR:** L – River Malše, Římovská water reservoir, Máchovo jezero pond system; D – Rivers Dyje, Morava and Svratka, South-Moravian ponds and pools; **SR:** D – Rivers Dunaj (= Danube), Latorica and Tisa.
Records: 1. Dyk 1952b; 2. Lucký 1953a; 3. Dyk 1954c; 4. Vojtek et al. 1954; 5. Vojtková 1959b; 6. Dyk 1961a; 7. Ergens 1961b; 8. Vojtek 1961; 9. Dyk and Lucký 1963; 10. Dyk and Lucký 1964; 11. Žitňan 1966b; 12. Žitňan 1967b; 13. Žitňan 1968c; 14. Ergens and Lom 1970; 15. Ergens et al. 1975; 16. Žitňan 1979e; 17. Willomitzer 1980a; 18. Moravec 1984e; 19. Moravec 1985a; 20. Žitňan 1985; 21. Žitňan 1986; 22. Scholz 1989a; 23. Gelnar et al. 1990; 24. Gelnar et al. 1994; 25. Moravec 1995a; 26. Hanzelová and Ryšavý 1999.

Paradilepis scolecina (Rudolphi, 1819) plerocerci
Site: abdominal cavity and surface of inner organs.
Hosts: *Aristichthys nobilis* (1)
Cyprinus carpio (1)
Distribution: **CR:** L – pond near Třeboň.
Record: 1. Scholz 1989c.

Valipora campylancristrota (Wedl, 1855) plerocerci
Syn. *Cysticercus Dilepis unilateralis* (Rudolphi, 1819); *Dilepis unilateralis* (Rudolphi, 1819); *Ophiovalipora unilateralis* (Rudolphi, 1819)

Site: gall bladder, rarely intestine.
Hosts: *Aristichthys nobilis* (22)
Carassius carassius (22, 24)
Ctenopharyngodon idella (22)
Cyprinus carpio (1, 3, 4, 5, 8, 9, 11, 12, 13, 14, 15, 16, 17, 19, 20, 21, 22, 25)
Scardinius erythrophthalmus (4, 23, 25)
Tinca tinca (1, 4, 5, 8, 17, 22, 23, 25)
Distribution: **CR:** L – River Malše, Římovská water reservoir, Máchovo jezero pond system, ponds near Třeboň and Jindřichův Hradec; D – South-Moravian ponds; **SR:** D – Rivers Dunaj (= Danube), Hron and Tisa, Domaša and Zemplínská Šírava water reservoirs, Slovak ponds, Tisa R. lowlands.
Records: 1. Ergens 1961b; 2. Vojtek 1961; 3. Žitňan 1964a; 4. Žitňan 1966b; 5. Žitňan 1966c; 6. Žitňan 1967b; 7. Žitňan 1968c; 8. Ergens and Lom 1970; 9. Žitňan 1971c; 10. Žitňan 1973b; 11. Žitňan 1977; 12. Moravec 1978d; 13. Žitňan 1978; 14. Žitňan 1979b; 15. Žitňan 1979c; 16. Žitňan 1979d; 17. Žitňan 1979e; 18. Žitňan 1981a; 19. Žitňan 1982a; 20. Moravec 1984e; 21. Žitňan 1985; 22. Scholz 1989a; 23. Gelnar et al. 1990; 24. Moravec 1995a; 25. Hanzelová and Ryšavý 1999.

TREMATODA

Fam. Bucephalidae Poche, 1907

Bucephalus polymorphus Baer, 1827
Syn. *Bucephalus markewitschi* Koval, 1949; *Bucephalus* sp. of Dyk and Lucký (1963)
Site: intestine.
Hosts: *Abramis sapa* (17)
Esox lucius (2, 3, 4, 8, 9, 12, 13, 16)
Perca fluviatilis (1, 3, 4, 6, 9, 10, 12, 13)
Silurus glanis (16)
Stizostedion lucioperca (5, 7, 14, 16)
Stizostedion volgense (13, 16, 18)
Zingel zingel (5, 7, 16, 17)
Distribution: **CR:** L – pond Velký Tisý; D – Rivers Dyje and its pools, Rokytná and Svratka, ponds and pools in southern Moravia; **SR:** D – Rivers Dunaj (= Danube) and Latorica.

Records: 1. Dyk 1952b; 2. Lucký 1953a; 3. Dyk 1954c; 4. Vojtek et al. 1954; 5. Dyk 1955b; 6. Lucký et al. 1955; 7. Vejnar 1956; 8. Lucký 1959a; 9. Vojtek 1959b; 10. Vojtková 1959b; 11. Dyk 1961a; 12. Dyk and Lucký 1963; 13. Lucký and Dyk 1964; 14. Ergens and Lom 1970; 15. Ergens et al. 1975; 16. Žitňan 1979e; 17. Žitňan 1985; 18. Žitňan 1986, 19. Vojtek 1989.

Rhipidocotyle campanula (Dujardin, 1845)
Syn. *Rhipidocotyle illense* (Ziegler, 1883)
Site: intestine.
Hosts: *Abramis sapa* (13)
Alburnus alburnus (13)
Esox lucius (6, 7, 8, 9, 10, 13)
Gobio kessleri (11)
Gobio uranoscopus (11)
Gymnocephalus schraetser (2)
Lota lota (10)
Perca fluviatilis (1, 2, 3, 5, 8, 10, 13, 14)
Rutilus rutilus (13)
Stizostedion lucioperca (5, 6, 7, 8, 14)
Stizostedion volgense (6)
Zingel streber (6, 10, 11)
Zingel zingel (2, 3, 6, 10, 11)
Distribution: **CR:** L – River Vltava, pond systems of Třeboň and Jindřichův Hradec; D –Rivers Dyje, Morava and Svratka, pools in southern Moravia, waterwork reservoir in Brno; O – localities not specified; **SR:** D – Rivers Dunaj (= Danube), Latorica and Tisa.
Records: 1. Dyk 1954c; 2. Dyk 1955b; 3. Vejnar 1956; 4. Dyk 1961a; 5. Ergens 1962a; 6. Žitňan 1966b; 7. Žitňan 1966c; 8. Ergens and Lom 1970; 9. Ergens et al. 1975; 10. Žiňan 1979e; 11. Žitňan 1985;

12. Vojtek 1989; 13. Gelnar et al. 1994; 14. Moravec et al. 1997.

Fam. Sanguinicolidae Graff, 1907

Sanguinicola armata Plehn, 1905
Site: blood system.
Host: *Chondrostoma nasus* (2)
Misgurnus fossilis (1)
Distribution: **CR:** D – River Dyje, surroundings of Pohořelice; **SR:** D – River Dunaj (= Danube).
Records: 1. Lucký and Dyk 1964; 2. Žitňan 1979e; 3. Žitňan 1981.

Sanguinicola inermis Plehn, 1905
Syn. *Sanguinicola* sp. of Volf (1946)
Site: blood system.
Hosts: *Cyprinus carpio* (1, 2, 3, 4, 6, 8, 9, 10, 11, 12, 13, 15, 16, 18, 19)
Misgurnus fossilis (7)
Scardinius erythrophthalmus (5)
Distribution: **CR:** L – Máchovo jezero pond system and other pond systems of the Labe (=Elbe) R. basin; D – drainage systems of Rivers Dyje and Jihlava; O – Odra (= Oder) R. basin; **SR:** D – Rivers Dunaj (= Danube) and Tisa.
Records: 1. Schäferna 1927; 2. Dyk 1944; 3. Volf 1946; 4. Dyk 1952a; 5. Vojtek et al. 1954; 6. Havlena and Pekář 1960; 7. Dyk 1961a; 8. Dyk and Lucký 1963; 9. Dyk 1964a; 10. Lucký 1964; 11. Lucký and Dyk 1964; 12. Lucký 1968a; 13. Řehulka and Tesarčík 1971; 14. Žitňan 1973b; 15. Moravec 1978d; 16. Žitňan 1979e; 17. Žitňan 1981a; 18. Moravec 1984e; 19. Žitňan 1985.

Sanguinicola intermedia Ejsmont, 1926
Site: blood system.
Host: *Misgurnus fossilis* (1, 2)
Distribution: **CR:** L – localities not mentioned; D – pools near Podivín; O – localities not specified; **SR:** D – localities not specified.
Records: 1. Dyk 1955a; 2. Baruš et al. 1997b.

Sanguinicola sp.
Site: blood system.
Host: *Barbus barbus* (1)
Distribution: **CR:** D – fish farm on Jihlava River.
Record: 1. Navrátil 1991.
Remark: This species is probably identical with *S. armata*.

Fam. Hemiuridae Lühe, 1901

Hemiurus appendiculatus (Rudolphi, 1802)
Syn. *Distomum appendiculatum* (Rudolphi, 1802)
Site: stomach, intestine.
Host: *Salmo salar* (1)
Distribution: **CR:** L – River Labe (= Elbe).
Record: 1. Frič 1893.

Fam. Halipegidae Poche, 1925

Derogenes varicus (Müller, 1784)
Syn. *Distomum varicum* Müller, 1784
Site: oesophagus, stomach, intestine.
Host: *Salmo salar* (1)
Distribution: **CR:** L – River Labe (= Elbe).
Record: 1. Frič 1893.

Fam. Monorchidae Odhner, 1911

Asymphylodora imitans (Mühling, 1898)
Site: intestine.
Hosts: *Abramis ballerus* (2, 11, 12)
Abramis bjoerkna (1, 2, 3, 4, 5, 6, 7, 11, 12, 13, 15)
Abramis brama (2, 6, 7, 8, 9, 11, 12, 13, 15)
Abramis sapa (7, 11, 12, 13, 14, 15)
Leuciscus cephalus (15, 16, 17)
**Perca fluviatilis* (15)
Scardinius erythrophthalmus (6, 7, 11)
Distribution: **CR:** L – localities not specified; D – Dunaj (= Danube) R. basin, Rivers Dyje and Morava, Kníničky water reservoir; O – localities not specified; **SR:** D – Rivers Dunaj (= Danube), Laborec and Tisa, surroundings of Komárno, Tisa R. lowlands.
Records: 1. Vojtek et al. 1954; 2. Vojtek 1959b; 3. Dyk 1961a; 4. Lucký and Dyk 1964; 5. Žitňan 1964f; 6. Žitňan 1965a; 7. Žitňan

1966b; 8. Žitňan 1966c; 9. Lucký and Jílek 1967; 10. Žitňan 1968c; 11. Ergens and Lom 1970; 12. Ergens et al. 1975; 13. Žitňan 1979e; 14. Žitňan 1986; 15. Gelnar et al. 1994; 16. Špakulová et al. 1995; 17. Gelnar et al. 1997.

Asymphylodora kubanicum Issaitschikoff, 1923
Site: intestine.
Host: *Scardinius erythrophthalmus* (1)
Distribution: **SR:** D – River Dunaj (= Danube).
Record: 1. Žitňan 1979e; 2. Žitňan 1981a.

Asymphylodora markewitschi Kulakovskaya, 1947
Syn. *Asymphylodora* sp. of Žitňan (1966b, 1967c, 1968c)
Site: intestine.
Hosts: *Barbus barbus* (2)
Carassius carassius (1)
Chondrostoma nasus (5)
Scardinius erythrophthalmus (2, 4, 5)
Distribution: **CR:** D – River Dyje; **SR:** D – Rivers Latorica and Tisa.
Records: 1. Lucký and Dyk 1964; 2. Žitňan 1966b; 3. Žitňan 1968c; 5. Ergens and Lom 1970; 6. Ergens et al. 1975.

Asymphylodora tincae (Modeer, 1790)
Site: intestine.
Hosts: *Rutilus rutilus* (18)
Tinca tinca (1, 2, 3, 4, 5, 6, 7, 8, 9, 10, 11, 12, 14, 17, 18, 19, 20, 21, 22)
Distribution: **CR:** L – River Malše, Máchovo jezero pond system, ponds near Chlumec nad Cidlinou and near Třeboň, River Labe (= Elbe) basin; D – Rivers Dyje and Morava, Kníničská water reservoir, Vranovská water reservoir, pools in southern Moravia, ponds near Lednice; O – localities not specified; **SR:** D – drainage systems of Rivers Dunaj (= Danube), Hornád, Hron, Latorica, Nitra, Váh and Tisa, surroundings of Komárno.
Records: 1. Dyk 1954c; 2. Vojtek et al. 1954; 3. Lucký 1955a; 4. Vojtek 1959b; 5. Ergens 1962a; 6. Lucký and Dyk 1964; 7. Žitňan 1966d; 8. Žitňan 1966c; 9. Žitňan 1966b; 10. Lucký and Jílek 1967; 11. Žitňan 1968c; 12. Ergens and Lom 1970; 13. Žitňan 1973b; 14. Moravec 1978d; 15. Žitňan 1979e; 16. Žitňan 1981a; 17. Žitňan 1981b; 18. Moravec 1985a; 19. Moravec 1986c; 20. Gelnar et al. 1990; 21. Gelnar et al.1994; 22. Našincová and Scholz 1994.

Palaeorchis incognitus Szidat, 1943
Site: intestine.
Hosts: *Abramis ballerus* (5, 6, 7)
Abramis bjoerkna (1, 2, 7)
Abramis brama (6, 7)
Abramis sapa (5, 6, 7, 9)
Leuciscus cephalus (3, 5, 6)
Leuciscus idus (7)
Leuciscus leuciscus (7)
Rutilus pigus (6, 7)
Rutilus rutilus (5, 6, 7, 8)
Scardinius erythrophthalmus (7)
Vimba vimba (7)
Distribution: **SR:** D – Rivers Dunaj (= Danube) Latorica and Tisa.
Records: 1. Vojtek 1959b; 2. Dyk 1961a; 3. Žitňan 1966b; 4. Žitňan 1968c; 5. Ergens and Lom 1970; 6. Ergens et al. 1975; 7. Žitňan 1979e; 8. Žitňan 1985; 9. Žitňan 1986.

Palaeorchis unicus Szidat, 1943
Site: intestine.
Hosts: *Abramis ballerus* (2)
Abramis bjoerkna (1, 3, 4, 5)
Distribution: **CR:** D – River Morava, southern Moravia (Přítluky); **SR:** D – Rivers Dunaj (= Danube) and Latorica, surroundings of Komárno.
Records: 1. Vojtek et al. 1954; 2. Vojtek 1959b; 3. Dyk 1961a; 4. Ergens et al. 1975; 5. Gelnar et al. 1994.

Fam. Bunoderidae Nicoll, 1914

Bunodera luciopercae (O. F. Müller, 1776)
Syn. *Distomum nodulosum* O. F. Müller, 1776
Site: intestine.

Hosts: *Esox lucius* (27, 28)
Gymnocephalus cernuus (3, 13, 21, 24, 25)
Gymnocephalus schraetser (21, 32)
Lota lota (17, 25)
Perca fluviatilis (1, 2, 3, 4, 5, 6, 7, 8, 9, 11, 13, 14, 15, 16, 17, 18, 19, 21, 22, 24, 25, 26, 27, 28, 30, 33, 34)
Silurus glanis (19, 20, 24)
Stizostedion lucioperca (6, 7, 8, 10, 11, 12, 19, 20, 21, 24, 25, 34)
Stizostedion volgense (19, 24, 25, 32)
Zingel streber (19, 30)
Zingel zingel (19, 21, 30)
Distribution: **CR:** L – River Labe (= Elbe), Máchovo jezero pond system, South-Bohemian ponds near Třeboň and Jindřichův Hradec; D – Rivers Jihlava, Morava and Svratka, Kníničska water reservoir, Vranovská water reservoir, South-Moravian ponds near Lednice and Pohořelice; O – localities not given; **SR:** D – Rivers Dunaj (= Danube), Hron, Latorica, Váh and Tisa, surroundings of Komárno, Tisa R. lowlands, lake Morské oko (Great Vihorlat Lake).
Records: 1. Šrámek 1900; 2. Šrámek 1901, 1903; 3. Frič 1908; 4. Dyk 1953a; 5. Dyk 1954c; 6. Dyk et al. 1954; 7. Vojtek et al. 1954; 8. Dyk 1955b; 9. Lucký et al. 1955; 10. Kašťák 1956d; 11. Vejnar 1956; 12. Kašťák 1957d; 13. Vojtek 1959b; 14. Vojtková 1959b; 15. Dyk 1961a; 16. Ergens 1962a; 17. Dyk and Lucký 1963; 18. Lucký and Dyk 1964; 19. Žitňan 1966b; 20. Žitňan 1966c; 21. Žitňan 1966d; 22. Přibyslavský and Lucký 1967; 23. Žitňan 1968c; 24. Ergens and Lom 1970; 25. Ergens et al. 1975; 26. Žitňan 1977; 27. Moravec 1978d; 28. Moravec 1979a; 29. Žitňan 1979e; 30. Žitňan 1985; 31. Scholz 1986; 32. Žitňan 1986; 33. Gelnar et al. 1994; 34. Moravec et al. 1997, 35. Špakulová 2000.

Crepidostomum auriculatum (Wedl, 1858)
Syn. *Acrolichanus auriculatus* (Wedl, 1858)
Site: intestine.
Host: *Acipenser ruthenus* (1, 2, 3, 4, 5)
Distribution: **SR:** D – Rivers Dunaj (= Danube) and Tisa.
Records: 1. Žitňan 1966b; 2. Žitňan 1968b; 3. Žitňan 1968c; 4. Žitňan 1979e; 5. Žitňan 1985.

Crepidostomum farionis (O. F. Müller, 1780)
Syn. *Distomum laureatum* Zeder, 1800; *Bunodera luciopercae* (O. F. Müller, 1776) of Dyk and Lucký (1953a)
Site: intestine.
Hosts: *Barbatula barbatula* (4, 7, 39)
Cottus gobio (7, 9, 12, 25, 26, 27)
Cottus poecilopus (17)
Lota lota (7)
Oncorhynchus mykiss (24, 33, 36, 37)
Salmo trutta m. *fario* (1, 2, 3, 4, 5, 6, 8, 9, 10, 11, 12, 13, 14, 15, 16, 17, 18, 19, 20, 21, 22, 23, 24, 25, 26, 27, 28, 29, 30, 31, 33, 36, 37, 38, 40, 41, 42, 43, 46)
Salmo trutta m. *lacustris* (5, 8, 12, 16, 22, 19, 37, 38)
Salvelinus fontinalis (12, 19, 22, 24, 38)
Thymallus baicalensis (34, 37)
Thymallus thymallus (2, 3, 4, 9, 13, 15, 17, 21, 22, 24, 25, 32, 33, 36, 37, 38, 44)
Distribution: **CR:** L – River Sázava and tributaries, Blanice R., brooks Černovický potok, Rokytka and Zlatý potok; D – Rivers Punkva and Bystřice near Olomouc, rivulets Lubina and Říčky, brook Adamovský potok; O – Rivers Moravice and Osoblaha and tributaries, surroundings of Malá Morávka and Karlov; **SR:** D – Rivers Dunaj (= Danube), Belá, Biely Váh, Bodva, Hnilec, Hornád, Hron, Revúca Váh and tributaries, Tisa R. basin, brooks Demänová and Mošnice, Dobšinská water reservoir, trout farm near Hor. Orešany; V – Rivers Dunajec and Poprad and tributaries, lakes Nové Štrbské pleso, Popradské pleso and Štrbské pleso.
Records: 1. Šrámek 1900; 2. Dyk and Lucký 1953a; 3. Dyk 1954c; 4. Dyk et al. 1954; 5. Dyk 1955b; 6. Kašťák 1955c; 7. Dyk 1956a; 8. Dyk 1956c; 9. Dyk and Lucký 1956b; 10. Kašťák 1956d; 11. Dyk 1957a; 12. Dyk 1957b; 13. Dyk and Lucký 1957; 14. Kašťák

1957d; 15. Pacák 1957; 16. Dyk 1958a; 17. Pacák 1958; 18. Kašťák and Žitňan 1959; 19. Kašťák and Žitňan 1960a, 20. Kašťák and Žitňan 1960b, 21. Žitňan 1960c, 22. Dyk 1961a, 23. Dyk and Dyková 1963; 24. Ergens 1963; 25. Dyk 1964a; 26. Dyk and Štědronský 1964; 27. Lucký and Dyk 1964; 28. Dyk 1965a; 29. Dyk 1965b; 30. Dyk 1965c; 31. Ergens 1965a; 32. Žitňan 1965c; 33. Žitňan 1966d; 34. Žitňan 1969b; 35. Ergens and Lom 1970; 36. Žitňan 1970–1971; 37. Žitňan 1973a; 38. Žitňan 1973b; 39. Moravec 1977c; 40. Tesarčík and Ivasik 1977; 41. Žitňan 1979e; 42. Moravec 1980b; 43. Žitňan 1980a; 44. Žitňan 1980b; 45. Žitňan 1981b; 46. Moravec 1982b.

Crepidostomum metoecus Braun, 1900
Site: intestine, pyloric caeca.
Hosts: *Cottus gobio* (4, 17, 20)
Oncorhynchus mykiss (3, 15)
Salmo trutta m. *fario* (1, 2, 3, 4, 5, 6, 7, 9, 12, 13, 16, 17, 18, 20)
Salmo trutta m. *lacustris* (13)
Salvelinus fontinalis (21)
Thymallus baicalensis (10, 13, 15)
Thymallus thymallus (3, 4, 8, 9, 12, 13, 19)
Distribution: **CR:** L – River Labe (= Elbe) basin, Černovický Brook, pond system of Jindřichův Hradec; D – River Dunaj (= Danube) basin, Bystřice River near Olomouc; O – River Osoblaha; **SR:** D – Rivers Hnilec, Hornád, Hron, Torysa and their drainage systems, Okna Brook, Dobšinská water reservoir; V – River Visla basin.
Records: 1. Žitňan 1960c; 2. Ergens 1962a; 3. Ergens 1963; 4. Dyk 1964a; 5. Hradil 1964; 6. Lucký and Dyk 1964; 7. Ergens 1965a; 8. Žitňan 1965c; 9. Žitňan 1966d; 10. Žitňan 1969b; 11. Ergens and Lom 1970; 12. Žitňan 1970–1971; 13. Žitňan 1973a; 14. Žitňan 1973b; 15. Žitňan 1974b; 16. Žitňan 1979e; 17. Moravec 1980b; 18. Žitňan 1980a; 19. Žitňan 1980b; 20. Moravec 1982b; 21. Moravec 1995a.

Fam. Gorgoderidae Looss, 1899

Phyllodistomum angulatum Linstow, 1907
Site: urinary bladder and kidneys.
Hosts: *Gymnocephalus cernuus* (1, 2)
Gymnocephalus schraetser (1, 2)
Distribution: **SR:** D – River Latorica.
Records: 1. Ergens and Lom 1970; 2. Ergens et al. 1975.

Phyllodistomum dogieli Pigulevsky, 1953
Site: urinary system.
Host: *Alburnus alburnus* (2)
Leuciscus cephalus (1)
Vimba vimba (1)
Distribution: **CR:** L – River Vltava; D – River Rokytná.
Records: 1. Moravec and Scholz 1991; 2. Moravec et al. 1997.

Phyllodistomum elongatum Nybelin, 1926
Site: ureters.
Hosts: *Abramis ballerus* (6, 8, 9)
Abramis bjoerkna (6)
Abramis brama (1, 2, 3, 6, 8, 11)
Abramis sapa (6, 15)
Alburnus alburnus (2, 5, 6, 8)
Aspius aspius (8, 9)
Cyprinus carpio (10, 12, 13, 14)
Leuciscus cephalus (6)
Rutilus rutilus (2, 6, 9)
Scardinius erythrophthalmus (8, 10, 12, 13, 14)
Tinca tinca (4, 8, 9, 10, 12, 13)
Vimba vimba (6)
Distribution: **CR:** L – Máchovo jezero pond system, ponds near Třeboň; D – River Svratka, Kníničská water reservoir, South-Moravian ponds; O – localities not given; **SR:** D – Rivers Dunaj (= Danube), Latorica and Tisa.
Records: 1. Vojtek et al. 1954; 2. Vojtková 1959b; 3. Dyk 1961a; 4. Ergens 1962a; 5. Žitňan 1965a; 6. Žitňan 1966b; 7. Žitňan 1968c; 9. Ergens and Lom 1970; 10. Ergens et al. 1975; 11. Moravec 1978d; 12. Žitňan 1979e; 13. Moravec 1984e; 14. Moravec 1985a; 15. Moravec 1986c; 16. Žitňan 1986.

Phyllodistomum folium (Olfers, 1926)
Site: urinary bladder and ureters.
Hosts: *Abramis ballerus* (17)
Abramis bjoerkna (7, 9, 10, 16)
Abramis brama (4, 5, 9, 10)
Abramis sapa (9, 13, 15)
Alburnus alburnus (8, 9, 10)
Esox lucius (6, 11, 12, 14)
**Gymnocephalus schraetser* (2, 3)
Leucaspius delineatus (9)
Leuciscus cephalus (1, 8, 9)
Rutilus rutilus (5, 9, 10)
Salmo trutta m. *fario* (8, 9)
Scardinius erythrophthalmus (4, 5, 9)
Tinca tinca (9)
Distribution: **CR:** L – Máchovo jezero pond system; D – Rivers Dyje, Morava and Svratka, Kníničská water reservoir; O – River Osoblaha; **SR:** D – Rivers Dunaj (= Danube) and Latorica, pools and irrigation canal of the Váh R.
Records: 1. Vojtek et al. 1954; 2.Dyk 1955b; 3. Vejnar 1956; 4. Vojtek 1959b; 5. Vojtková 1959b; 6. Dyk 1961a; 7. Lucký and Dyk 1964; 8. Ergens 1965a; 9. Ergens and Lom 1970; 10. Ergens et al. 1975; 11. Moravec 1978d; 12. Moravec 1979a; 13. Žitňan 1979e; 14. Bakke and Žďárská 1985; 15. Žitňan 1986; 16. Gelnar et al. 1994; 17. Moravec 1995a.
Remark: According to Moravec (1979a), records of this species from cyprinids may concern *Phyllodistomum dogieli* Pigulevsky, 1953.

Phyllodistomum megalorchis Nybelin, 1926
Site: urinary bladder.
Host: *Salmo trutta* m. *fario* (1, 2, 3, 4, 5)
Distribution: **CR:** O – localities not given; **SR:** D – River Dunaj (= Danube) basin, River Hron; V – Visla R. basin.
Records: 1. Žitňan 1966d; 2. Ergens and Lom 1970; 3. Žitňan 1979e; 4. Žitňan 1980a; 5. Žitňan 1981b.

Phyllodistomum simile Nybelin, 1926
Site: ureters, urinary bladder.
Hosts: *Cottus gobio* (1)
Salmo trutta m. *fario* (1)
Distribution: **CR** and **SR** (localities not given).
Record: 1. Moravec 1995a.

Phyllodistomum sp.
Site: urinary system.
Hosts: *Alburnus alburnus* (7)
Barbatula barbatula (1)
Carassius carassius (6)
Salmo trutta m. *fario* (2, 3, 4, 5)
Distribution: **CR:** L – Brooks Černovický potok and Chotčínský potok (Lužnice R. basin); D – River Morava, Bečva R. basin, ponds near Pohořelice; O – River Osoblaha, brook Krásný potok; **SR:** D – River Bodva.
Recors: 1. Vojtek et al. 1954; 2. Dyk and Valenta 1955; 3. Žitňan 1960c; 4. Dyk 1961a; 5. Lucký and Dyk 1964; 6. Ergens 1965a; 7. Gelnar et al. 1994.

Fam. Acanthocolpidae Lühe, 1909

Skrjabinopsolus semiarmatus (Molin, 1858)
Syn. *Skrjabinopsolus acipenseris* Ivanov, 1934
Site: intestine.
Host: *Acipenser ruthenus* (1, 2, 4, 5, 6)
Distribution: **SR:** D – Rivers Dunaj (= Danube), Hron and Tisa.
Records: 1. Žitňan 1966b; 2. Žitňan 1966d; 3. Žitňan 1968c; 4. Žitňan 1979e; 5. Žitňan 1985; 6. Moravec et al. 1989.

Fam. Azygiidae Odhner, 1911

Azygia lucii (O. F. Müller, 1776)
Syn. *Distomum tereticolle* (Rudolphi, 1802)
Site: stomach and oesophagus.
Hosts: *Anguilla anguilla* (28, 29, 33)
Esox lucius (1, 2, 3, 4, 5, 6, 7, 9, 12, 14, 15, 16, 17, 18, 19, 21, 22, 23, 24, 27, 28, 29, 30, 32, 33, 37)
Lota lota (10, 11, 12, 17, 24, 35)
Perca fluviatilis (8, 17, 18, 19, 29, 33, 36)
Silurus glanis (20, 22, 23)
Stizostedion lucioperca (4, 18, 19, 20, 22, 23, 28, 29, 33)
Zingel streber (34)
Zingel zingel (22, 24, 34)

Distribution: **CR:** L – River Labe (= Elbe), Máchovo jezero pond system, ponds near Třeboň and Jindřichův Hradec, Lipno water reservoir; D – pools of Dyje River, ponds near Podivín; O – River Moravice and its drainage system; **SR:** D – Rivers Dunaj (= Danube), Bodva, Hron, Latorica and Tisa, surroundings of Komárno, Zemplínská Šírava water reservoir.
Records: 1. Šrámek 1900, 2. Šrámek 1901; 1903, 3. Frič 1908; 4. Dyk 1954b; 5. Dyk 1954c; 6. Vojtek et al. 1954; 7. Dyk 1955b; 8. Lucký et al. 1955; 9. Dyk and Lucký 1956b; 10. Kašťák 1955b; 11. Kašťák 1956d; 12. Dyk and Lucký 1957; 13. Kašťák 1957d; 14. Vojtek 1959b; 15. Žitňan 1960c; 16. Dyk 1961a; 17. Ergens 1962a; 18. Dyk and Lucký 1963; 19. Lucký and Dyk 1964; 20. Žitňan 1965a; 21. Ergens 1966a; 22. Žitňan 1966b; 23. Žitňan 1966c; 24. Žitňan 1966d; 25. Žitňan 1968c; 26. Žitňan 1973b; 27. Žitňan 1977; 28. Moravec 1978d; 29. Moravec 1979a; 30. Žitňan 1979e; 31. Žitňan 1981a; 32. Žitňan 1981b; 33. Moravec 1985b; 34. Žitňan 1985; 35. Žitňan 1986; 36. Scholz 1987b; 37. Namuleno and Scholz 1994.

Fam. Orientocreadiidae Skryabin et Koval, 1960

Orientocreadium siluri (Bykhovskiy et Dubinina, 1954)
Site: intestine.
Host: *Silurus glanis* (1, 2, 3, 4)
Distribution: **SR:** D – Rivers Dunaj (= Danube) and Tisa.
Records: 1. Žitňan 1966b; 2. Žitňan 1966c; 3. Žitňan 1968b; 4. Žitňan 1968c; 5. Žitňan 1979e.

Fam. Allocreadiidae Looss, 1902

Allocreadium carparum Odening, 1959
Site: intestine.
Host: *Cyprinus carpio* (1, 2, 3, 4, 5, 6, 7)
Distribution: **CR:** L – Máchovo jezero pond system, pond near Chlumec nad Cidlinou; D – pond systems of central Moravia, Kníničská water reservoir.
Records: 1. Dyk 1961a; 2. Lucký and Kazda 1961; 3. Lucký and Dyk 1964; 4. Lucký and Jílek 1967; 5. Moravec 1978d; 6. Moravec 1984a; 7. Moravec 1984e.

Allocreadium dogieli Koval, 1950
Site: intestine.
Host: *Abramis bjoerkna* (1, 2, 3)
Distribution: **SR:** D – River Latorica.
Records: 1. Ergens 1964; 2. Ergens and Lom 1970; 3. Ergens et al. 1975.

Allocreadium isoporum (Looss, 1894)
Syn. *Allocreadium isoporum dubium* Koval, 1957; *Allocreadium isoporum macrorchis* Koval et Kulakovskaya, 1957
Site: intestine.
Hosts: *Alburnoides bipunctatus* (8, 14, 17, 24, 26, 27)
Alburnus alburnus (11, 14, 18, 25, 26, 27)
Barbatula barbatula (11, 22, 23, 26, 27)
Barbus barbus (1, 2, 3, 7, 10, 11, 12, 13, 14, 20, 21, 26, 27, 29)
Barbus petenyi (14, 17)
Chondrostoma nasus (7, 11)
Cyprinus carpio (7)
**Esox lucius* (18)
Gobio albipinnatus (26, 27)
Gobio gobio (4, 7, 9, 11, 13, 14, 18, 21, 26, 27)
Leuciscus cephalus (4, 5, 7, 9, 11, 13, 14, 17, 21, 23, 26, 27)
Leuciscus leuciscus (7, 11, 26, 27)
**Perca fluviatilis* (28)
Phoxinus phoxinus (4, 5, 9, 14, 15, 19, 24)
Rutilus rutilus (5, 7)
Distribution: **CR:** L – River Malše, Bystřička Brook near Hradec Králové, brooks Běchovický potok and Černovický potok, ponds near Jindřichův Hradec, ponds near Střížkov (central Bohemia); D – Rivers Bystřice near Olomouc, Jihlava, Rokytná and Želetavka, Vranovská water reservoir; O – River Osoblaha, Hořina Brook; **SR:** D – Rivers Dunaj (= Danube), Hornád,

Hron, Laborec, Latorica and Tisa, Dobšinská water reservoir; V – River Dunajec.
Records: 1. Vojtek et al. 1954; 2. Kašťák 1956d; 3. Kašťák 1957d; 4. Lucký 1959a; 5. Žitňan 1960c; 6. Ergens 1962a; 7. Ergens 1964; 8. Hradil 1964; 9. Lucký and Dyk 1964; 10. Žitňan 1964f; 11. Ergens 1965a; 12. Žitňan 1965a; 13. Žitňan 1966b; 14. Žitňan 1966d; 15. Řehulka 1968; 16. Žitňan 1968c; 17. Žitňan 1970–1971; 18. Žitňan 1971c; 19. Žitňan 1973a; 20. Žitňan 1974c; 21. Žitňan 1979e; 22. Moravec 1980b; 23. Moravec 1984a; 24. Žitňan 1986; 25. Gelnar et al. 1990; 26. Moravec and Scholz 1991b; 27. Moravec 1992a; 28. Moravec 1995a; 29. Moravec et al. 1997.
Remark: Records of *A. isoporum* from *P. phoxinus* probably concern the species *Allocreadium papilligerum* (see Moravec 1984a).

Allocreadium markewitschi Koval, 1949
Site: intestine.
Hosts: *Chondrostoma nasus* (1, 2, 3, 4, 5, 6, 7, 8, 9, 12, 13, 14, 15, 16, 19)
**Gobio gobio* (18)
**Phoxinus phoxinus* (4, 6, 10, 12, 15, 17, 19)
Distribution: **CR:** L – Lužiště Brook (northeastern Bohemia); D – Rivers Rokytná and Svratka; O – River Osoblaha; Hořina Brook; **SR:** – Rivers Dunaj (= Danube), Hornád, Hron, Latorica, Tisa and Toplá.
Records: 1. Žitňan 1960c; 2. Dyk 1961a; 3. Lucký and Kazda 1961; 4. Ergens 1964; 5. Dyk and Lucký 1964; 6. Ergens 1965a; 7. Žitňan 1965a; 8. Žitňan 1966b; 9. Žitňan 1966d;10. Řehulka 1968; 11. Žitňan 1968c; 12. Ergens and Lom 1970; 13. Žitňan 1971c; 14. Ergens et al. 1975; 15. Žitňan 1979e; 16. Moravec 1984a; 17. Žitňan 1986; 18. Moravec 1995a; 19. Baruš et al. 1997b.
Remark: Records of *A. markewitschi* from *P. phoxinus* probably concern the species *Allocreadium papilligerum* (see Moravec 1984a).

Allocreadium papilligerum (Rees, 1968)
Syn. *Allocreadium isoporum* (Looss, 1894) of Moravec (1977c)
Site: intestine.
Host: *Phoxinus phoxinus* (1, 2, 3)
Distribution: **CR:** L – Rokytka Brook near Prague; D – localities not given; O – localities not given; **SR:** – localities not given.
Records: 1. Moravec 1977c; 2. Moravec 1984a; 3. Baruš et al. 1997b.

Allocreadium transversale (Rudolphi, 1802)
Site: intestine.
Hosts: *Cobitis taenia* (3, 4, 7, 8, 10, 11)
Misgurnus fossilis (1, 2, 3, 4, 6, 7, 9, 11)
Distribution: **CR:** L – localities not given; D – River Morava; O – localities not given; **SR:** D – Rivers Dunaj (= Danube), Hron, Latorica and Tisa, ponds in eastern Slovakia.
Records: 1. Žitňan 1960c; 2. Ergens 1964; 3. Žitňan 1966b; 4. Žitňan 1966d; 5. Žitňan 1968c; 6. Ergens and Lom 1970; 7. Žitňan 1979e; 8. Žitňan 1985; 9. Žitňan 1986; 10. Gelnar et al. 1994; 11. Baruš et al. 1997b.

Allocreadium sp.
Site: intestine.
Hosts: *Gymnocephalus cernuus* (2)
Leuciscus cephalus (1)
Phoxinus phoxinus (1)
Distribution: **SR:** D – Rivers Dunaj and Hornád.
Record: 1. Žitňan 1970–1971; 2. Žitňan 1979e.
Remark: The trematodes from *L. cephalus* and *P. phoxinus* probably belonged to *A. isoporum* and *A. papilligerum*, respectively.

Fam. Opecoelidae Ozaki, 1925

Nicolla skrjabini (Iwanitzky, 1928)
Syn. *Coitocaecum skrjabini* Iwanitzky, 1928; *Crowcrocaecum skrjabini* (Iwanitzky, 1928)
Site: intestine.
Hosts: *Abramis ballerus* (7, 16)
Abramis brama (13)
Abramis sapa (7, 13, 16, 20)

Acipenser ruthenus (7, 19)
Alburnoides bipunctatus (16, 20)
Alburnus alburnus (6, 7, 13, 16)
Aspius aspius (7, 16)
Barbus barbus (7, 14)
Barbus petenyi (7, 17)
Chondrostoma nasus (7, 16)
Cobitis taenia (16)
Cottus gobio (13)
Cottus poecilopus (5, 13)
Cyprinus carpio (16, 18, 19)
Esox lucius (16,18)
Gobio albipinnatus (7, 14)
Gobio gobio (7, 16)
Gobio kessleri (7, 19)
Gobio uranoscopus (16, 17, 19)
Gymnocephalus cernuus (1, 2, 3, 4, 6, 7, 9, 13, 14, 15, 16)
Gymnocephalus schraetser (4, 6, 7, 9, 13, 14, 16, 20)
Leuciscus cephalus (6, 7, 11, 16)
Leuciscus idus (7, 16)
Leuciscus leuciscus (7)
Lota lota (7, 16, 20)
Pelecus cultratus (13, 16, 19)
Perca fluviatilis (7, 16)
Rutilus pigus (14)
Salmo trutta m. *fario* (5, 13)
Scardinius erythrophthalmus (7)
Silurus glanis (6, 7, 8, 16, 18)
Stizostedion lucioperca (4, 6, 7, 8, 14, 16, 18)
Stizostedion volgense (7, 16, 20)
Vimba vimba (16)
Zingel streber (7, 16, 19)
Zingel zingel (4, 7, 9, 16, 19)

Distribution: **CR:** O – River Osoblaha; **SR:** D – Rivers Dunaj (= Danube), Hron, Laborec, Latorica and Tisa, pools of Váh R., Zemplínská Šírava water reservoir, localities Senné and Čierna Voda, Tisa R. lowlands.
Records: 1. Kašťák 1956d; 2. Kašťák 1957d; 3. Vojtek 1959b; 4. Žitňan 1964f; 5. Ergens 1965a; 6. Žitňan 1965a; 7. Žitňan 1966b; 8. Žitňan 1966c; 9. Žitňan 1966d; 10. Žitňan 1968c; 11. Žitňan 1971c; 12. Žitňan 1973b; 13. Ergens and Lom 1975; 14. Ergens et al. 1975; 15. Žitňan 1977; 16. Žitňan 1979e; 17. Žitňan 1981a; 18. Žitňan 1981b; 19. Žitňan 1985; 20. Žitňan 1986.

Nicolla testiobliquum (Wisniewski, 1932)
Syn. *Coitocaecum testiobliquum* Wisniewski, 1932, *Crowcrocaecum testiobliquum* (Wisniewski, 1932)
Site: intestine.
Hosts: *Perca fluviatilis* (1, 3)
Gymnocephalus cernuus (1)
Gymnocephalus schraetser (3)
Salmo trutta m. *fario* (2, 4, 5)
Zingel zingel (1, 3)

Distribution: **SR:** D – Rivers Dunaj (= Danube), Hornád and Torysa.
Records: 1. Dyk 1955b; 2. Kašťák 1956d; 3. Vejnar 1956; 4. Kašťák 1957d; 5. Hradil 1964.

Nicolla wisniewskii (Ślusarski, 1958)
Syn. *Crowcrocaecum wisniewskii* Ślusarski, 1958; *Excoitocaecum wisniewskii* (Ślusarski, 1958)
Site: intestine.
Hosts: *Cottus poecilopus* (3, 5, 10)
Oncorhynchus mykiss (3, 5, 7)
Salmo trutta m. *fario* (1, 3, 4, 5, 8, 9, 11)
Salvelinus fontinalis (7)
Thymallus baicalensis (5, 7)
Thymallus thymallus (2, 3, 4, 9)

Distribution: **SR:** D – Rivers Dunaj (= Danube), Hnilec, Hornád, Hron and Váh, Tisa R. basin, Okna Brook, Dobšínská water reservoir; V – River Dunajec.
Records: 1. Žitňan 1960c; 2. Žitňan 1965c; 3. Žitňan 1966d; 4. Žitňan 1970–1971; 5. Žitňan 1973a; 6. Žitňan 1973b; 7. Žitňan 1974b; 8. Žitňan 1979e; 9. Žitňan 1980a; 10. Žitňan 1986; 11. Moravec 1995a.

Plagioporus sp.
Site: intestine.
Host: *Vimba vimba* (1, 2)
Distribution: **SR:** D – Lake Ököri.
Records: 1. Kašťák 1956d; 2. Kašťák 1957d.
Remark: Probably a misidentified *Allocreadium* sp.

Sphaerostomum bramae (Müller, 1776)
Syn. *Sphaerostoma bramae* (Müller, 1776)
Site: intestine.
Hosts: *Abramis bjoerkna* (1, 15, 16)
Abramis brama (5, 6, 7, 8, 9, 12, 13, 15)
Alburnus alburnus (16)
Aspius aspius (2, 3, 4, 17)
Leuciscus cephalus (6, 10, 11, 12, 13, 19, 20)
Leuciscus leuciscus (15)
**Perca fluviatilis* (18)
Rutilus rutilus (5, 6, 15)
Silurus glanis (5)
Vimba vimba (15)
Distribution: **CR:** L – localities not given; D – Rivers Dyje, Jihlava, Morava, Svitava and Svratka, ponds in southern Moravia; O – Oder R. basin; **SR:** D – Rivers Dunaj (= Danube), Laborec, Latorica and Tisa, surroundings of Komárno, localities Čierna Voda and Senné, Tisa R. lowlands.
Records: 1. Vojtek et al. 1954; 2. Kašťák 1956d; 3. Kašťák 1957b; 4. Kašťák 1957d; 5. Vojtek 1959b; 6. Vojtková 1959b; 7. Pacák 1962; 8. Dyk and Lucký 1963; 9. Lucký and Dyk 1964; 10. Žitňan 1964f; 11. Žitňan 1965a; 12. Žitňan 1966b; 13. Žitňan 1966c; 14. Žitňan 1968c; 15. Ergens and Lom 1970; 16. Žitňan 1979e; 17. Gelnar et al. 1994; 18. Moravec 1995a; 19. Šebelová et al. 1995; 20. Gelnar et al. 1997.

Sphaerostomum globiporum (Rudolphi, 1802)
Syn. *Distomum globiporum* Rudolphi, 1802
Site: intestine.
Hosts: *Abramis ballerus* (5)
Abramis bjoerkna (3)
Abramis brama (2, 3, 4, 5)
Aspius aspius (6)
Cyprinus carpio (4)
Leuciscus cephalus (2, 3, 4, 5)
Leuciscus idus (4)
Leuciscus leuciscus (3, 4)
Perca fluviatilis (1, 2)
Rutilus rutilus (3, 4)
Scardinius erythrophthalmus (4)
Tinca tinca (4)
Vimba vimba (3, 4)
Distribution: **CR:** L – River Labe (= Elbe), Bohemia; **SR:** D – Rivers Dunaj (= Danube) and Latorica.
Records: 1. John 1877; 2. Šrámek 1900; 3. Šrámek 1901, 1903; 4. Frič 1908; 5. Ergens et al. 1975; 6. Žitňan 1979e.
Remark: *Sphaerostomum globiporum* may be a junior synononym of *S. bramae*. Probably both these forms were confused in the Czech and Slovak literature.

LARVAE

Fam. Bucephalidae Poche, 1907

Bucephalus polymorphus Baer, 1827 metacercariae
Syn. *Bucephalus* sp. of Dyk and Lucký (1963) and Lucký and Dyk (1964)
Site: gills and subcutaneous tissue
Hosts: *Abramis ballerus* (2)
Abramis bjoerkna (2, 5, 6)
Abramis brama (2, 3, 5, 6)
Alburnus alburnus (3, 5, 6)
Gobio gobio (1, 6)
Leuciscus cephalus (1, 5, 6)
Pelecus cultratus (2)
Rutilus rutilus (1, 2, 5, 6)
Scardinius erythrophthalmus (2)
Stizostedion lucioperca (5, 6)
Tinca tinca (5, 6)
Distribution: **CR:** D – Rivers Dyje, Rokytná and Svratka; **SR:** D – surroundings of Komárno.
Records: 1. Lucký 1959a; 2. Vojtek 1959b; 3. Vojtková 1959b; 4. Vojtek 1961; 5. Dyk and Lucký 1963; 6. Lucký and Dyk 1964; 7. Vojtek in Ergens and Lom 1970.

Rhipidocotyle campanula (Dujardin, 1845) metacercariae
Syn. *Rhipidocotyle illense* (Ziegler, 1883)
Site: gills, fins, musculature, eyes, brain, subcutaneous connective tissue.
Hosts: *Abramis bjoerkna* (4)
Abramis brama (2, 4)
Abramis sapa (4)
Alburnus alburnus (2, 4, 13)
Aristichthys nobilis (5, 6, 7)
Aspius aspius (2)
Barbus petenyi (10)
Chondrostoma nasus (4)
Ctenopharyngodon idella (5, 6, 7)
Cyprinus carpio (2, 8, 9, 11)
Gobio albipinnatus (2)
Gobio kessleri (2)
Gobio uranoscopus (10)
Leucaspius delineatus (2, 12)
Leuciscus cephalus (1)
Leuciscus idus (4)
Leuciscus leuciscus (1, 2)
Lota lota (12)
Perca fluviatilis (14)
Phoxinus phoxinus (12)
Rhodeus sericeus (2)
Rutilus rutilus (1, 2, 3, 4, 11, 13, 15)
Scardinius erythrophthalmus (1)
Stizostedion volgense (12)
Vimba vimba (2)
Distribution: **CR:** D – River Morava; **SR:** D – Rivers Dunaj (= Danube), Latorica and Tisa, southern Slovakia, Slovak ponds and water reservoirs.
Records: 1. Ergens 1962a; 2. Žitňan 1966b; 3. Vojtek in Ergens and Lom 1970; 4. Pár and Vojtek 1972; 5. Žitňan 1972a; 6. Žitňan 1974a; 7. Žitňan 1974b; 8. Žitňan 1978; 9. Žitňan 1979d; 10. Žitňan 1981a; 11. Žitňan 1985; 12. Žitňan 1986; 13. Gelnar et al. 1994; 14. Moravec 1995a; 15. Ondračková and Jurajda 2000.

Fam. Echinostomatidae Dietz, 1909

Echinostomatidae gen. sp. metacercariae
Site: gills.
Hosts: *Gymnocephalus cernuus* (1, 2, 3, 4)
Umbra krameri (5, 8)
Distribution: **SR:** D – Rivers Dunaj (= Danube) and Tisa, Čierna voda in Senné.
Records: 1. Kašťák 1956d; 2. Kašťák 1956e; 3. Kašťák 1957b; 4. Kašťák 1957d; 5. Žitňan 1966b; 6. Žitňan 1968c; 7. Vojtek 1974; 8. Žitňan 1985.

Fam. Diplostomidae Poirier, 1886

Diplostomum indistinctum (Guberlet, 1923) metacercariae
Site: lens of eye.
Host: *Hypophthalmichthys molitrix* (1)
Distribution: **CR:** D – ponds near Pohořelice.
Record: 1. Lucký 1981b.

Diplostomum spathaceum (Rudolphi, 1819) metacercariae
Syn. *Diplostomulum spathaceum* (Rudolphi, 1819); *Proalaria spathacea* (Rudolphi, 1819)
Site: lens of eye.
Hosts: *Abramis ballerus* (23, 28, 37, 44, 50)
Abramis bjoerkna (1, 2, 7, 13, 16, 18, 21, 23, 25, 36, 37, 44)
Abramis brama (1, 2, 7, 13, 15, 16, 20, 21, 23, 24, 25, 26, 36, 37, 44, 45, 46, 47, 50, 51, 52, 53, 57)
Abramis sapa (23, 25, 37, 44, 50, 55)
Acipenser gueldenstadti (53)
Acipenser ruthenus (37, 50, 53)
Alburnoides bipunctatus (55)
Alburnus alburnus (1, 15, 16, 18, 20, 21, 23, 25, 36, 44, 50, 57)
Aristichthys nobilis (38, 42, 43, 49)
Aspius aspius (2, 7, 16, 25, 36, 37, 44, 50, 57)
Barbatula barbatula (16, 25, 50, 55)
Barbus barbus (25, 39)
Carassius auratus gibelio (43)
Carassius carassius (13, 16, 23, 25, 51, 52)
Chondrostoma nasus (15, 20, 21, 23, 25, 37, 44, 50, 58)
Cottus gobio (55)
Ctenopharyngodon idella (29, 34, 38, 42, 43, 49)

Cyprinus carpio (1, 2, 7, 10, 13, 14, 16, 17, 18, 20, 21, 23, 24, 25, 26, 36, 37, 45, 46, 47, 51, 52, 53)
Esox lucius (16, 22, 23, 24, 45, 46, 47, 50, 51, 52)
Gobio albipinnatus (18, 21, 23, 25, 50)
Gobio gobio (1, 13, 15, 20, 23, 25, 36, 39, 46, 47, 50, 51, 52)
Gobio kessleri (23, 50)
Gymnocephalus cernuus (1, 2, 5, 9, 13, 16, 18, 21, 23, 25, 27, 44, 45, 50, 51, 52, 57)
Gymnocephalus schraetser (3, 6, 18, 21, 23, 25, 37, 50, 55)
Hypophthalmichthys molitrix (49)
Ictalurus nebulosus (18, 21, 23, 25, 43, 44, 50)
Lepomis gibbosus (23, 25, 43, 50)
Leucaspius delineatus (13, 51, 52, 55)
Leuciscus cephalus (1, 16, 18, 20, 21, 23, 25, 36, 37, 44, 46, 47, 51, 52, 57, 58, 59, 60)
Leuciscus idus (16, 25, 36, 37, 44)
Leuciscus leuciscus (1, 13, 15, 20, 23, 57)
Lota lota (2, 5, 13, 16, 23, 25, 50, 55)
Misgurnus fossilis (55)
Oncorhynchus mykiss (8, 19, 39, 43, 50)
Pelecus cultratus (23, 25, 37, 50)
Perca fluviatilis (4, 13, 15, 16, 20, 21, 23, 25, 27, 37, 45, 50)
Phoxinus phoxinus (30, 39, 55)
Proterorhinus marmoratus (7, 50, 55, 61)
Rhodeus sericeus (15, 16, 20, 23, 25, 50, 55)
Rutilus pigus (21, 23, 36, 44, 50)
Rutilus rutilus (1, 2, 5, 7, 13, 15, 16, 18, 20, 21, 23, 25, 26, 36, 37, 44, 45, 46, 47, 48, 50, 51, 52, 53, 54, 57, 62)
Scardinius erythrophtahlmus (1, 7, 15, 16, 20, 25, 37, 44, 45, 50, 51, 52, 54, 57)
Salmo trutta m. *fario* (8)
Silurus glanis (16, 21, 23, 25, 50)
Stizostedion lucioperca (2, 9, 16, 18, 21, 23, 24, 25, 37, 45, 50, 57)
Stizostedion volgense (23, 25, 50, 55)
Tinca tinca (1, 16, 24, 25, 46, 47, 50, 51, 52, 57)
Vimba vimba (1, 23, 25, 37, 50, 57)
Zingel streber (23, 50)
Zingel zingel (18, 23, 25, 50)

Distribution: **CR:** L – Rivers Blanice, Malše, Ohře and Vltava, brook Černovický potok, South-Bohemian ponds (regions of Vodňany and Jindřichův Hradec), Lipno water reservoir, Máchovo jezero pond system, trout farm in Liběchov; D – Rivers Dřevnice and Morava, Kníničská and Mušov water reservoirs, South-Moravian ponds and pools, fish farm on Jihlava R., fish hatchery Třebíč; O – River Opava, brook Hořině; **SR:** D – Rivers Dunaj (= Danube), Hron, Laborec, Latorica and Tisa, Dobšinská and Zemplínská Šírava water reservoirs, surroundings of Komárno, Slovak fishponds.

Records: 1. Romanovský 1954b; 2. Dyk 1955a; 3. Dyk 1955b; 4. Lucký et al. 1955; 5. Dyk 1956a; 6. Vejnar 1956; 7. Vojtek 1959b; 8. Havlena and Pekář 1960; 9. Dyk 1961a; 10. Lucký 1961; 11. Vojtek 1961; 12. Žitňan 1961; 13. Ergens 1962a; 14. Dyk and Lucký 1963; 15. Dyk 1964a; 16. Lucký and Dyk 1964; 17. Žitňan 1964a; 18. Žitňan 1964f; 19. Ergens et al. 1965; 20. Rodák et al. 1965; 21. Žitňan 1965a; 22. Ergens 1966a; 23. Žitňan 1966b; 24. Žitňan 1966c; 25. Žitňan 1966d; 26. Lucký and Jílek 1967; 27. Přibyslavský and Lucký 1967; 28. Žitňan 1967b; 29. Lucký 1968b; 30. Řehulka 1968; 31. Žitňan 1968a; 32. Žitňan 1968c; 33. Vojtek in Ergens and Lom 1970; 34. Žitňan 1970b; 35. Vojtek and Vojtková 1971; 36. Žitňan 1971c; 37. Pár and Vojtek 1972; 38. Žitňan 1972a; 39. Žitňan 1973a; 40. Žitňan 1973b; 41. Vojtek 1974; 42. Žitňan 1974a; 43. Žitňan 1974b; 44. Ergens et al. 1975; 45. Žitňan 1977; 46. Moravec 1978d; 47. Moravec 1979a; 48. Moravec 1980b; 49. Lucký 1981b; 50. Vojtek 1981; 51. Moravec 1984e; 52. Moravec 1985a; 53. Žitňan 1985; 54. Moravec 1986c; 55. Žitňan 1986; 56. Vojtek 1989; 57. Gelnar et al. 1990; 58. Navrátil 1991; 59. Gelnar et al. 1997; 60. Divílek et al. 1998; 61. Koubková and Baruš 2000; 62. Ondračková and Jurajda 2000.

Diplostomum volvens (Nordmann, 1832) metacercariae
Syn. *Diplostomum baeri* Dubois, 1937
Site: eye.
Hosts: *Gymnocephalus cernuus* (1)
Hypophthalmichthys molitrix (2)
Distribution: **CR:** L – Prague; D – ponds near Pohořelice.
Records: 1. John 1877, 2. Lucký 1981b.

Diplostomum sp. metacercariae
Syn. *Diplostomulum* sp.
Hosts: *Abramis ballerus* (8)
Abramis bjoerkna (8, 14, 17)
Abramis brama (8, 14, 17)
Abramis sapa (8, 14)
Acipenser gueldenstaedti (8)
Acipenser ruthenus (8)
Alburnoides bipunctatus (8, 13)
Alburnus alburnus (8, 14)
Aristichthys nobilis (3)
Aspius aspius (8, 14, 17)
Barbatula barbatula (1, 8, 14)
Barbus barbus (8, 18)
Barbus petenyi (8)
Carassius auratus gibelio (8, 14)
Carassius carassius (8, 14)
Chondrostoma nasus (8, 14)
Cobitis taenia (2)
Cottus gobio (8)
Ctenopharyngodon idella (3, 8)
Cyprinus carpio (1, 4, 5, 6, 7, 8, 12, 14, 17)
Esox lucius (8, 14)
Gobio albipinnatus (8, 14)
Gobio gobio (8, 13, 14)
Gobio uranoscopus (8)
Gymnocephalus cernuus (8)
Gymnocephalus schraetser (8)
Ictalurus nebulosus (8)
Lepomis gibbosus (8)
Leucaspius delineatus (8)
Leuciscus cephalus (1, 8, 13, 14, 15, 16, 17)
Leuciscus idus (8, 14)
Leuciscus leuciscus (8, 14, 17)
Lota lota (8)
Misgurnus fossilis (8)
Pelecus cultratus (8)
Perca fluviatilis (3, 8, 14)
Phoxinus phoxinus (8)
Proterorhinus marmoratus (8)
Rhodeus sericeus (8, 14)
Rutilus pigus (8)
Rutilus rutilus (1, 8, 14, 17, 19)
Salmo trutta m. *fario* (9)
Scardinius erythophthalmus (8, 17)
Silurus glanis (8, 14)
Stizostedion lucioperca (8)
Stizostedion volgense (8)
Thymallus thymallus (10)
Tinca tinca (8, 14)
Vimba vimba (8, 14)
Zingel streber (8)
Zingel zingel (8)
Distribution: **CR:** L – Rivers Malše and Vltava; D – Rivers Jihlava, Morava and Rokytná, Morava R. basin, South-Moravian ponds; **SR:** D – River Dunaj (= Danube), Váh River basin, Slovak ponds.
Records: 1. Lucký 1959a; 2. Vojtek 1959b; 3. Tesarčík and Ehl 1968; 4. Žitňan 1978; 5. Žitňan 1979b; 6. Žitňan 1979c; 7. Žitňan 1979d; 8. Žitňan 1979e; 9. Žitňan 1980a; 10. Žitňan 1980b; 11. Žitňan 1981a; 12. Žitňan 1982b; 13. Moravec and Scholz 1991; 14. Gelnar et al. 1994; 15. Koníček and Gelnar 1994; 16. Šebelová et al. 1995; 17. Moravec et al. 1997a; 18. Moravec et al. 1997a; 19. Kadlec et al. 1998.

Tylodelphys clavata (Nordmann, 1832) metacercariae
Syn. *Diplostomum clavatum* Nordmann, 1832; *Diplostomulum clavatum* (Nordmann, 1832)
Site: humour of eye.
Hosts: *Abramis ballerus* (23, 32)
Abramis bjoerkna (9, 10, 23, 32)
Abramis brama (5, 10, 23, 29, 30, 31, 32, 37)
Abramis sapa (23, 32)
Alburnoides bipunctatus (5)
Alburnus alburnus (10)
Aspius aspius (10)

Barbatula barbatula (5, 10)
Barbus barbus (13, 14, 32)
Carassius carassius (10)
Chondrostoma nasus (5, 14, 16, 32)
Cobitis taenia (31, 36)
Cyprinus carpio (10, 11, 16, 32, 36)
Esox lucius (9, 10, 12, 13, 14, 15, 16, 29, 30, 31, 32)
Gobio gobio (2, 5, 9, 32)
Gymnocephalus cernuus (1, 4, 9, 10, 12, 13, 14, 16, 17, 27, 28, 32, 35)
Gymnocephalus schraetser (12, 13, 14, 16, 31, 32, 38)
Lepomis gibbosus (16, 26)
Leuciscus cephalus (5, 6, 9, 10, 23, 27, 29, 30, 31, 32, 40, 41, 43, 44, 45, 46)
Leuciscus idus (10, 16, 23, 31, 32)
Leuciscus leuciscus (9, 29)
Lota lota (7, 9, 10, 16, 32, 38)
Misgurnus fossilis (38)
Perca fluviatilis (1, 3, 4, 9, 10, 16, 23, 27, 29, 31, 32, 33, 34, 35, 39)
Phoxinus phoxinus (5)
Proterorhinus marmoratus (47)
Rhodeus sericeus (10)
Rutilus pigus (23, 31, 32)
Rutilus rutilus (2, 5, 9, 10, 13, 14, 23, 27, 29, 30, 31, 32, 36, 37)
Scardinius erythrophthalmus (2, 10, 23, 29, 30, 32, 37)
Silurus glanis (7, 10, 16)
Stizostedion lucioperca (10, 13, 14, 15, 16, 19, 28, 31, 33)
Stizostedion volgense (16, 38)
Tinca tinca (5, 10, 14, 15, 32)
Vimba vimba (31)

Distribution: **CR:** L – River Malše, Máchovo jezero pond system; D – Rivers Dřevnice, Dyje, Jihlava, Morava, Rokytná, Svitava and Svratka, Kníničská and Mušov water reservoirs, small water reservoirs in Morava R. system, South-Moravian ponds, fish hatchery Třebíč; O – localities not specified; **SR:** D – Rivers Dunaj (= Danube), Bodva, Hron, Laborec, Latorica and Tisa, Zemplínská Šírava water reservoir, southern Slovakia, ponds near Levice, Tisa R. lowlands.

Records: 1. Dyk 1954c; 2. Vojtek et al. 1954; 3. Dyk 1955b; 4. Vejnar 1956; 5. Vojtková 1959b; 6. Žitňan 1960c; 7. Dyk 1961a; 8. Vojtek 1961; 9. Ergens 1962a; 10. Dyk and Lucký 1963; 11. Lucký and Dyk 1964; 12. Žitňan 1964f; 13. Žitňan 1965a; 14. Žitňan 1966b; 15. Žitňan 1966c; 16. Žitňan 1966d; 17. Přibyslavský and Lucký 1967; 18. Žitňan 1967b; 19. Žitňan 1968a; 20. Žitňan 1968c; 21. Vojtek in Ergens and Lom 1970; 22. Vojtek and Vojtková 1971; 23. Pár and Vojtek 1972; 24. Žitňan 1973b; 25. Vojtek 1974; 26. Žitňan 1974b; 27. Ergens et al. 1975; 28. Žitňan 1977; 29. Moravec 1978d; 30. Moravec 1979a; 31. Žitňan 1979e, 32. Vojtek 1981; 33. Lucký and Navrátil 1983; 34. Lucký and Navrátil 1984; 35. Moravec and Scholz 1984; 36. Žitňan 1985; 37. Moravec 1986c; 38. Žitňan 1986; 39. Scholz 1987b, 40. Gelnar et al. 1990; 41. Moravec and Scholz 1991; 42. Navrátil 1991; 43. Gelnar et al. 1994; 44. Koníček and Gelnar 1994; 45. Divílek et al. 1998; 46. Kadlec et al. 1998; 47. Koubková and Baruš 2000.

Tylodelphys podicipina Kozicka et Niewiadomska, 1960 metacercariae
Site: humour of eye.
Hosts: *Gymnocephalus cernuus* (1)
Perca fluviatilis (1, 2)
Distribution: **CR:** L – Máchovo jezero pond system.
Records: 1. Moravec and Scholz 1984; 2. Scholz 1987b.

Hysteromorpha triloba (Rudolphi, 1819) metacercariae
Site: musculature, less often under skin.
Hosts: *Abramis ballerus* (1, 2, 4)
Abramis bjoerkna (1, 2, 4)
Abramis brama (1, 2, 4)
Abramis sapa (1, 2, 4)
Rutilus rutilus (1, 2, 4)
Distribution: **SR:** D – Rivers Dunaj (= Danube) and Morava, irrigation canals in southern Slovakia.
Records: 1. Vojtek and Vojtková 1971; 2. Pár and Vojtek 1972; 3. Vojtek 1974; 4. Vojtek 1981.

Posthodiplostomum brevicaudatum (Nordmann, 1832) metacercariae
Syn. *Neascus brevicaudatus* Nordmann, 1832; *Distomum retroconstrictum* Šrámek, 1901
Site: eye.
Hosts: *Abramis brama* (1)
Cyprinus carpio (2, 4)
Distribution: **CR:** L – River Labe (= Elbe); D – River Dyje.
Records: 1. Šrámek 1901, 1903; 2. Lucký and Dyk 1964; 3. Vojtek 1974; 4. Vojtek 1981.

Posthodiplostomum cuticola (Nordmann, 1832) metacercariae
Syn. *Neascus cuticola* Nordmann, 1832; *Neodiplostomum cuticola* (Nordmann, 1832)
Site: skin, subcutaneous connective tissue, musculature, brain.
Hosts: *Abramis ballerus* (16, 27, 39)
Abramis bjoerkna (3, 7, 12, 13, 16, 18, 23, 26, 27, 38, 39, 47, 52)
Abramis brama (7, 11, 12, 13, 16, 17, 18, 23, 26, 27, 33, 38, 39, 47, 52)
Abramis sapa (18, 27, 38, 39, 45, 47)
Alburnus alburnus (2, 4, 8, 12, 13, 16, 18, 23, 39, 47)
Aspius aspius (7, 15, 16, 38, 39)
Barbatula barbatula (3, 39)
Barbus barbus (27, 39)
Carassius auratus gibelio (38)
Carassius carassius (12, 13, 16, 18, 23, 39)
Chondrostoma nasus (16, 26, 27, 38, 39)
Cobitis elongatoides (53)
Cobitis elongatoides × *Cobitis* sp. (53)
Cobitis taenia (16, 38, 39, 44)
Ctenopharyngodon idella (24, 28, 31, 32, 38)
Cyprinus carpio (7, 12, 13, 14, 15, 16, 17, 18, 23, 26, 33, 34, 35, 36, 37, 38, 39, 41, 42, 44)
Gobio gobio (38)
Leucaspius delineatus (16, 38, 39, 45)
Leuciscus cephalus (16, 26, 27, 38, 39, 47, 48, 49, 50)
Leuciscus idus (9, 16, 18, 27, 38, 39)
Leuciscus leuciscus (38)
Misgurnus fossilis (12, 13, 16, 18, 23, 38, 39, 45)
Pelecus cultratus (16, 18, 27, 38, 39, 43)
Phoxinus phoxinus (1, 6, 8)
Rhodeus sericeus (16, 39, 45)
Rutilus pigus (15, 16, 18, 27, 38, 39)
Rutilus rutilus (2, 3, 4, 7, 8, 12, 13, 16, 18, 23, 27, 38, 39, 44, 47, 51, 52)
Scardinius erythrophthalmus (1, 5, 6, 7, 13, 16, 18, 23, 27, 38, 39, 52)
Vimba vimba (27, 38, 39, 47)
Distribution: **CR:** L – localities not specified; D – Rivers Dyje and Morava, brook near Frenštát, Morava R. basin, Vranovská water reservoir, South-Moravian pools and ponds; O – localities not specified; **SR:** D – Rivers Dunaj (= Danube), Hron, Váh, Laborec, Latorica, Ondava and Tisa, pools and irrigation canal near Komárno, southern Slovakia, Zemplínská Šírava water reservoir, Tisa R. lowlands, Slovak ponds.
Records: 1. Dyk 1954c; 2. Michalovič 1954; 3. Vojtek et al. 1954; 4. Dyk 1955b; 5. Lucký 1955d; 6. Dyk 1956a; 7. Vojtek 1959b; 8. Dyk 1961a; 9. Pacák 1961; 10. Vojtek 1961; 11. Pacák 1962; 12. Dyk and Lucký 1963; 13. Lucký and Dyk 1964; 14. Žitňan 1964a; 15. Žitňan 1965a; 16. Žitňan 1966b; 17. Žitňan 1966c; 18. Žitňan 1966d; 19. Žitňan 1967b; 20. Žitňan 1968a; 21. Žitňan 1968c; 22. Vojtek in Ergens and Lom 1970; 23. Lucký 1970; 24. Žitňan 1970b; 25. Vojtek and Vojtková 1971; 26. Žitňan 1971c; 27. Pár and Vojtek 1972; 28. Žitňan 1972a; 29. Žitňan 1973b; 30. Vojtek 1974; 31. Žitňan 1974a; 32. Žitňan 1974b; 33. Žitňan 1977; 34. Žitňan 1978; 35. Žitňan 1979b; 36. Žitňan 1979c; 37. Žitňan 1979d; 38. Žitňan 1979e; 39. Vojtek 1981; 40. Žitňan 1981a; 41. Žitňan et al. 1981; 42. Žitňan 1982a; 43. Žitňan 1985; 44. Žitňan 1985; 45. Žitňan 1986; 46. Ondračková et al. 1993; 47. Gelnar et al. 1994; 48. Koníček and Gelnar 1994; 49. Šebelová et al. 1995; 50. Gelnar et al. 1997; 51. Kadlec et al. 1998; 52. Bartošová et al. 1999; 53. Halačka et al. 2000.

Fam. Strigeidae Railliet, 1919

Apharyngostrigea cornu (Zeder, 1800) metacercariae
Syn. *Apharyngostrigea* sp. of Vojtek (1959b), Dyk (1961a) and Lucký and Jílek (1967)
Site: musculature.
Hosts: *Abramis bjoerkna* (4, 5, 11, 12, 19)
Abramis brama (1, 2, 5, 10, 11, 12, 19)
Abramis sapa (11, 12, 19, 21)
Ctenopharyngodon idella (13, 15, 16, 18)
Cyprinus carpio (11, 12, 19)
Leuciscus cephalus (21, 22)
Leuciscus idus (11, 12, 19)
Rutilus rutilus (6, 10, 18, 19)
Scardinius erythrophthalmus (1, 2, 10, 11, 12, 17, 18, 19)
Vimba vimba (11, 12, 19)
Distribution: **CR:** L – localities not specified, D – River Morava, Kníničská water reservoir, South Moravia, O – localities not specified, **SR:** D – Rivers Dunaj (= Danube), Laborec and Latorica, Tisa R. basin, Zemplínská Šírava water reservoir, surrounding of Komárno, ponds near Hroboňovo, southern Slovakia.
Records: 1. Vojtek 1959b; 2. Dyk 1961a; 3. Vojtek 1961; 4. Žitňan 1965a; 5. Žitňan 1966b; 6. Lucký and Jílek 1967; 7. Žitňan 1967c; 8. Žitňan 1968a; 9. Žitňan 1968c; 10. Vojtek in Ergens and Lom 1970; 11. Vojtek and Vojtková 1971; 12. Pár and Vojtek 1972; 13. Žitňan 1972a; 14. Vojtek 1974; 15. Žitňan 1974a; 16. Žitňan 1974b; 17. Žitňan 1977; 18. Žitňan 1979e; 19. Vojtek 1981; 20. Žitňan 1981a; 21. Gelnar et al. 1994; 22. Koníček and Gelnar 1994.

Ichthyocotylurus pileatus (Rudolphi, 1802) metacercariae
Syn. *Tetracotyle ovata* Linstow, 1877; *T. variegata* (Creplin, 1825); *Tetracotyle* sp. of Vojtek et al. (1954)
Site: swimbladder wall, region of heart, kidneys.
Hosts: *Abramis ballerus* (26)
Abramis bjoerkna (12, 13, 20)
Abramis brama (2, 11, 15, 16, 20, 23)
Aspius aspius (25, 26)
Carassius carassius (1, 2, 12, 13, 20, 23)
Chondrostoma nasus (19)
Cyprinus carpio (12, 13, 20, 22, 23)
Esox lucius (23)
Gobio gobio (15, 16, 19, 20, 23)
Gymnocephalus cernuus (3, 9, 10, 18, 19, 20, 23)
Gymnocephalus schraetser (26)
Leucaspius delineatus (23)
Leuciscus cephalus (6, 10, 15, 20, 23)
Leuciscus idus (15, 16, 19, 20)
Perca fluviatilis (19)
Rutilus rutilus (2, 3, 20, 23, 27)
Scardinius erythrophthalmus (1, 2, 3, 4, 5, 7, 8, 10, 12, 13, 20, 23, 24)
Stizostedion lucioperca (26)
Tinca tinca (1, 2, 10, 23)
Vimba vimba (26)
Distribution: **CR:** L – Máchovo jezero pond system, pond Velký Tisý near Třeboň; D – River Dyje, Morava R. basin, South-Moravian ponds near Lednice and Pohořelice, fish farm on Jihlava R.; **SR:** D – River Dunaj (= Danube) and surroundings, Latorica River.
Records: 1. Dyk 1953a; 2. Lucký 1953a; 3. Dyk 1954c; 4. Dyk and Lucký 1954a; 5. Dyk and Lucký 1954e; 6. Vojtek et al. 1954; 7. Lucký et al. 1955; 8. Dyk 1956a; 9. Vejnar 1956; 10. Dyk 1961a; 11. Pacák 1962; 12. Dyk and Lucký 1963; 13. Lucký and Dyk 1964; 14. Vojtek in Ergens and Lom 1970; 15. Vojtek and Vojtková 1971; 16. Pár and Vojtek 1972; 17. Vojtek 1974; 18. Ergens et al. 1975; 19. Žitňan 1979e; 20. Vojtek 1981; 21. Žitňan 1981a; 22. Moravec 1984e; 23. Moravec 1985; 24. Moravec 1986c; 25. Navrátil 1991; 26. Moravec 1995a; 27. Kadlec et al. 1998.

Ichthyocotylurus platycephalus (Creplin, 1852) metacercariae
Syn. *Tetracotyle percae fluviatilis* Moulinié, 1856; *T. percae* Zschokke, 1884; *T. musculorum percae* Maturé, 1910

Site: musculature.
Hosts: *Abramis sapa* (10)
Gymnocephalus cernuus (3, 6, 9)
Gymnocephalus schraetser (1, 2, 9)
Perca fluviatilis (1, 2, 9)
Stizostedion lucioperca (1, 2, 9)
Distribution: **CR:** D – River Morava; **SR:** D – Rivers Dunaj (= Danube) and Hron, pools and irrigation canal of Váh R. near Komárno, southern Slovakia.
Records: 1. Dyk 1955b; 2. Vejnar 1956; 3. Vojtek 1959b; 4. Dyk 1961a; 5. Vojtek 1961; 6. Žitňan 1966d; 7. Vojtek in Lom and Ergens 1970; 8. Vojtek 1974; 9. Vojtek 1981; Gelnar 1994.

Apatemon cobitidis (Linstow, 1890) metacercariae
Syn. *Tetracotyle cobitidis* (Linstow, 1890); *Tetracotyle* sp. of Vojtková (1959b); *Apatemon cobitidis proterorhini* Vojtek, 1964; *Opisthorchis felineus* (Ribolta, 1884) of Vojtek et al. (1954)
Site: abdominal cavity, musculature, nervous system, liver, kidneys, gonads.
Hosts: *Barbatula barbatula* (1, 3, 6, 7, 8, 9, 10, 11, 13, 14, 15)
Cottus gobio (6, 7, 9, 10, 11, 13, 14, 15)
**Phoxinus phoxinus* (16)
Proterorhinus marmoratus (2, 4, 6, 7, 9, 10, 11, 13, 14, 15, 17)
Distribution: **CR:** D – River Svitava, South-Moravian brooks and ponds, Mušov water reservoir; O – River Osoblaha; **SR:** D – River Dunaj (= Danube), pools and irrigation canal of Váh R., southern Slovakia.
Records: 1. Vojtek et al. 1954; 2. Vojtek 1959b; 3. Vojtková 1959b; 4. Dyk 1961a; 5. Vojtek 1961; 6. Vojtek 1964a; 7. Vojtek 1964b; 8. Ergens 1965a; 9. Vojtek in Ergens and Lom 1970; 10. Vojtek and Vojtková 1971; 11. Vojtek 1972; 12. Vojtek 1974; 13. Žitňan 1979e; 14. Vojtek 1981; 15. Žitňan 1986; 16. Moravec 1995a; 17. Koubková and Baruš 2000b.

Fam. Cyathocotylidae Poche, 1925

Holostephanus cobitidis Opravilová, 1968 metacercariae
Site: musculature, abdominal cavity, humour of eye.
Host: *Cobitis taenia* (1, 2, 3, 4, 5, 6, 7)
Distribution: **SR:** D – River Dunaj (= Danube), southern and southwestern Slovakia.
Records: 1. Opravilová 1968; 2. Vojtek in Ergens and Lom 1970; 3. Vojtek and Vojtková 1971; 4. Pár and Vojtek 1972; 5. Vojtek 1974; 6. Žitňan 1979e; 7. Vojtek 1981; 8. Žitňan 1985.

Cyathocotylidae gen. sp. metacercariae
Site: not given.
Hosts: fishes (1).
Distribution: Czechoslovakia (present **CR** and **SR**) (localities not given).
Record: 1. Vojtek 1961.
Remark: Probably identical with *Holostephanus cobitidis* Opravilová, 1968.

Fam. Prohemistomatidae Sudarikov, 1961

Mesostephanus appendiculatus (Price, 1934) metacercariae
Site: musculature, fins.
Hosts: *Abramis brama* (3)
cyprinid fishes (1, 2)
Distribution: Czechoslovakia (present **CR** and **SR**): D – localities not specified; **SR:** D – River Dunaj (= Danube).
Records: 1. Vojtek in Ergens and Lom 1970; 2. Vojtek 1974; 3. Žitňan 1979e.

Paracoenogonimus ovatus Katsurada, 1914 metacercariae
Syn. *Diplostomum hughesi* Markevich, 1934
Site: musculature, gills, brain, liver, kidneys, gonads.
Hosts: *Abramis ballerus* (6, 8, 9)
Abramis bjoerkna (1, 6, 8, 9)
Abramis brama (1, 6, 8, 9)
Abramis sapa (6, 8, 9, 12)
Alburnoides bipunctatus (8, 10, 12)
Alburnus alburnus (8)

Aspius aspius (6, 9)
Barbus barbus (6, 9)
Carassius carassius (1, 9)
Chondrostoma nasus (6, 8, 9)
Esox lucius (4, 6, 8, 9)
Gobio gobio (8)
Gobio uranoscopus (8, 10)
Leucaspius delineatus (8, 12)
Leuciscus cephalus (6, 8, 9)
Leuciscus idus (6, 8, 9)
Leuciscus leuciscus (8)
Pelecus cultratus (8)
Perca fluviatilis (1, 2, 4, 9)
Phoxinus phoxinus (8, 10, 12)
Rhodeus sericeus (8, 12)
Rutilus pigus (6, 8, 9)
Rutilus rutilus (1, 6, 8, 9, 11)
Scardinius erythrophthalmus (1, 6, 8, 9)
Stizostedion lucioperca (6, 8, 9)
Stizostedion volgense (8, 12)
Tinca tinca (1, 9)
Vimba vimba (6, 8)

Distribution: **CR:** L – localities not specified; **SR:** D – River Dunaj (= Danube), southern Slovakia.
Records: 1. Vojtek 1959b; 2. Dyk 1961a; 3. Vojtek 1961; 4. Vojtek in Ergens and Lom 1970; 5. Vojtek and Vojtková 1971; 6. Pár and Vojtek 1972; 7. Vojtek 1974; 8. Žitňan 1979e; 9. Vojtek 1981; 10. Žitňan 1981a; 11. Žitňan 1985; 12. Žitňan 1986.

Prohemistomum sp. metacercariae
Site: musculature, fins, gills, mesenteries.
Host: *Leuciscus idus* (1)
Distribution: **SR:** D – southern Slovakia.
Records: 1. Pár and Vojtek 1972; 2. Vojtek 1974.

Fam. Clinostomatidae Lühe, 1901

Clinostomum complanatum (Rudolphi, 1819) metacercariae
Site: subcutaneous musculature, abdominal cavity.
Hosts: *Cobitis taenia* (1, 3, 6, 7)
Perca fluviatilis (3, 7)
Rhodeus sericeus (9)
Scardinius erythrophthalmus (3, 5, 7)

Distribution: **CR:** L – localities not given; D – River Morava; **SR:** D – Rivers Dunaj (= Danube) and Latorica, surroundings of Komárno.
Records: 1. Vojtek 1959b; 2. Vojtek 1961; 3. Vojtek in Ergens and Lom 1970; 4. Vojtek 1974; 5. Ergens et al. 1975; 6. Žitňan 1979e; 7. Vojtek 1981; 8. Žitňan 1985; 9. Gelnar et al. 1994.

Fam. Opisthorchidae Lühe, 1911

Metorchis xanthosomus (Creplin, 1846) metacercariae
Syn. *Metorchis intermedius* Heinemann, 1937; *Opisthorchis* sp. of Řehulka (1968)
Site: musculature, gills, liver.
Host: *Cobitis taenia* (3)
Phoxinus phoxinus (1)

Distribution: **CR:** O – Hořina Brook; **SR:** D – surroundings of Komárno, southern Slovakia.
Records: 1. Řehulka 1968; 2. Vojtek in Ergens and Lom 1970; 3. Pár and Vojtek 1972; 4. Vojtek 1974; 5. Vojtek 1981.

Fam. Heterophyidae Odhner, 1914

Apophallus donicus (Skryabin et Lindtrop, 1919) metacercariae
Syn. *Rossicotrema donica* Skryabin et Lindtrop, 1919
Site: skin, under scales, fins.
Hosts: *Alburnoides bipunctatus* (1, 2)
Alburnus alburnus (1, 2)
Gymnocephalus cernuus (3, 7, 16, 17)
Gymnocephalus schraetser (7, 13, 16, 17, 19)
Lepomis gibbosus (15)
Leuciscus cephalus (20, 21)
Perca fluviatilis (2, 3, 7, 16, 17)
Stizostedion lucioperca (7, 8, 16, 17)
Stizostedion volgense (7, 8, 16, 17, 19)
Zingel zingel (7, 8, 16, 17, 18)

Distribution: **CR:** L – localities not given; D – River Morava; O – localities not given; **SR:** D – Rivers Dunaj (= Danube), Hron, Ipeľ and tributories, Laborec, Latorica and

Tisa, surroundings of Komárno, southern Slovakia, Tisa R. lowlands.
Records: 1. Gotthardt 1954; 2. Dyk 1955b; 3. Vojtek 1959a; 4. Vojtek 1959c; 5. Vojtek 1961; 6. Žitňan 1964g; 7. Žitňan 1966d; 8. Žitňan 1966b; 9. Žitňan 1967b; 10. Žitňan 1968a; 11. Žitňan 1968c; 12. Vojtek in Ergens and Lom 1970; 13. Pár and Vojtek 1972; 14. Vojtek 1974; 15. Žitňan 1974b; 16. Žitňan 1979e; 17. Vojtek 1981; 18. Žitňan 1985; 19. Žitňan 1986; 20. Gelnar et al. 1994; 21. Koníček and Gelnar 1994.

Apophallus muehlingi (Jägerskiöld, 1898) metacercariae
Site: fins, gills
Hosts: *Abramis ballerus* (16, 19, 20)
Abramis bjoerkna (1, 2, 4, 6, 7, 9, 10, 16, 18, 20, 23)
Abramis brama (2, 9, 10, 16, 18, 20)
Abramis sapa (10, 16, 19, 20, 22)
Alburnoides bipunctatus (11, 19, 20, 22)
Alburnus alburnus (7, 9, 10, 11, 16, 19, 20, 23)
Aspius aspius (18, 19)
Chondrostoma nasus (16, 18, 19, 20)
Esox lucius (18)
Gobio gobio (19)
Leucaspius delineatus (19, 22)
Leuciscus cephalus (10, 16, 18, 19, 20)
Leuciscus idus (10, 16, 19, 20)
Leuciscus leuciscus (11, 19)
Rhodeus sericeus (1, 2, 6, 11, 18, 19, 20, 22)
Rutilus pigus (18, 19)
Rutilus rutilus (2, 6, 9, 10, 16, 18, 20, 21)
Scardinius erythrophthalmus (2, 9, 10, 16, 18, 19, 20)
Tinca tinca (2, 19, 20)
Vimba vimba (16, 19, 20)
Distribution: **CR:** D – River Dyje and nearby pools, Morava River; **SR:** D – Rivers Dunaj (= Danube), Hron, Loborec, Latorica and Tisa, surroundings of Komárno, southern Slovakia, Tisa R. lowlands.
Records: 1. Lucký 1957a; 2. Vojtek 1959b; 3. Vojtek 1959c; 4. Dyk 1961a; 5. Vojtek 1961; 6. Lucký and Dyk 1964; 7. Žitňan 1964f; 8. Žitňan 1964g; 9. Žitňan 1965a; 10. Žitňan 1966b; 11. Žitňan 1966d; 12. Žitňan 1967b; 13. Žitňan 1968a; 15. Žitňan 1968c; 16. Vojtek in Ergens and Lom 1970; 17. Pár and Vojtek 1972; 18. Vojtek 1974; 19. Ergens et al. 1975; 20. Žitňan 1979e; 21. Vojtek 1981; 22. Žitňan 1985; 23. Žitňan 1986; 24. Gelnar et al. 1994.

Metagonimus yokogawai Katsurada, 1912 metacercariae
Site: scales, fins.
Hosts: *Abramis ballerus* (4, 6, 14, 15)
Abramis bjoerkna (2, 4, 6, 9, 14, 15)
Abramis brama (1, 2, 4, 6, 9, 14, 15, 17)
Abramis sapa (4, 6, 14, 15)
Acipenser ruthenus (6, 14, 16, 18)
Alburnoides bipunctatus (6, 14, 15)
Alburnus alburnus (4, 5, 6, 9, 14, 15)
Anguilla anguilla (14)
Aspius aspius (1, 2, 4, 5, 6, 9, 14, 15, 17)
Barbatula barbatula (6)
Barbus barbus (6, 14, 17)
Barbus petenyi (6)
Carassius auratus (14)
Carassius carassius (6, 12)
Chondrostoma nasus (4, 6, 9, 14, 17)
Cobitis taenia (6)
Ctenopharyngodon idella (10, 12, 13, 14, 17, 19)
Cyprinus carpio (2, 6, 9, 14, 15, 17)
Esox lucius (6, 14, 17)
Gobio albipinnatus (6, 14, 15)
Gobio gobio (6, 9, 14)
Gobio kessleri (6)
Gobio uranoscopus (14, 16)
Gymnocephalus cernuus (6, 14, 15)
Gymnocephalus schraetser (6, 14, 15)
Ictalurus nebulosus (6)
Lepomis gibbosus (6)
Leucaspius delineatus (14)
Leuciscus cephalus (4, 5, 6, 9, 14, 17)
Leuciscus idus (4, 5, 6, 9, 14)
Leuciscus leuciscus (6, 14, 16)
Lota lota (6)
Misgurnus fossilis (6)

Pelecus cultratus (4, 6, 14)
Perca fluviatilis (6, 14, 15)
Phoxinus phoxinus (14, 16)
Rhodeus sericeus (6, 14)
Rutilus pigus (4, 5, 6, 14, 15)
Rutilus rutilus (2, 4, 5, 6, 9, 14, 15)
Scardinius erythrophthalmus (2, 4, 5, 6, 14, 15)
Silurus glanis (6, 14, 15, 17)
Stizostedion lucioperca (4, 6, 14, 15, 17)
Stizostedion volgense (4, 6, 14, 15)
Tinca tinca (6)
Vimba vimba (2, 4, 5, 6, 14, 15)
Zingel zingel (6, 14, 15, 18)
Distribution: **SR:** D – Rivers Dunaj (= Danube), Hron, Ipeľ, Nitra, Váh, Žitava, surroundings of Komárno, southern Slovakia.
Records: 1. Vojtek 1959a; 2. Vojtek 1959b; 3. Vojtek 1959c; 4. Žitňan 1959; 5. Žitňan 1960b; 6. Žitňan 1960d; 7. Vojtek 1961; 8. Žitňan 1964g; 9. Žitňan 1971c; 10. Žitňan 1972a; 11. Vojtek 1974; 12. Žitňan 1974a; 13. Žitňan 1974b; 14. Žitňan 1979e; 15. Vojtek 1981; 16. Žitňan 1981a; 17. Žitňan 1981b; 18. Žitňan 1985; 19. Moravec 1995a.

NEMATODA

Fam. Capillariidae Railliet, 1915

Pseudocapillaria salvelini (Polyansky, 1952)
Syn. *Capillaria salvelini* Polyansky, 1952
Site: intestine.
Host: *Salmo trutta* m. *fario* (1, 2, 3, 4)
Distribution: **CR**: L – Kamenice R. near Hřensko.
Records: 1. Moravec 1980d; 2. Moravec 1980e; 3. Moravec 1984d; 4. Moravec 1987; 5. Moravec 1994.

Pseudocapillaria tomentosa (Dujardin, 1843)
Syn. *Capillaria brevispicula* (Linstow, 1873)
Site: intestine.
Hosts: *Alburnoides bipunctatus* (13, 17)
Alburnus alburnus (14)
Barbatula barbatula (4, 7, 11, 14)
Barbus barbus (6)
Carassius auratus gibelio (14)
Carassius carassius (14)
Ctenopharyngodon idella (17)
Cyprinus carpio (4, 5, 7, 8, 9, 10, 11, 18)
Gobio albipinnatus (6, 14)
Gobio gobio (2, 3, 4, 6, 7, 11, 13, 14)
Leuciscus cephalus (4, 7, 11, 12, 13, 14, 15)
Leuciscus leuciscus (2, 11, 12)
Phoxinus phoxinus (4, 7, 11)
Proterorhinus marmoratus (19)
Pseudorasbora parva (14)
Rutilus rutilus (1, 2, 7, 11)
Tinca tinca (3, 4, 5, 7, 8, 10, 11)
Distribution: **CR**: L – River Malše and Třeboň, Jindřichův Hradec and Máchovo jezero pond systems; D – Rivers Jihlava, Morava, Rokytná, Svratka, Bystřice near Olomouc, Mušov water reservoir; O – locality not given; **SR:** D – River Dunaj (= Danube).
Records: 1. Vojtková 1959b; 2. Ergens 1962a; 3. Moravec in Ergens and Lom 1970; 4. Moravec 1971a; 5. Moravec 1978d; 6. Žitňan 1979e; 7. Moravec 1980e; 8. Moravec 1983b; 9. Moravec 1984e; 10. Moravec 1985a; 11. Moravec 1987; 12. Gelnar et al. 1990; 13. Moravec and Scholz 1991b; 14. Gelnar et al. 1994; 15. Koníček and Gelnar 1994; 16. Moravec 1994; 17. Moravec 1995a; 18. Moravec et al. 1997; 19. Koubková and Baruš 2000b.

Schulmanela petruschewskii (Shulman, 1948)
Syn. *Hepaticola petruschewskii* Shulman, 1948; *Capillaria petruschewskii* (Schulman, 1948)
Site: liver.
Hosts: Abramis bjoerkna (6, 7)
Abramis brama (5)
Carassius auratus gibelio (9)
Lepomis gibbosus (8)

Leuciscus cephalus (11)
Perca fluviatilis (3)
Silurus glanis (9)
Tinca tinca (1, 2, 4, 6, 7)
Distribution: **CR:** D – Rivers Dřevnice, Dyje and Morava, periodical pools of Dyje R.; **SR:** D – Dunaj (= Danube) R., pools and irrigation canal of Váh R. (lower reaches).
Records: 1. Dyk 1955a; 2. Lucký 1955c; 3. Vojtek 1959b; 4. Dyk 1961a; 5. Pacák 1962; 6. Dyk and Lucký 1963; 7. Lucký and Dyk 1964; 8. Žitňan 1979e; 9. Gelnar et al. 1994; 10. Moravec 1994; 11. Divílek et al. 1998.

Fam. Cystoopsidae Skrjabin, 1923

Cystoopsis acipenseris Wagner, 1867
Site: skin (mainly between bone scales).
Host: *Acipenser ruthenus* (2, 3)
Distribution: **SR:** D – River Dunaj (= Danube).
Records: 1. Rašín 1930b; 2. Rašín 1930c; 3. Jedlička and Matis 1966; 4. Moravec 1994.

Fam. Anisakidae Railliet et Henry, 1915

Hysterothylacium bidentatum (Linstow, 1899)
Syn. *Contracaecum bidentatum* (Linstow, 1899)
Site: oesophagus, stomach, intestine.
Host: *Acipenser ruthenus* (1, 2, 3)
Distribution: **SR:** D – River Dunaj (= Danube).
Records: 1. Dyk 1952a, 2. Zubrický 1981, 3. Moravec et al. 1989; 4. Moravec 1994.

Hysterothylacium gadi (Müller, 1776)
Syn. *Ascaris clavata* Rudolphi, 1809
Site: stomach, intestine.
Host: *Salmo salar* (1)
Distribution: **CR:** L – River Labe (= Elbe).
Record: 1. Frič 1893.

Raphidascaris acus (Bloch, 1779)
Syn. *Ascaris acus* Bloch, 1779; *A. mucronata* Schrank, 1790; *A. labiata* Rudolphi, 1809; *A. cristata* Linstow, 1872; *Raphidascaris* sp. of Pacák (1960)
Site: intestine, pyloric caeca.
Hosts: *Anguilla anguilla* (11, 54, 55, 56, 60, 62)
Esox lucius (1, 2, 3, 4, 13, 15, 18, 19, 20, 22, 23, 25, 26, 27, 28, 32, 33, 36, 37, 38, 39, 44, 45, 50, 52, 54, 55, 56, 60, 61, 62, 64, 65, 68)
Gymnocephalus cernuus (15, 20)
Hucho hucho (51, 60, 63)
**Leuciscus cephalus* (69)
Lota lota (1, 9, 12, 13, 44)
Perca fluviatilis (2, 35, 38, 39, 44, 52)
Oncorhynchus mykiss (14, 22, 29, 46, 49, 60)
Salmo trutta m. *fario* (5, 6, 7, 8, 10, 13, 14, 16, 17, 21, 22, 24, 27, 29, 30, 33, 34, 41, 42, 46, 49, 52, 56, 57, 58, 59, 60, 61, 64, 67)
Salmo trutta m. *lacustris* (46)
Salvelinus fontinalis (48)
Thymallus baicalensis (40, 46, 48)
Thymallus thymallus (7, 8, 14)
Umbra krameri (63)
Distribution: **CR:** L – Rivers Blanice, Kamenice, Lužnice, Malše, Sázava, Teplá, Teplá Vltava, Vltava, Želivka, some small streams (Černovický, Chotčínský and Bílinský Brooks), Lipno water reservoir, Třeboň, Jindřichův Hradec and Máchovo jezero pond systems; D – Rivers Bečva, Bystřice near Olomouc, Jihlava, Morava, Punkva, Svratka and some small streams (Adamovský, Bílý, Chvojnice and Říčky Brooks); O – Rivers Moravice and Osoblaha and some small streams near Karlov, Karlovec and Malá Morávka; **SR:** D – Rivers Bodrog, Bodva, Čierna Voda, Dunaj (= Danube), Hnilec, Hornád, Hron, Laborec, Latorica, Malý Dunaj, Tisa, Váh (including periodical pools and irrigation canal), some small streams (Demänová, Revúca, Selec Brooks, streams near Senné), Dobšiská, Oravská and Zemplínská Šírava water reservoirs; V – Rivers Poprad and Dunajec, lake Štrbské pleso.
Records: 1. John 1877; 2. Šrámek 1900; 3. Šrámek 1901, 1903; 4. Frič 1908; 5. Dyk 1953b; 6. Dyk 1954b; 7. Dyk 1954c; 8. Dyk

and Lucký 1954e; 9. Dyk et al. 1955; 10. Dyk and Valenta 1955; 11. Volf and Smíšek 1955; 12. Dyk 1956a; 13. Dyk 1956b; 14. Dyk and Lucký 1956b; 15. Kašťák 1956d; 16. Dyk 1957a; 17 Dyk and Lucký 1957; 18. Kašťák 1957b; 19. Kašťák 1957c; 20. Kašťák 1957d; 21. Dyk and Štědronský 1958; 22. Pacák 1958; 23. Vojtek 1959b; 24. Kašťák and Žitňan 1960a; 25. Pacák 1960; 26. Žitňan 1960c; 27. Dyk 1961a; 28. Ergens 1962a; 29. Dyk 1964a; 30. Lucký and Dyk 1964; 31. Dyk and Štědronský 1964; 32. Žitňan 1964f; 33. Dyk 1965a; 34. Ergens 1965a; 35. Žitňan 1965a; 36. Ergens 1966a; 37. Žitňan 1966c; 38. Žitňan 1967c; 39. Žiňan 1968c; 40. Žitňan 1969b; 41. Moravec 1970; 42. Žitňan 1970–1971; 43. Moravec 1971a; 44. Žitňan 1971a; 45. Žitňan 1971c; 46. Žitňan 1973a; 47. Žitňan 1973b; 48. Žitňan 1974b; 49. Žitňan 1974c; 50. Ergens et al. 1975; 51. Žitňan 1976; 52. Tesarčík and Ivasik 1977; 53. Žitňan 1977; 54. Moravec 1978d; 55. Moravec 1979a; 56. Žitňan 1979e; 57. Moravec 1980c; 58. Žitňan 1980a; 59. Žitňan 1981a; 60. Žitňan 1982b; 61. Moravec 1984d; 62. Moravec 1985b; 63. Žitňan 1985; 64. Gelnar et al. 1990; 65. Gelnar et al. 1994; 66. Moravec 1994; 67. Moravec 1996a; 68. Moravec et al. 1997; 69. Divílek et al. 1998.

Fam. Quimperiidae Gendre, 1928

Paraquimperia tenerrima (Linstow, 1878)
Syn. *Filaria conoura* Linstow, 1885
Site: intestine.
Hosts: *Anguilla anguilla* (1, 2, 3, 4, 5, 7, 8)
**Aspius aspius* (1, 2)
**Leuciscus cephalus* (1, 2)
**Perca fluviatilis* (1, 2)
Distribution: **CR:** L – Rivers Labe (= Elbe), Teplá and Želivka; D – River Rokytná.
Records: 1. Šrámek 1901, 1903; 2. Frič 1908; 3. Moravec 1966a; 4. Moravec 1966b; 5. Moravec 1971; 6. Moravec 1974; 7. Moravec and Scholz 1991; 8. Moravec 1994.

Fam. Cucullanidae Cobbold, 1864

Cucullanus dogieli Krotas, 1959
Syn. *Cucullanellus minutus* of Kašťák (1956d, 1957d), nec *Cucullanus minutus* Rudolphi, 1819
Site: intestine.
Hosts: *Abramis bjoerkna* (1, 2, 4, 6, 9)
Abramis brama (5, 6)
Abramis sapa (6, 12)
Alburnus alburnus (6)
Aspius aspius (6, 9, 10)
Leuciscus cephalus (3, 4, 6, 9)
Rutilus pigus (6, 9)
Rutilus rutilus (6, 11)
Distribution: **SR:** D – Rivers Dunaj (= Danube), Laborec, Latorica and Tisa, lake Ököri.
Records: 1. Kašťák 1956d; 2. Kašťák 1957d; 3. Žitňan 1964f; 4. Žitňan 1965a; 5. Žitňan 1966b; 6. Žitňan 1967c; 7. Žitňan 1968c; 8. Moravec 1971a; 9. Ergens et al. 1975; 10. Žitňan 1979e; 11. Žitňan 1985; 12. Žitňan 1986; 13. Moravec 1994.

Cucullanus truttae Fabricius, 1794
Syn. *Bulbodacnitis globosa* (Zeder, 1800)
Site: intestine, pyloric caeca.
Host: *Salmo trutta* m. *fario* (1, 2, 3, 4, 5, 6, 7, 8, 9)
Distribution: **CR:** L – Rivers Teplá and Úhlava, Černovický Brook near Tábor.
Records: 1. Dyk and Dyková 1957; 2. Dyk 1961a; 3. Lucký and Dyk 1964; 4. Moravec 1971a; 5. Moravec 1976b; 6. Moravec 1979c; 7. Moravec 1980b; 8. Moravec 1983a; 9. Moravec 1984d; 10. Moravec 1994.

Fam. Camallanidae Railliet et Henry, 1915

Camallanus lacustris (Zoega, 1776)
Syn. *Cucullanus elegans* Zeder, 1800
Site: pyloric caeca, intestine.
Hosts: *Anguilla anguilla* (1, 2, 3, 23, 28, 29, 33, 34)
Cyprinus carpio (33, 34)
Esox lucius (1, 2, 3, 5, 6, 8, 14, 15, 17, 19, 23, 27, 28, 29, 32, 33, 34)

Gymnocephalus cernuus (14, 15, 23)
Leuciscus cephalus (23, 27, 41)
Lota lota (1, 5, 6, 14, 15, 17, 23)
Oncorhynchus mykiss (40)
Pelecus cultratus (40)
Perca fluviatilis (2, 3, 4, 5, 6, 7, 8, 9, 10, 11, 12, 14, 15, 16, 17, 18, 21, 23, 27, 28, 29, 30, 31, 33, 34, 35, 36, 38, 39)
Proterorhinus marmoratus (42)
Salmo trutta m. *fario* (23, 24)
Silurus glanis (23)
Stizostedion lucioperca (4, 6, 7, 8, 11, 12, 15, 17, 20, 21, 23, 28, 29, 33, 34, 39)
Zingel zingel (23)

Distribution: **CR:** L – Rivers Labe (= Elbe), Vltava, Teplá and Želivka, Třeboň (Horusický, Rožmberk, Velký Tisý), Jindřichův Hradec and Máchovo jezero pond systems, Lipno water reservoir; D – Rivers Dřevnice, Dyje, Jihlava and Želetavka, Mušov water reservoir, retention water reservoirs and waterworks in Morava R. basin, South-Moravian ponds and pools near Lednice, Podivín, Pohořelice and Přítluky; O – Rivers Opava and Opavice; **SR:** D – Rivers Dunaj (= Danube), Tisa and Latorica.

Records: 1. John 1877; 2. Šrámek 1901, 1903; 3. Frič 1908; 4. Dyk 1954b; 5. Dyk 1954c; 6. Dyk and Lucký 1954b; 7. Dyk and Lucký 1954c; 8. Vojtek et al. 1954; 9. Dyk 1955b; 10. Lucký et al. 1955; 11. Dyk 1956a; 12. Vejnar 1956; 13. Dyk 1961a; 14. Ergens 1962a; 15. Dyk and Lucký 1963; 16. Dyk 1964a; 17. Lucký and Dyk 1964; 18. Rodák et al. 1965; 19. Ergens 1966a; 20. Žitňan 1966c; 21. Žitňan 1967c; 22. Žitňan 1968c; 23. Moravec in Ergens and Lom 1970; 24. Moravec 1971a; 25. Moravec 1971c; 26. Žitňan 1973b; 27. Ergens et al. 1975; 28. Moravec 1978d; 29. Moravec 1979a; 30. Lucký and Navrátil 1983; 31. Lucký and Navrátil 1984; 32. Moravec 1984d; 33. Moravec 1984e; 34. Moravec 1985b; 35. Scholz 1986; 36. Scholz 1987b; 37. Gelnar et al. 1990; 38. Moravec 1994; 39. Moravec et al. 1994; 40. Moravec 1995a; 41. Divílek et al. 1998; 42. Koubková and Baruš 2000b.

Camallanus truncatus (Rudolphi, 1814)
Syn. *Camallanus lacustris* of Dyk and Lucký (1953), nec *C. lacustris* (Zoega, 1776)
Site: pyloric caeca, intestine.
Hosts: *Abramis brama* (8)
Anguilla anguilla (8, 12, 15, 16, 23, 29)
Aspius aspius (18, 22)
Cottus gobio (26)
Esox lucius (7, 18, 22, 24)
Gobio albipinnatus (18, 22)
Gymnocephalus cernuus (7, 12, 14, 15, 16, 18, 22)
Gymnocephalus schraetser (15, 16, 18, 23, 26)
Lota lota (15, 16, 20, 26)
Perca fluviatilis (2, 7, 8, 9, 10, 15, 16, 18, 22, 27)
Silurus glanis (13, 15, 16, 18, 22)
Stizostedion lucioperca (1, 2, 3, 4, 5, 6, 8, 9, 10, 11, 12, 13, 15, 16, 18, 19, 20, 22, 23, 27)
Stizostedion volgense (15, 16, 18, 19, 20, 22, 23, 26)
Zingel zingel (15, 16, 20, 23, 25)

Distribution: **CR:** D – Rivers Morava and Svratka, Kníničská water reservoir, South--Moravian ponds near Pohořelice and Lednice (Nesyt); **SR:** D – Rivers Dunaj (= Danube), Hron, Váh, Laborec, Latorica and Tisa, surroundings of Komárno, lake Morské oko (Great Vihorlat Lake).

Records: 1. Dyk 1954b; 2. Dyk 1954c; 3. Dyk 1955b; 4. Dyk and Lucký 1954b; 5. Kašťák 1956d; 6. Kašťák 1957d; 7. Vojtek 1959; 8. Vojtková 1959; 9. Dyk and Lucký 1963; 10. Lucký and Dyk 1964; 11. Žitňan 1964; 12. Žitňan 1965; 13. Žitňan 1966; 14. Přibyslavský and Lucký 1967; 15. Žitňan 1967a; 16. Žitňan 1967b; 17. Žitňan 1968; 18. Moravec in Ergens and Lom 1970; 19. Moravec 1971; 20. Žitňan 1971; 21. Žitňan 1973; 22. Ergens et al. 1975; 23. Žitňan 1979, 24. Moravec 1984; 25. Žitňan 1985; 26. Žitňan 1986; 27. Gelnar et al. 1994; 28. Moravec 1995a; 29. Moravec et al. 1996; 30. Špakulová 2000.

Fam. Anguillicolidae Yamaguti, 1935

Anguillicola crassus Kuwahara, Niimi et Itagaki, 1974
Site: swimbladder.
Host: *Anguilla anguilla* (1, 2, 4, 5, 6, 7, 8, 9, 10, 11, 12, 13, 14, 15, 16, 17, 18, 19, 20, 21, 22, 23, 24)
Distribution: **CR:** L – River Labe (= Elbe) near Hřensko, Orlická water reservoir; D – Rivers Dyje, Hloučela, Kyjovka, Morava and Svratka, Vranovská, Mušov and Koryčany water reservoirs, small reservoir near Podivín, Morava R. basin.
Records: 1. Moravec 1992b; 2. Moravec et al. 1993; 3. Baruš 1994, 4. Moravec 1996b; 5. Baruš 1995; 6. Moravec 1995a; 7. Baruš et al. 1996; 8. Baruš and Prokeš 1996; 9. Moravec et al. 1996; 10. Baruš 1997; 11. Moravec 1997; 12. Baruš et al. 1998a; 13. Baruš et al. 1998b; 14. Baruš et al. 1998d; 15. Kráčmar et al. 1998; 16. Moravec and Škoríková 1998; 17. Vostradovský 1998; 18. Baruš et al. 1999a; 19. Baruš et al. 1999b; 20. Baruš et al. 1999c; 21. Baruš et al. 1999d; 22. Navrátil et al. 1999; 23. Tenora et al. 1999; 24. Koubková and Baruš 2000a; 25. Koubková and Baruš 2000b.

Fam. Skrjabillanidae Shigin et Shigina, 1958

Esocinema bohemicum Moravec, 1977
Site: serosa of swimbladder.
Host: *Esox lucius* (1, 2, 3, 4, 5, 6)
Distribution: **CR:** L – Máchovo jezero pond system.
Records: 1. Moravec 1977d; 2. Moravec 1978d; 3. Moravec 1979a; 4. Moravec 1984d; 5. Moravec 1990b; 6. Moravec 1994.

Molnaria intestinalis (Dogiel et Bychowsky, 1934)
Syn. *Philometra intestinalis* Dogiel et Bychowsky, 1934; *Skrjabillanus erythrophthalmi* Molnár, 1966
Site: serosa covers of intestine, kidneys, swimbladder and other organs, abdominal cavity.
Hosts: *Abramis brama* (2)
Leuciscus cephalus (1, 8)
Scardinius erythrophthalmus (3, 4, 5, 6, 7)
Distribution: **CR:** L – Máchovo jezero pond system; D – River Rokytná; **SR:** D – Rivers Dunaj (= Danube) and Nitra.
Records: 1. Dyk 1961a; 2. Pacák 1962; 3. Moravec 1968b; 4. Moravec 1971a; 5. Moravec 1978d; 6. Žitňan 1979e; 7. Moravec 1986c; 8. Moravec and Scholz 1991; 9. Moravec 1994.

Skrjabillanus scardinii Molnár, 1966
Syn. *Agrachanus scardinii* (Molnár, 1966)
Site: serosa covers of swimbladder, kidneys and gonads, urinary bladder, orbits, humour of eyes, abdominal cavity.
Host: *Scardinius erythrophthalmus* (1, 2, 3, 4, 5, 6)
Distribution: **CR:** L – Máchovo jezero pond system; **SR:** D – Dunaj (= Danube) River.
Records: 1. Moravec 1968b; 2. Moravec 1971a; 3. Moravec 1978d; 4. Žitňan 1979e; 5. Moravec 1986c; 6. Moravec 1994.

Skrjabillanus tincae Shigin et Shigina, 1958
Site: serosa of swimbladder, urinary system, surface of heart and intestine.
Host: *Tinca tinca* (1, 2, 3, 4, 5, 6)
Distribution: **CR:** L – Máchovo jezero pond system; D – River Svratka.
Records: 1. Ergens 1960a; 2. Moravec 1968b; 3. Moravec 1971a; 4. Moravec 1978d; 5. Moravec 1985a; 6. Moravec 1994.

Fam. Daniconematidae Moravec et Køie, 1987

Daniconema anguillae Moravec et Køie, 1987
Site: serosa of swimbladder and intestine.
Host: *Anguilla anguilla* (1, 2)
Distribution: **CR:** L – Máchovo jezero pond system.
Records: 1. Moravec and Køie 1987, 2. Moravec 1994.

Fam. Philometridae Baylis et Daubney, 1926

Philometra abdominalis Nybelin, 1928
Syn. *Filaria ovata* Zeder, 1803 of John (1877), nec Zeder, 1803; *Ichthyonema ovatum* (Zeder, 1803) of Šrámek (1900, 1901, 1903); *Philometra* sp. of Žitňan (1964f, 1967a)
Site: abdominal cavity, serosa of swimbladder.
Hosts: *Gobio gobio* (1, 6, 7, 9, 11, 13, 14, 15, 17, 23, 24)
Gobio uranoscopus (20, 21)
Leuciscus cephalus (2, 3, 4, 5, 8, 9, 10, 11, 12, 13, 14, 15, 16, 20, 22, 23, 24, 25, 27, 28, 29, 30)
Leuciscus leuciscus (29)
Phoxinus phoxinus (11, 13, 14, 17, 18, 19)
Distribution: **CR:** L – Rivers Labe (= Elbe) and Vltava, Rokytka Brook near Prague; D – Rivers Dřevnice, Morava, Rokytná, Svratka and Bystřice near Olomouc, Klanečnica Brook; O – Rivers Moravice and Osoblaha; **SR:** D – Rivers Dunaj (= Danube), Hron, Laborec, Latorica and Nitra, lake Morské oko (Great Vihorlat Lake).
Records: 1. John 1877; 2. Šrámek 1900; 3. Šrámek 1901, 1903; 4. Dyk 1954c; 5. Čákay 1957; 6. Lucký 1959a; 7. Vojtková 1959b; 8. Dyk 1961a; 9. Lucký a Dyk 1964; 10. Žiňan 1964f; 11. Ergens 1965a; 12. Žitňan 1967c; 13. Moravec in Ergens and Lom 1970; 14. Moravec 1971; 15. Žitňan 1971; 16. Ergens et al. 1975; 17. Moravec 1977a; 18. Moravec 1977b; 19. Moravec 1977c; 20. Žitňan 1979; 21. Žitňan 1985; 22. Gelnar et al. 1990; 23. Moravec and Scholz 1991; 24. Gelnar et al. 1994; 25. Koníček and Gelnar 1994; 26. Moravec 1994; 27. Šebelová et al. 1995; 28. Gelnar et al. 1997; 29. Moravec et al. 1997; 30. Divílek et al. 1998; 31. Špakulová 2000.

Philometra kotlani (Molnár, 1969)
Site: abdominal cavity, serosa of swimbladder.
Host: *Aspius aspius* (1, 2, 3, 4, 5, 6)
Distribution: **CR:** L – River Vltava; **SR:** D – Rivers Dunaj (= Danube) and Latorica.
Records: 1. Moravec 1971a; 2. Ergens et al. 1975; 3. Žitňan 1979e; 4. Žitňan 1981a; 5. Moravec 1994; 6. Moravec et al. 1997.

Philometra obturans (Prenant, 1886)
Site: gill arteries, ventral aorta, abdominal cavity, vitreous body of eyes.
Host: *Esox lucius* (1, 2, 3, 4, 5, 6, 7)
Distribution: **CR:** L – Máchovo jezero pond system; **SR:** D – River Latorica.
Records: 1. Moravec in Ergens and Lom 1970; 2. Moravec 1971a; 3. Moravec 1978b; 4. Moravec 1978c; 5. Moravec 1978d; 6. Moravec and Dyková 1978; 7. Moravec 1979a; 8. Moravec 1984d; 9. Moravec 1994.

Philometra ovata (Zeder, 1800)
Syn. *Agamonema ovatum* (Zeder, 1800); *Ichthyonema ovatum* (Zeder, 1800); *I. sanguineum* (Rudolphi, 1819) of Šrámek (1901, 1903) and of Frič (1908), nec Rudolphi 1819
Site: abdominal cavity and serosa of swimbladder.
Hosts: *Abramis ballerus* (13, 14, 15, 22, 33)
Abramis bjoerkna (4, 9, 17, 30, 33)
Abramis brama (15, 17, 18, 19, 20, 21, 22, 25, 27, 30, 31)
Aspius aspius (15)
**Gobio gobio* (12)
**Leuciscus cephalus* (1, 6, 7, 9, 10, 12, 19, 24, 28, 29, 30)
**Leuciscus leuciscus* (27)
**Phoxinus phoxinus* (27)
**Proterorhinus marmoratus* (8, 9, 32)
Rutilus rutilus (5, 6, 7, 9, 11, 13, 14, 15, 16, 17, 18, 19, 22, 24, 30, 33)
**Tinca tinca* (19, 22, 24)
Vimba vimba (2, 3)
Distribution: **CR:** L – Rivers Labe (= Elbe) and Vltava, Máchovo jezero pond system; D – Rivers Morava and Svratka, Kníničská and Mušov water reservoirs; O – Rivers Moravice, Opava and Opavice; **SR:** D – Rivers Dunaj (= Danube), Hornád, Latorica, Nitra and Váh, pools and irrigation canal of Váh R.

Records: 1. Šrámek 1900; 2. Šrámek 1901, 1903; 3. Frič 1908; 4. Volf and Poustka 1933; 5. Kašťák 1956b; 6. Kašťák 1956d; 7. Kašťák 1957d; 8. Vojtek 1959b; 9. Dyk 1961a; 10. Hradil 1964; 11. Lucký and Dyk 1964; 12. Rodák et al. 1965; 13. Žitňan 1965a; 14. Lucký and Jílek 1967; 15. Žitňan 1967c; 16. Moravec in Lom and Ergens 1970; 17. Moravec 1971a; 18. Žitňan 1971c; 19. Moravec 1978d; 20. Žitňan 1979e; 21. Moravec 1980a; 22. Moravec 1985a; 23. Moravec 1986b; 24. Moravec 1986c; 25. Gelnar et al. 1994; 26. Moravec 1994; 27. Moravec 1995a; 28. Šebelová et al. 1995; 29. Gelnar et al. 1997; 30. Moravec et al. 1997; 31. Baruš et al. 1998a; 32. Koubková and Baruš 2000b; 33. Tenora et al. 2000.

Philometra rischta Skrjabin, 1923
Syn. *Ichthyonema ovatum* (Zeder, 1803) of Volf and Poustka (1933) (in part), nec Zeder, 1803
Site: under skin of head, gill covers, buccal cavity, fins, serosa of swimbladder.
Hosts: *Abramis bjoerkna* (1, 4, 9)
Abramis brama (2, 3, 5)
Alburnus alburnus (3, 4, 6, 8, 9, 10)
Aspius aspius (10, 11)
Chondrostoma nasus (8, 9)
Leuciscus idus (10)
Leuciscus leuciscus (10, 11)
Rutilus rutilus (3)
Distribution: **CR:** L – River Labe (= Elbe), D – Rivers Dyje and Svratka, pools near Svratka R., **SR:** D – Rivers Dunaj (= Danube), Hron and Tisa.
Records: 1. Volf and Poustka 1933; 2. Vojtková 1959b; 3. Dyk 1961a; 4. Lucký and Dyk 1964; 5. Lucký and Jílek 1967; 6. Žitňan 1967c; 7. Žitňan 1968c; 8. Žitňan 1971a; 9. Žitňan 1971c; 10. Žitňan 1979e; 11. Žitňan 1981a; 12. Moravec 1994.

Philometra sp.
Site: serosa of swimbladder.
Hosts: *Tinca tinca* (1)
Distribution (finding): **CR:** D – River Morava.
Records: 1. Gelnar et al. 1994.

Philometroides sanguinea (Rudolphi, 1819)
Syn. *Philometra sanguinea* (Rudolphi, 1819)
Site: fins, abdominal cavity, serosa of swimbladder.
Hosts: **Barbus petenyi* (2, 3)
Carassius auratus gibelio (7)
Carassius carassius (1, 4, 5, 6)
Distribution: **CR:** L – Pools near Labe (= Elbe) River; D – Morava River; **SR:** D – Rivers Latorica and Váh, Myslava Brook.
References: 1. Čákay 1957a; 2. Kašťák 1956d; 3. Kašťák 1957d; 4. Moravec in Ergens and Lom 1970; 5. Moravec 1971a; 6. Ergens et al. 1975; 7. Gelnar et al. 1994; 8. Moravec 1994.

Fam. Rhabdochonidae Travassos, Artigas et Pereira, 1928

Rhabdochona denudata (Dujardin, 1845)
Syn.: *Ascaris dentata* (Zeder, 1800) of Šrámek (1900, 1901, 1903)
Site: intestine.
Hosts: *Abramis bjoerkna* (40)
Abramis brama (38)
Alburnoides bipunctatus (34, 36)
Alburnus alburnus (3, 11, 16, 17, 18, 21, 25, 26, 34, 36, 40)
Aspius aspius (24, 33)
**Barbatula barbatula* (4)
**Barbus barbus* (3, 7, 8, 11, 15, 16)
**Barbus petenyi* (15)
**Gobio gobio* (18)
Leuciscus cephalus (1, 2, 3, 5, 6, 7, 8, 9, 12, 15, 16, 18, 21, 22, 25, 27, 29, 34, 35, 36, 37, 40)
Leuciscus leuciscus (11, 13, 14, 18, 19, 22, 25, 27, 34, 35, 36)
**Oncorhynchus mykiss* (10)
Pelecus cultratus (25, 32)
**Perca fluviatilis* (19)
**Phoxinus phoxinus* (23, 25, 33)
Rutilus rutilus (1, 22)
**Salmo trutta* m. *fario* (14, 18)
**Salvelinus fontinalis* (15)
**Thymallus thymallus* (4, 12, 13, 16, 20, 25, 30)

Distribution: **CR:** L – Rivers Labe (= Elbe), Blanice, Malše, Vltava, Želivka, Žernonický Brook; D – Rivers Dyje, Rokytná and Svitava, Vranovská water reservoir, pools near Přítluky, fish farm on Jihlava R.; O – Rivers Moravice, Opava, Opavice and Osoblaha, Hořina Brook; **SR:** D – Rivers Dunaj (= Danube), Hornád, Hron, Muráňka, Laborec, Latorica, Tisa, Torysa R. basin, Myslava Brook, lake Morské oko (Great Vihorlat Lake), pool near Nové Město nad Váhom; V – Dunajec River.
Records: 1. Šrámek 1900; 2. Šrámek 1901, 1903; 3. Vojtek et al. 1954; 4. Dyk and Lucký 1956b; 5. Kašťák 1956d; 6. Kašťák 1957d; 7. Dyk and Štědronský 1958; 8. Lucký 1959a; 9. Vojtková 1959b; 10. Kašťák and Žitňan 1960a; 11. Žitňan 1960c; 12. Dyk 1961a; 13. Dyk 1964a; 14. Dyk and Štědronský 1964; 15. Hradil 1964; 16. Lucký and Dyk 1964; 17. Žitňan 1964f; 18. Ergens 1965a; 19. Rodák et al. 1965; 20. Žitňan 1965a; 21. Žitňan 1965c; 22. Moravec 1968a; 23. Řehulka 1968; 24. Moravec 1971a; 25. Žitňan 1971a; 26. Žitňan 1971c; 27. Žitňan 1974c; 28. Moravec 1978e; 29. Žitňan 1979e; 30. Žitňan 1980b; 31. Mészáros 1983; 32. Žitňan 1985; 33. Žitňan 1986; 34. Moravec 1989; 35. Gelnar et al. 1990; 36. Moravec and Scholz 1991b; 37. Navrátil 1991; 38. Gelnar et al. 1994; 39. Moravec 1994; 40. Moravec et al. 1997; 41. Špakulová 2000.

Rhabdochona ergensi Moravec, 1968
Syn. *Rhabdochona acuminata* (Molin, 1860) of Ergens (1965a) and of Žitňan (1971a) (partly), nec Molin, 1860
Site: intestine.
Host: *Barbatula barbatula* (1, 2, 3, 4, 5, 6, 7, 8, 9, 10, 11, 12, 13)
Distribution: **CR:** D – River Morava, Klanečnica Brook; O – River Osoblaha; **SR:** D – River Hron, rivers of East Slovakia.
Records: 1. Ergens 1965a; 2. Moravec 1968a; 3. Moravec 1971a; 4. Žitňan 1971a; 5. Moravec 1972; 6. Moravec 1975; 7. Moravec 1978e; 8. Mészáros 1983; 9. Moravec 1989; 10. Moravec 1990b; 11. Moravec and Scholz 1991b; 12. Gelnar et al. 1994; 13. Moravec 1994.

Rhabdochona hellichi (Šrámek, 1901)
Syn. *Filaria hellichi* Šrámek, 1901; *R. acuminata* (Molin, 1860) of Žitňan (1967c, 1968c, 1971a) (partly), nec Molin, 1860
Site: intestine.
Host: *Barbus barbus* (1, 2, 3, 5, 6, 7, 8, 9, 10, 15, 16, 17, 18, 22, 23, 24)
Barbus barbus × *Barbus petenyi* (reported incorrectly as *B. plebejus*) (13)
Barbus petenyi (2, 7, 8, 16, 23)
**Leuciscus cephalus* (1)
**Salmo trutta* m. *fario* (12)
**Zingel zingel* (19)
Distribution: **CR:** L – Rivers Labe (= Elbe), Malše, Teplá and Želivka; D – Rivers Jihlava, Morava and Rokytná, fish farm on Jihlava R.; **SR:** D – Rivers Dunaj (= Danube), Hron, Latorica and Tisa; V – River Dunajec.
Records: 1. Šrámek 1901, 1903; 2. Žitňan 1967c; 3. Moravec 1968a; 4. Žitňan 1968c; 5. Moravec in Ergens and Lom 1970; 6. Moravec 1971a; 7. Žitňan 1971a; 8. Žitňan 1974c; 9. Ergens et al. 1975; 10. Moravec 1975; 11. Moravec 1978e; 12. Žitňan 1979e; 13. Kirka et al. 1982; 14. Mészáros 1983; 15. Moravec 1989; 16. Gelnar et al. 1990; 17. Moravec and Schoz 1991b; 18. Navrátil 1991; 19. Gelnar et al. 1994; 20. Jurajda et al. 1994; 21. Moravec 1994; 22. Moravec 1995; 23. Baruš et al. 1997b; 24. Moravec et al. 1997a.

Rhabdochona phoxini Moravec, 1968
Site: intestine.
Host: *Phoxinus phoxinus* (1, 2, 3, 4, 5, 6, 7)
Distribution: **CR:** L – Rokytka Brook; D – Klanečnica Brook.
Records: 1. Moravec 1968a; 2. Moravec 1971a; 3. Moravec 1976a; 4. Moravec 1977c; 5. Moravec 1978e; 6. Moravec 1990b; 7. Moravec 1994.

Rhabdochona sp.
Site: intestine.
Hosts: *Gymnocephalus schraetser* (1)
Thymallus thymallus (2)
Zingel zingel (1)
Distribution: **SR:** D – Rivers Dunaj (= Danube) and Hornád.
Records: 1. Vejnar 1956; 2. Žitňan 1970–1971.

Fam. Cystidicolidae Skrjabin, 1946

Cystidicola farionis Fischer, 1798
Site: swimbladder.
Hosts: *Hucho hucho* (11, 17, 22, 24)
Oncorhynchus mykiss (3, 4, 6, 7, 13, 14, 22)
Salmo trutta m. *fario* (1, 2, 3, 4, 5, 7, 8, 11, 13, 14, 15, 16, 18, 19, 22, 23)
Salmo trutta m. *lacustris* (13, 14)
Salvelinus fontinalis (16)
Thymallus baicalensis (10, 14, 16)
Thymallus thymallus (4, 5, 9, 11, 13, 14, 20, 22)
Distribution: **SR:** D – Rivers Dunaj (= Danube), Belá, Hnilec, Hornád, Hron, Muráňka, Revúca, Váh, brooks Demänová, Mošnice, Okna and Selec, lake Vrbické pleso, Dobšinská reservoir, trout farm Stupava; V – Rivers Dunajec and Poprad
Records: 1. Kašťák 1956d; 2. Kašťák 1957d; 3. Pacák 1957; 4. Pacák 1958; 5. Pacák 1960; 6. Žitňan 1960c; 7. Dyk 1961a; 8. Dyk and Dyková 1963; 9. Žitňan 1965c; 10. Žitňan 1969b; 11. Žitňan 1970–1971; 12. Moravec 1971a; 13. Žitňan 1971a; 14. Žitňan 1973a; 15. Žitňan 1974c; 16. Žitňan 1974b; 17. Žitňan 1976; 18. Žitňan 1979e; 19. Žitňan 1980a; 20. Žitňan 1980b; 21. Žitňan 1981a; 22. Žitňan 1982b; 23. Moravec 1984d; 24. Žitňan 1985; 25. Moravec 1994.

Cystidicoloides ephemeridarum (Linstow, 1872)
Syn. *Sterliadochona ssavini* Skrjabin, 1946; *Metabronema truttae* Baylis, 1935; *Cystidicoloides tenuissima* (Rudolphi, 1809), nec Zeder, 1800
Site: stomach, intestine, pyloric caeca.
Hosts: **Anguilla anguilla* (15, 16)
**Barbatula barbatula* (5)
**Esox lucius* (15, 16)
Hucho hucho (24)
Oncorhynchus mykiss (2, 6, 7, 11, 18, 20, 23)
**Perca fluviatilis* (14, 15, 16)
**Phoxinus phoxinus* (5)
Salmo trutta m. *fario* (1, 2, 4, 6, 7, 8, 9, 10, 11, 12, 13, 14, 15, 16, 17, 18, 19, 20, 22, 25, 26, 27, 28, 30, 31, 32)
Thymallus thymallus (15, 16, 19, 28)
Distribution: **CR:** L – Rivers Blanice, Malše, Teplá and Želivka, Bílinský, Černovický, Ondřejovický and Žernovický Brooks, pond system of Jindřichův Hradec, trout farm at Liběchov; D – Rivers Bečva, Bystřice near Olomouc, Klanečnica, Olešnice, Punkva, Senice and Svratka, brooks Adamovská říčka, Bílý and Chvojenec; O – Rivers Moravice, Osoblaha and Zlatá Opavice; **SR:** D – Rivers Bodva, Dunaj (= Danube), Hnilec, Hornád, Hron, Toplá, Torysa and Váh, brooks Okna and Ždiarský, tributaries of Nitra R. and Tisa R., streams in southern Slovakia, Dobšinská water reservoir, lake Morské oko (Great Vihorlat Lake); V – Dunajec River.
Records: 1. Kašťák 1956d; 2. Čákay 1957b; 3. Kašťák 1957d; 4. Dyk and Štědronský 1958; 5. Kašťák and Žitňan 1960a; 6. Žitňan 1960c; 7. Dyk 1961a; 8. Ergens 1962a; 9. Dyk and Dyková 1963; 10. Dyk and Štědronský 1964; 11. Hradil 1964; 12. Lucký and Dyk 1964; 13. Dyk 1965a; 14. Ergens 1965a; 15. Moravec 1967; 16. Moravec in Ergens and Lom 1970; 17. Žitňan 1970–1971; 18. Moravec 1971a; 19. Moravec 1971d; 20. Žitňan 1971a; 21. Žitňan 1973b; 22. Žitňan 1974c; 23. Žitňan 1974b; 24. Žitňan 1976; 25. De and Moravec 1979; 26. Žitňan 1979e; 27. Moravec 1980b; 28. Žitňan 1980a; 29. Žitňan 1980b; 30. Moravec and De 1982; 31. Moravec 1984e; 32. Gelnar et al. 1990; 33. Moravec 1994; 34. Špakulová 2000.

Spinitectus inermis (Zeder, 1800)
Site: stomach, intestine.
Host: *Anguilla anguilla* (1, 2, 3, 4, 5, 6)
Distribution: **CR:** L – Máchovo jezero pond system.
Records: 1. Moravec 1977e; 2. Moravec 1978d; 3. Moravec 1979b; 4. Moravec 1985b; 5. Moravec 1994; 6. Moravec 1996c.

Nematodes of uncertain systematic status and accidental findings in fish

Ichthyobronema sp.
Site: not given.
Hosts: *Gymnocephalus cernuus* (1)
Zingel zingel (1)
Distribution: **SR:** D – River Dunaj (= Danube).
Record: 1. Dyk 1955b.
Remark: The systematic status of these nematodes is unclear.

Oswaldocruzia filiformis (Goeze, 1782) (Trichostrongylidae)
Syn. *Oswaldocruzia* sp. of Kašťák and Žitňan (1960a) and of Žitňan (1974c)
Site: intestine.
Hosts (facultative): **Leuciscus cephalus* (3)
**Salmo trutta* m. *fario* (1, 2)
**Salmo trutta* m. *lacustris* (1, 2)
Distribution: **CR:** D – River Rokytná; **SR:** V – Lakes Štrbské pleso and Nové Štrbské pleso.
Records: 1. Kašťák and Žitňan 1960a; 2. Žitňan 1974c; 3. Moravec and Scholz 1991b.
Remark: Nematodes of this genus are common parasites of amphibians and reptiles. These records represent, undoubtedly, accidental findings where the nematodes survived for some time in the digestive tract of fishes which had ingested their original amphibian or reptilian definitive hosts.

Nematoda gen. sp.
Hosts: *Salmo trutta* m. *fario* (1, 2)
Ctenopharyngodon idella (3)
Distribution: **CR:** D – Punkva River, ponds near Pohořelice; **SR:** V – Poprad River and streams Biela Voda and Javorinka.
Records: 1. Dyk 1957a; 2. Dyk and Dyková 1963; 3. Willomitzer 1980a.

LARVAE

Fam. Dioctophymatidae Railliet, 1915

Eustrongylides mergorum (Rudolphi, 1809) larva
Syn. *Eustrongylides* sp. sensu Moravec in Ergens and Lom (1970) and Moravec (1971a, 1978d)
Site: stomach wall.
Host: *Anguilla anguilla* (1, 2, 3, 4, 5)
Distribution: **CR:** L – Máchovo jezero pond system.
Records: 1. Moravec in Ergens and Lom 1970; 2. Moravec 1971a; 3. Moravec 1978d; 4. Moravec 1985b; 5. Moravec 1994.

Fam. Anisakidae Railliet et Henry, 1912

Anisakis simplex (Rudolphi, 1809) larvae
Syn. *Agamonema capsularia* of Frič (1893)
Site: abdominal cavity (surface of intestine).
Host: *Alosa alosa* (3)
Salmo salar (1, 2)
Distribution: **CR:** L – River Labe (= Elbe), Bohemia.
Records: 1. Frič 1893; 2. Frič 1908; 3. Moravec 1995a.
Remark: The record from *A. alosa* based on a specimen from the National Museum in Prague.

Contracaecum microcephalum (Stossich, 1890) larvae
Site: abdominal cavity.
Hosts: *Cyprinus carpio* (1)
Perca fluviatilis (1, 2)
Distribution: **CR:** L – Máchovo jezero pond system.

Record: 1. Moravec 1984e; 2. Scholz 1987b.

Contracaecum sp. larvae
Site: not given.
Host: *Rutilus rutilus* (1, 2)
Distribution: **SR:** D – River Dunaj (= Danube).
Record: 1. Moravec 1971a; 2. Moravec 1995a.

Raphidascaris acus (Bloch, 1799) larvae
Syn. *Ascaris gracillima* Linstow, 1890
Site: liver and other internal organs, abdominal cavity.
Hosts: *Abramis bjoerkna* (28)
Abramis brama (15, 17, 18, 22, 23, 24, 25, 29, 32)
Alburnus alburnus (7, 28)
Anguilla anguilla (17, 18, 22, 23, 24)
Aspius aspius (29)
Barbatula barbatula (1, 2, 3, 4, 5, 6, 7, 9, 12, 13, 14, 19, 22, 26, 29)
Barbus barbus (31)
Cobitis taenia (7, 30)
Cottus gobio (12, 19, 26, 28)
Cottus poecilopus (7, 12, 13, 14, 26)
Cyprinus carpio (17, 18, 22)
Esox lucius (17, 18, 19, 22, 23, 24)
Gobio gobio (7, 14)
Hucho hucho (16)
Ictalurus nebulosus (19)
Lampetra planeri (20)
Leuciscus cephalus (12, 14, 28)
Leuciscus leuciscus (10, 31)
Lota lota (6, 7, 19, 26, 28)
Misgurnus fossilis (19, 26)
Oncorhynchus mykiss (5, 14)
Perca fluviatilis (7, 22, 23, 24, 27)
Phoxinus phoxinus (6, 10, 12, 14, 19, 26)
Proterorhinus marmoratus (33)
Rutilus rutilus (7, 22, 23, 24, 25, 31)
Salmo trutta m. *fario* (3, 5, 6, 12, 14, 19)
Salmo trutta m. *lacustris* (14)
Scardinius erythrophthalmus (17, 18, 22, 24, 25, 28)
Thymallus baicalensis (11, 14)
Thymallus thymallus (6, 12, 14, 21, 28)
Tinca tinca (7, 17, 18, 22, 23, 24)

Distribution: **CR:** L – Rivers Kamenice, Malše and Vltava, Máchovo jezero pond system, pond systems of Třeboň and Jindřichův Hradec, surroundings of Karlovy Vary; D – Rivers Bystřice near Olomouc, Jihlava R. and Morava R.; O – Rivers Moravice and Osoblaha, Hořina Brook; **SR:** D – Rivers Dunaj (= Danube), Hnilec, Hornád, Latorica, Váh and Tisa, Dobšinská water reservoir; V – Rivers Dunajec and Poprad.
Records: 1. Kohn 1929; 2. Dyk 1954c; 3. Dyk and Lucký 1954e; 4. Dyk and Lucký 1956b; 5. Dyk and Lucký 1957; 6. Dyk 1961a; 7. Ergens 1962a, 8. Ergens 1965a; 9. Žitňan 1967a; 10. Řehulka 1968; 11. Žitňan 1969b; 12. Moravec 1970; 13. Žitňan 1970–1971; 14. Žitňan 1973a; 15. Ergens et al. 1975; 16. Žitňan 1976; 17. Moravec 1978d; 18. Moravec 1979a; 19. Žitňan 1979e; 20. Moravec 1980c; 21. Žitňan 1980b; 22. Moravec 1984e; 23. Moravec 1985a; 24. Moravec 1985b; 25. Moravec 1986c; 26. Žitňan 1986; 27. Scholz 1987b; 28. Gelnar et al. 1990; 29. Gelnar et al. 1994; 30. Moravec 1995a; 31. Moravec et al. 1997a; 32. Moravec et al. 1997b; 33. Koubková and Baruš 2000b.

Fam. Ascarididae Baird, 1853

Porrocaecum crassum (Deslongchamps, 1824) larva
Syn. *Porrocaecum* sp. of Moravec in Ergens and Lom (1970)
Site: intestine.
Host (facultative): *Gymnocephalus cernuus* (1, 2, 3)
Distribution: **SR:** – River Latorica.
Records: 1. Moravec in Ergens and Lom 1970; 2. Moravec 1971a; 3. Ergens et al. 1975.

Porrocaecum ensicaudatum (Zeder, 1800) larvae
Site: intestine, pyloric caeca.
Host (facultative): *Salmo trutta* m. *fario* (1)
Distribution: **CR:** D – River Bystřice near Olomouc.
Record: 1. Moravec 1971a.

Porrocaecum semiteres (Zeder, 1800) larvae
Site: intestine.
Host (facultative): *Perca fluviatilis* (1)
Distribution: **CR:** L – Máchovo jezero pond system.
Record: 1. Scholz 1987b.

Fam. Camallanidae Railliet et Henry, 1915

Camallanus lacustris (Zoega, 1776) larvae
Site: intestine.
Host: *Tinca tinca* (1)
Distribution (finding): **CR** and **SR:** locality not given.
Record: 1. Moravec 1995a.

Fam. Gnathostomatidae Railliet, 1895

Spiroxys contortus (Rudolphi, 1819) larvae
Site: intestinal wall, mesentery.
Host: *Leuciscus idus* (1, 2, 3)
Distribution: **SR:** D – River Latorica.
Records: 1. Moravec and Vojtková 1970; 2. Moravec 1971a; 3. Ergens et al. 1975; 4. Moravec 1994.

Fam. Anguillicolidae Yamaguti, 1935

Anguillicola crassus Kuwahara, Niimi et Itagaki, 1974 larvae
Site: abdominal cavity.
Host: *Proterorhinus marmoratus* (1, 2)
Distribution (finding): **CR:** D – Mušov water reservoir.
Record: 1. Koubková and Baruš 2000a; 2. Koubková and Baruš 2000b.

Fam. Philometridae Baylis et Daubney, 1926

Philometra obturans (Prenant, 1886) larvae
Syn. Nematoda gen. sp. larv. of Žitňan (1965, 1967) and of Moravec (1969, 1971)
Site: humour of eye.
Hosts: *Perca fluviatilis* (1, 2, 3, 4, 6)
Scardinius erythrophthalmus (4, 5)
Distribution: **CR:** L – Máchovo jezero pond system, Bystřice Brook near Hradec Králové; **SR:** D – River Latorica.
Records: 1. Žitňan 1965a; 2. Žiňan 1967c; 3. Moravec 1971c; 5. Moravec 1978d; 6. Moravec 1986c; 7. Scholz 1987b.

Philometra sp. larvae
Site: swimbladder serosa.
Hosts: *Barbus barbus* (2)
Tinca tinca (1)
Distribution: **CR:** D – Rivers Jihlava and Morava.
Records: 1. Gelnar et al. 1994; 2. Moravec et al. 1997a.

Fam. Rhabdochonidae Travassos, Artigas et Pereira, 1928

Rhabdochona phoxini Moravec, 1968 larvae
Site: intestine.
Host: **Barbatula barbatula* (1)
Distribution: **CR:** L – Rokytka Brook.
Record: 1. Moravec 1977c.

Fam. Cystidicolidae Skrjabin, 1946

Cystidicoloides ephemeridarum (Linstow, 1872) larvae
Syn. *Cystidicoloides tenuissima* Zeder, 1800
Site: stomach.
Hosts: *Barbatula barbatula* (1)
Barbus barbus (4)
Cottus gobio (1, 2, 3)
Distribution: **CR:** L – Černovický Brook; D – River Jihlava.
Records: 1. Moravec 1971a; 2. De and Moravec 1979; 3. Moravec and De 1982; 4. Moravec et al. 1997a.

Fam. Acuariidae Railliet, Henry et Sisoff, 1912

Streptocara crassicauda (Creplin, 1829) larvae
Site: stomach and intestinal wall, liver.
Hosts: *Barbatula barbatula* (1, 2)
Phoxinus phoxinus (1, 2, 4, 6)
Rutilus rutilus (2)
Salmo trutta m. *fario* (2)
Scardinius erythrophthalmus (3, 5)
Distribution (findings): **CR:** L – Máchovo jezero pond system; D – River Klanečnica; **SR:** River Dunaj (= Danube).

Records: 1. Moravec in Ergens and Lom 1970; 2. Moravec 1971a; 3. Moravec 1978d; 4. Žitňan 1979e; 5. Moravec 1986c; 6. Žitňan 1986.

Syncuaria squamata (Linstow, 1883) larvae
Site: intestinal serosa.
Host: *Tinca tinca* (1)
Distribution (finding): **CR:** L – Pond Ženich near Třeboň.
Record: 1. Moravec and Scholz 1994a.

Agamospirura sp. larvae
Site: surface of intestine.
Host: *Tinca tinca* (1)
Distribution (finding): **CR:** L – Máchovo jezero pond system.
Record: 1. Moravec 1985a.
Remark: *Agamospirura* Henry et Sisoff, 1913 is a collective name for unidentified spiruroid larvae.

Nematoda gen sp. larvae
Site: intestine.
Host: *Aspius aspius* (1)
Distribution (finding): **CR:** D – Fish farm on Jihlava River.
Record: 1. Navrátil 1991.

NEMATOMORPHA

LARVAE

Gordius aquaticus Linnaeus, 1766 larvae
Site: gut surface.
Host: *Lampetra planeri* (1)
Distribution: **CR:** L – Černovický Brook (South Bohemia).
Record: 1. Moravec 1980b.

ACANTHOCEPHALA

Fam. Neoechinorhynchidae Ward, 1918

Neoechinorhynchus rutili (Müller, 1780)
Site: intestine.
Hosts: *Abramis bjoerkna* (42)
Abramis brama (23, 37, 54, 56, 62, 64, 66, 69, 74)
Alburnoides bipunctatus (41, 42, 67, 70, 71)
Anguilla anguilla (62, 69, 70)
Aspius aspius (59)
Barbatula barbatula (9, 13, 25, 41, 42, 46, 67)
Barbus barbus (15, 26, 38, 70, 71, 73)
Carassius carassius (62)
Chondrostoma nasus (25, 26)
Cobitis taenia (34, 51, 65)
Cottus gobio (41, 42, 59, 67)
Cottus poecilopus (42, 46, 67)
Ctenopharyngodon idella (40, 45, 48, 49)
Cyprinus carpio (1, 2, 3, 4, 5, 22, 25, 26, 32, 33, 34, 38, 39, 42, 44, 53, 55, 57, 58, 62, 63, 64, 65)
Esox lucius (23, 31, 37, 54, 62, 63, 64, 66, 74)
Gobio albipinnatus (27, 34)
Gobio gobio (17, 20, 23, 25, 26, 37, 42, 43, 46, 59, 62, 63, 64, 66, 70, 71)
Gymnocephalus cernuus (23, 37, 62, 64)
Hucho hucho (41, 52)
Ictalurus nebulosus (42, 49)
Leucaspius delineatus (62)
Leuciscus cephalus (42, 43, 46, 69, 70, 71)
Leuciscus idus (8, 11, 16, 18, 19, 21, 43, 50)
Oncorhynchus mykiss (41, 46, 49)

Perca fluviatilis (25, 26, 62, 63, 64, 66, 67)
Phoxinus phoxinus (42, 46, 59, 67)
Proterorhinus marmoratus (72)
Rutilus pigus (42)
Rutilus rutilus (23, 25, 26, 37, 38, 42, 62, 65, 66, 71)
Salmo trutta m. *fario* (7, 8, 10, 11, 12, 14, 16, 18, 19, 20, 21, 22, 23, 24, 26, 28, 29, 37, 41, 42, 46, 50, 60, 71)
Salmo trutta m. *lacustris* (7, 8, 11, 18, 21, 46, 50)
Salvelinus fontinalis (8, 11, 18, 21, 49, 50)
Scardinius erythrophthalmus (6, 21, 54, 56, 62, 66)
Thymallus baicalensis (36, 46, 49)
Thymallus thymallus (9, 13, 25, 30, 41, 42, 46, 61)
Tinca tinca (62, 63, 64, 66)

Distribution: **CR:** L – Rivers Malše and Vltava, brook Černovický potok, Lipno and Římovská water reservoirs, Máchovo jezero pond system, South-Bohemian ponds near Jindřichův Hradec, Třeboň and Vodňany; D – Rivers Jihlava, Punkva and Rokytná, Kníničská water reservoir, South-Moravian ponds; O – River Moravice; **SR:** D – Rivers Dunaj (= Danube), Bodva, Hnilec, Hornád, Hron, Laborec, Latorica, Muráňka, Tisa and Váh, lake Morské oko (Great Vihorlat Lake), Dobšinská and Zemplínská Šírava water reservoirs, Slovak ponds; V – Rivers Dunajec and Poprad and tributaries, lakes Nové Štrbské pleso, Popradské pleso and Štrbské pleso.

Records: 1. Dyk 1953a; 2. Dyk 1954c; 3. Dyk and Lucký 1954a; 4. Dyk and Lucký 1954c; 5. Štědronský and Pekař 1954; 6. Vojtek et al. 1954; 7. Dyk 1955b; 8. Dyk 1956c; 9. Dyk and Lucký 1956b; 10. Kašťák 1956d; 11. Dyk 1957b; 12. Dyk and Dyková 1957; 13. Dyk and Lucký 1957; 14. Kašťák 1957d; 15. Dyk and Štědronský 1958; 16. Kašťák and Žitňan 1959; 17. Lucký 1959a; 18. Kašťák and Žitňan 1960a; 19. Kašťák and Žitňan 1960b; 20. Žitňan 1960c; 21. Dyk 1961a; 22. Dyk 1961b; 23. Ergens 1962a; 24. Dyk and Dyková 1963; 25. Dyk and Lucký 1963; 25. Dyk 1964a; 26. Lucký and Dyk 1964; 27. Žitňan 1964f; 28. Dyk 1965a; 29. Dyk 1965c; 30. Žitňan 1965c; 31. Ergens 1966a; 32. Žitňan 1966c; 33. Lucký and Jílek 1967; 34. Žitňan 1967c; 35. Žitňan 1968c; 36. Žitňan 1969b; 37. Ergens and Lom 1970; 38. Tesarčík 1970a; 39. Tesarčík 1970b; 40. Žitňan 1970b; 41. Žitňan 1970–1971; 42. Žitňan 1971a; 43. Žitňan 1971c; 44. Tesarčík 1972; 45. Žitňan 1972a; 46. Žitňan 1973a; 47. Žitňan 1973b; 48. Žitňan 1974a; 49. Žitňan 1974b; 50. Žitňan 1974c; 51. Ergens et al. 1975; 52. Žitňan 1976; 53. Žitňan 1977; 54. Moravec 1978d; 55. Žitňan 1978; 56. Moravec 1979a; 57. Žitňan 1979b; 58. Žitňan 1979d; 59. Žitňan 1979e; 60. Žitňan 1980a; 61. Žitňan 1980b; 62. Moravec 1984c; 63. Moravec 1984e; 64. Moravec 1985a; 65. Žitňan 1985; 66. Moravec 1986c; 67. Žitňan 1986; 68. Scholz 1987b; 69. Gelnar et al. 1990; 70. Moravec and Scholz 1991b; 71. Moravec and Scholz 1994b; 72. Moravec 1995a; 73. Moravec et al. 1997a; 74. Moravec et al. 1997b; 75. Špakulová 2000.

Fam Leptorhynchoididae Witenberg, 1932

Leptorhynchoides plagicephalus (Westrumb, 1821)
Site: intestine.
Hosts: *Acipenser ruthenus* (1, 2, 3, 4, 5, 6)
Acipenser sp. (2)
Distribution: **SR:** D – River Dunaj (= Danube).
Records: 1. Rašín 1936; 2. Rašín 1949a; 3. Žitňan 1979e; 4. Žitňan 1981a; 5. Žitňan 1982b; 6. Žitňan 1985.

Fam. Echinorhynchidae Cobbold, 1876

Acanthocephalus anguillae (Müller, 1780)
Syn. *Echinorhynchus globulosus* Rudolphi, 1819
Site: intestine.

Hosts: *Abramis ballerus* (40)
Abramis bjoerkna (3, 4, 9, 11, 40, 47, 53)
Abramis bjoerkna x *Rutilus rutilus* (3)
Abramis brama (2, 3, 4, 6, 7, 8, 9, 10, 11, 19, 22, 23, 26, 30, 36, 40, 53)
Abramis sapa (40, 45)
Acipenser gueldenstaedti (40, 44)
Acipenser ruthenus (9, 11, 40, 44)
Alburnoides bipunctatus (40, 45, 46)
Alburnus alburnus (15, 40, 47, 53)
Anguilla anguilla (3, 4, 12, 40)
Aspius aspius (9, 11, 23, 26, 31, 36, 40)
Barbatula barbatula (19, 28, 35, 40, 45, 48)
Barbus barbus (1, 3, 4, 5, 6, 7, 8, 9, 11, 13, 23, 24, 26, 40, 43, 46, 47, 52, 53)
Barbus petenyi (40)
Carassius carassius (3)
Chondrostoma nasus (40)
Cobitis taenia (40, 44)
Cyprinus carpio (6, 7, 8, 9, 11, 23, 24, 26, 29, 38, 40, 44)
Esox lucius (3, 4, 34)
Gymnocephalus cernuus (3, 40)
Gymnocephalus schraetser (40, 45)
Leuciscus cephalus (3, 9, 10, 11, 13, 14, 16, 17, 19, 20, 24, 25, 27, 28, 29, 31, 35, 38, 40, 46, 47, 48, 49, 51, 53)
Leuciscus idus (3, 4, 17, 21, 31, 35, 38, 40, 48)
Leuciscus leuciscus (3, 4, 13, 14, 16, 24, 29, 40, 46)
Lota lota (2, 3, 40, 45)
Oncorhynchus mykiss (8)
Pelecus cultratus (40, 44)
Perca fluviatilis (3, 34, 46)
Phoxinus phoxinus (15, 28, 32, 45)
Proterorhinus marmoratus (54)
Rutilus pigus (40)
Rutilus rutilus (3, 4, 6, 7, 8, 23, 24, 26, 29, 30, 40, 44)
Salmo trutta m. *fario* (13, 15, 24, 28, 29, 39, 40, 41)
Scardinius erythrophthalmus (3, 4, 6, 7, 10, 11, 18, 29, 35)
Silurus glanis (2, 3, 9, 11, 34, 40, 48)
Stizostedion lucioperca (34)
Tinca tinca (3, 4, 8, 19, 23, 26)
Vimba vimba (3, 4, 8, 40)
Zingel streber (40)
Zingel zingel (40, 44)

Distribution: **CR:** L – Rivers Labe (= Elbe), Sázava and Vltava; D – Rivers Dyje, Morava, Rokytná, Svitava and Svratka, Kníničská and Mušov water reservoirs, South-Moravian ponds, fish farm on Jihlava R.; O – Rivers Moravice, Opava and Osoblaha, Hořina Brook; **SR:** D – Rivers Dunaj (= Danube), Hornád, Hron, Laborec, Latorica, Morava, Ondava, Tisa, Torysa and Váh, localities Čierna Voda and Senné.

Records: 1. John 1877; 2. Šrámek 1900; 3. Šrámek 1901, 1903; 4. Frič 1908; 5. Dyk 1952b; 6. Dyk 1953a; 7. Lucký 1953a; 8. Dyk 1954c; 9. Michalovič 1954; 10. Vojtek et al. 1954; 11. Dyk 1955b; 12. Volf and Smíšek 1955; 13. Dyk and Lucký 1956b; 14. Kašťák 1956d; 15. Dyk and Lucký 1957; 16. Kašťák 1957d; 17. Kašťák 1957b; 18. Vojtek 1959b; 19. Vojtková 1959b; 20. Žitňan 1960c; 21. Pacák 1961; 22. Pacák 1962; 23. Dyk and Lucký 1963; 24. Dyk 1964a; 25. Hradil 1964; 26. Lucký and Dyk 1964; 27. Žitňan 1964f; 28. Ergens 1965a; 29. Rodák et al. 1965; 30. Lucký and Jílek 1967; 31. Žitňan 1967c; 32. Řehulka 1968; 33. Žitňan 1968c; 34. Ergens and Lom 1970; 35. Žitňan 1971a; 36. Žitňan 1971c; 37. Žitňan 1973b; 38. Ergens et al. 1975; 39. Tesarčík and Ivasik 1977; 40. Žitňan 1979e; 41. Žitňan 1980a; 42. Žitňan 1981a; 43. Kirka et al. 1982; 44. Žitňan 1985; 45. Žitňan 1986; 46. Moravec and Scholz 1991; 47. Navrátil 1991; 48. Gelnar et al. 1994; 49. Koníček and Gelnar 1994; 50. Šebelová et al. 1995; 51. Gelnar et al. 1997; 52. Moravec et al. 1997a; 53. Moravec et al. 1997b; 54. Koubková and Baruš 2000b.

Acanthocephalus lucii (Müller, 1776)
Syn. *Echinorhynchus angustatus* Rudolphi, 1809
Site: intestine.
Hosts: *Abramis bjoerkna* (56)

Abramis brama (2, 8, 10, 24, 47, 48, 50, 54)
Acipenser ruthenus (43)
Anguilla anguilla (12, 15, 20, 26, 36, 41, 42, 43, 47, 49)
Barbatula barbatula (18, 25, 27)
Barbus barbus (8, 10, 43)
Cyprinus carpio (25, 27, 41, 42, 47, 48)
Esox lucius (2, 3, 5, 8, 10, 21, 23, 25, 27, 31, 34, 36, 37, 39, 40, 41, 42, 43, 47, 48, 54)
Gobio gobio (47, 48)
Gymnocephalus cernuus (2, 6, 9, 13, 16, 17, 23, 33, 36, 39, 47, 48, 54)
Ictalurus nebulosus (43)
Lepomis gibbosus (37)
Leuciscus cephalus (9, 13, 16, 17, 36, 54, 55)
Leuciscus idus (2)
Leuciscus leuciscus (36)
Lota lota (23, 43, 52)
Oncorhynchus mykiss (6, 12)
Perca fluviatilis (2, 4, 6, 7, 9, 10, 11, 14, 19, 20, 21, 23, 25, 26, 27, 29, 30, 33, 34, 36, 37, 39, 41, 42, 43, 45, 46, 47, 48, 51, 53, 54, 57)
Proterorhinus marmoratus (58)
Rutilus rutilus (43)
Salmo trutta m. *fario* (6, 12, 15, 26, 27, 28)
Silurus glanis (6, 8, 10, 25, 27, 32, 34, 36, 43)
Stizostedion lucioperca (10, 20, 32, 34, 36, 37, 43, 57)
Stizostedion volgense (34, 36, 37, 43, 52)
Tinca tinca (41, 47, 48)
Vimba vimba (6, 22)

Distribution: **CR:** L – River Vltava, Máchovo jezero pond system, South-Bohemian ponds near Třeboň and Jindřichův Hradec, ponds in central Bohemia; D – Rivers Dyje, Morava, Punkva, Rokytná, Svitava, Svratka, brook Bystřička near Bělá, Kníničská, Mušov and Vranovská water reservoirs, small reservoirs in Morava R. basin, South-Moravia ponds and pools; O – Rivers Moravice, Opava and Osoblaha, Karlovec; **SR:** D – Rivers Dunaj (= Danube), Bodva, Hornád, Hron, Latorica, Nitra, Tisa and Váh, Zemplíská širava water reservoir, localities Čierna Voda and Senné (Laborec R. basin).

Records: 1. John 1877; 2. Šrámek 1900; 3. Jírovec 1940; 4. Dyk 1953a; 5. Lucký 1953a; 6. Dyk 1954ac 7. Dyk and Lucký 1954d; 8. Michalovič 1954; 9. Vojtek et al. 1954; 10. Dyk 1955b; 11. Lucký et al. 1955; 12. Dyk and Lucký 1956b; 13. Kašťák 1956d; 14. Vejnar 1956; 15. Dyk and Lucký 1957; 16. Kašťák 1957d; 17. Kašťák 1957b; 18. Lucký 1959a; 19. Vojtek 1959b; 20. Vojtková 1959b; 21. Žitňan 1960c; 22. Dyk 1961a; 23. Ergens 1962a; 24. Pacák 1962; 25. Dyk and Lucký 1963; 26. Dyk 1964a; 27. Lucký and Dyk 1964; 28. Dyk 1965a; 29. Ergens 1965a; 30. Rodák et al. 1965; 31. Žitňan 1965a; 32. Žitňan 1966c; 33. Přibyslavský and Lucký 1967; 34. Žitňan 1967c; 35. Žitňan 1968c; 36. Ergens and Lom 1970; 37. Žitňan 1971a; 38. Žitňan 1973b; 39. Ergens et al. 1975; 40. Žitňan 1977; 41. Moravec 1978d; 42. Moravec 1979a; 43. Žitňan 1979e; 44. Žitňan 1981a; 45. Lucký and Navrátil 1983; 46. Lucký and Navrátil 1984; 47. Moravec 1984e; 48. Moravec 1985a; 49. Moravec 1985b; 50. Moravec 1986c; 51. Scholz 1986; 52. Žitňan 1986; 53. Scholz 1987b; 54. Gelnar 1994; 55. Koníček and Gelnar 1994; 56. Moravec 1995a; 57. Moravec et al. 1997b; 58. Koubková and Baruš 2000b.

Acanthocephalus sp.
Site: intestine.
Host: *Cyprinus carpio* (1)
Distribution: **CR:** L – ponds near Vodňany.
Record: 1. Štědronský and Pekař 1954.

Echinorhynchus salmonis Müller, 1780
Site: intestine.
Host: *Salmo salar* (1)
Distribution: **CR:** L – River Labe (= Elbe).
Record: 1. Frič 1893.

Echinorhynchus truttae Schrank, 1788
Echinorhynchus salmonis Müller, 1780, auctores; *E. clavula* Dujardin, 1845, auctores
Site: intestine.
Hosts: *Cottus gobio* (6, 16, 17, 24)
Cottus poecilopus (6, 16, 17, 18, 24)
Esox lucius (13, 25)
Gobio gobio (9)
Leuciscus cephalus (17)
Lota lota (14, 16, 17, 24)
Oncorhynchus mykiss (5, 8, 17, 18, 20)
Salmo trutta m. *fario* (1, 2, 3, 4, 6, 8, 9, 10, 11, 12, 16, 17, 18, 21, 23)
Thymallus baicalensis (18, 20)
Thymallus thymallus (17, 18, 22)
Distribution: **CR:** L – River Malše, Lipno and Římovská water reservoirs; D – River Punkva; O – localities not specified;
SR: D – Rivers Dunaj (= Danube), Belá, Hron, Revúca, Tisa, Torysa and Váh, brooks Demänovský potok and Hnilec, Dobšinská water reservoir.
Records: 1. Kašťák 1955a; 2. Kašťák 1956d; 3. Dyk 1957a; 4. Kašťák 1957d; 5. Pacák 1957; 6. Pacák 1958; 7. Pacák 1960; 8. Dyk 1961a; 9. Hradil 1964; 10. Lucký and Dyk 1964; 11. Dyk 1965a; 12. Žitňan 1965c; 13. Ergens 1966a; 14. Žitňan 1967a; 15. Žitňan 1968c; 16. Ergens and Lom 1970; 17. Žitňan 1971a; 18. Žitňan 1973a; 19. Žitňan 1973b; 20. Žitňan 1974b; 21. Žitňan 1979e; 22. Žitňan 1980b; 23. Žitňan 1980a; 24. Žitňan 1986; 25. Gelnar et al. 1990.

Fam. Pomphorhynchidae Yamaguti, 1939

Pomphorhynchus laevis (Müller, 1776)
Syn. *Echinorhynchus proteus* Westrumb, 1821
Site: intestine.
Hosts: *Abramis ballerus* (35)
Abramis bjoerkna (35, 36, 46)
Abramis brama (21, 35)
Abramis sapa (35)
Acipenser ruthenus (46, 50)
Alburnoides bipunctatus (2, 23, 35, 51, 52, 53)
Alburnus alburnus (17, 23, 35, 36, 41, 46, 49, 52, 53)
Anguilla anguilla (46, 52, 53, 59)
Aspius aspius (5, 7, 9, 13, 17, 23, 31, 35, 51)
Barbatula barbatula (2, 3, 6, 16, 33, 35, 51, 52, 53)
Barbus barbus (2, 3, 5, 6, 8, 11, 17, 22, 23, 24, 26, 27, 31, 33, 35, 46, 49, 52, 53, 55)
Barbus petenyi (9, 13, 23, 31, 33, 34, 35, 46, 49)
Carassius carassius (7, 16, 35)
Chondrostoma nasus (17, 23, 27, 31, 35, 36)
Cottus gobio (14, 34, 35, 46, 51)
Ctenopharyngodon idella (37, 39, 40)
Cyprinus carpio (5, 7, 23, 25, 26, 29, 31, 35, 36, 43, 44, 45, 46, 50)
Esox lucius (5, 7, 35, 36)
Gobio albipinnatus (52, 53)
Gobio gobio (2, 23, 33, 35, 36, 52, 53)
Gobio uranoscopus (46, 50)
Gymnocephalus cernuus (10, 35, 46, 52, 53)
Gymnocephalus schraetser (7, 35, 46, 51)
Ictalurus nebulosus (26, 31, 35, 40, 46)
Lepomis gibbosus (35, 40)
Leucaspius delineatus (51)
Leuciscus cephalus (2, 3, 4, 5, 7, 9, 12, 13, 15, 17, 23, 24, 27, 31, 33, 34, 35, 36, 41, 46, 49, 51, 52, 53, 56, 58, 60, 61)
Leuciscus idus (12, 19, 35, 36, 42, 46, 51, 52, 53)
Leuciscus leuciscus (8, 9, 11, 13, 17, 22, 24, 33, 35, 46, 49, 51, 52, 53)
Lota lota (35, 46, 51)
Oncorhynchus mykiss (3, 16, 17, 23, 35, 40)
Pelecus cultratus (35, 50)
Perca fluviatilis (3, 10, 35)
Phoxinus phoxinus (16, 23, 24, 35, 46, 51, 62)
Proterorhinus marmoratus (51)
Pseudorasbora parva (57)

Rhodeus sericeus* (6, 33)
Rutilus pigus (35)
Rutilus rutilus (30, 35, 36, 46, 50, 51, 52, 53)
Salmo trutta m. *fario* (16, 17, 23, 35, 46, 47)
Scardinius erythrophthalmus (46)
Silurus glanis (5, 7, 35, 46)
Stizostedion lucioperca (5, 7, 35)
Thymallus thymallus (28, 35, 48)
Tinca tinca (35, 52, 53)
Vimba vimba (17, 23, 35, 46, 52, 53)
Zingel streber (46, 50)
Zingel zingel (35, 46, 50)

Distribution: **CR:** L – Bohemia (localities not specified); D – Rivers Dřevnice, Dyje, Morava, Rokytná, Svitava and Svratka, Obrava Brook, Kníničská and Vranovská water reservoirs, fish farm on Jihlava R.; O – River Moravice; **SR:** D – Rivers Dunaj (= Danube), Bodva, Hron, Hornád, Ipeľ, Laborec, Latorica, Ondava, Tisa, Torysa and Váh, brooks Myslava,Okna and Olšava, lake Morské oko (Great Vihorlat Lake), Domaša water reservoir, localities Čierna Voda and Senné, Slovak ponds; V – River Dunajec.

Records: 1. John 1877; 2. Rašín 1928; 3. Dyk 1954c; 4. Gotthardt 1954; 5. Michalovič 1954; 6. Vojtek et al. 1954; 7. Dyk 1955b; 8. Dyk and Lucký 1956b; 9. Kašťák 1956d; 10. Vejnar 1956; 11. Dyk and Lucký 1957; 12. Kašťák 1957b; 13. Kašťák 1957d; 14. Pacák 1958; 15. Vojtková 1959b; 16. Kašťák and Žitňan 1960a; 17. Žitňan 1960c; 18. Dyk 1961a; 19. Pacák 1961; 20. Vojtek 1961; 21. Pacák 1962; 22. Dyk 1964a; 23. Hradil 1964; 24. Lucký and Dyk 1964; 25. Žitňan 1964a; 26. Žitňan 1964f; 27. Žitňan 1965a; 28. Žitňan 1965c; 29. Žitňan 1966c; 30. Lucký and Jílek 1967; 31. Žitňan 1967c; 32. Žitňan 1968c; 33. Ergens and Lom 1970; 34. Žitňan 1970–1971; 35. Žitňan 1971a; 36. Žitňan 1971c; 37. Žitňan 1972a; 38. Žitňan 1973b; 39. Žitňan 1974a; 40. Žitňan 1974b; 41. Žitňan 1974c; 42. Ergens et al. 1975; 43. Žitňan 1978; 44. Žitňan 1979b; 45. Žitňan 1979d; 46. Žitňan 1979e; 47. Žitňan 1980a; 48. Žitňan 1980b; 49. Kirka et al. 1982; 50. Žitňan 1985; 51. Žitňan 1986; 52. Moravec and Scholz 1991a; 53. Moravec and Scholz 1991b; 54. Navrátil 1991; 55.Gelnar et al. 1994; 56. Koníček and Gelnar 1994; 57. Moravec 1995a; 58. Šebelová et al. 1995; 59. Moravec et al. 1996; 60. Gelnar et al. 1997; 61. Divílek et al. 1998; 62. Dudiňák 2000; 63. Špakulová 2000.

HIRUDINEA

Fam. Piscicolidae Johnston, 1865

Piscicola fasciata Kollar, 1842
Syn. *Cystobranchus fasciatus* (Kollar, 1842)
Site: mainly surface of head.
Host: *Silurus glanis* (1, 2, 3)
Distribution: **CR:** D – River Dyje;
SR: – Slovakia (localities not specified).
Records: 1. Dyk and Lucký 1963; 2. Lucký and Dyk 1964; 3. Žitňan 1982b.

Piscicola geometra (Linnaeus, 1761)
Site: body surface, fins, mouth and gill cavities.
Hosts: *Abramis ballerus* (38)
Abramis bjoerkna (28, 31)
Abramis brama (14, 16, 27, 46, 54, 57)
Abramis sapa (38)
Anguilla anguilla (53)
Aristichthys nobilis (43)
Aspius aspius (14, 16, 28, 30, 31)
Barbus barbus (20, 21, 30)
Barbus petenyi (30, 51)
Carassius auratus gibelio (38)
Carassius carassius (14, 16, 26, 28, 31)
Chondrostoma nasus (30)
Ctenopharyngodon idella (39, 43, 48)
Cyprinus carpio (3, 4, 5, 6, 8, 9, 11, 12, 13, 14, 16, 18, 20, 21, 24, 26, 28, 30, 31, 32, 34, 36, 37, 38, 41, 44, 52)

Esox lucius (7, 9, 12, 14, 16, 28, 31, 35, 44, 47)
Gobio gobio (30, 57)
Gymnocephalus cernuus (18, 26)
Hypophthalmichthys molitrix (50)
Leucaspius delineatus (26)
Leuciscus cephalus (14, 20, 21, 30, 41, 51, 56, 57, 58, 59, 60, 62)
Leuciscus idus (25, 42)
Lota lota (26)
Micropterus salmoides (9, 12)
Perca fluviatilis (12, 18, 26, 28, 31, 38, 46, 55, 61)
Phoxinus phoxinus (33)
Rutilus rutilus (18, 26, 34, 54, 63)
Salmo trutta m. *fario* (30)
Scardinius erythrophthalmus (54)
Silurus glanis (14, 16, 19, 28, 31, 44)
Stizostedion lucioperca (9, 18, 28, 31, 36, 38, 46)
Thymallus thymallus (49)
Tinca tinca (5, 9, 11, 17, 18, 26, 28, 31, 36, 37, 38)
Vimba vimba (30)

Distribution: **CR:** L – Rivers Labe (= Elbe) and Vltava, Máchovo jezero pond system, Lipno water reservoir, South-Bohemian ponds near Jindřichův Hradec, Vodňany and Třeboň; D – Rivers Dyje and Morava, Kníničská water reservoir, Morava R. basin, South-Moravian ponds and pools, fish farm on Jihlava R.; O – Rivers Opava and Osoblaha, ponds of Odra R. basin; **SR:** D – Rivers Dunaj (= Danube), Hron, Latorica, Morava, Ondava, Tisa, Torysa and Váh, locality Senné, Slovak ponds.

Records: 1. Šrámek 1900; 2. Frič and Vávra 1903; 3. Frič 1908; 4. Kohn 1929; 5. Poustka 1934; 6. Dyk 1939; 7. Jírovec 1940; 8. Dyk 1941; 9. Podubský 1952; 10. Dyk 1953a; 11. Lucký 1953a; 12. Dyk 1954c; 13. Gotthardt 1954; 14. Michalovič 1954; 15. Podubský and Štědronský 1954; 16. Dyk 1955b; 17. Lucký 1955a; 18. Lucký et al. 1955; 19. Dyk and Lucký 1956a; 20. Kašťák 1956d; 21. Kašťák 1957b; 22. Onderiková 1958; 23. Dyk 1961a; 24. Dyk 1961b; 25. Pacák 1961; 26. Ergens 1962a; 27. Pacák 1962; 28. Dyk and Lucký 1963; 29. Dyk 1964a; 30. Hradil 1964; 31. Lucký and Dyk 1964; 32. Žitňan 1964a; 33. Ergens 1965a; 34. Rodák et al. 1965; 35. Ergens 1966a; 36. Žitňan 1966c; 37. Lucký and Jílek 1967; 38. Žitňan 1967c; 39. Tesarčík and Ehl 1968; 40. Žitňan 1968c; 41. Žitňan 1971a; 42. Žitňan 1971c; 43. Žitňan 1972a; 44. Koubková and Vojtková 1973; 45. Žitňan 1973b; 46. Ergens et al. 1975; 47. Moravec 1979a; 48. Willomitzer 1980; 49. Žitňan 1980b; 50. Lucký 1981b; 51. Kirka et al. 1982; 52. Moravec 1984e; 53. Moravec 1985b; 54. Moravec 1986c; 55. Scholz 1987b; 56. Navrátil 1991; 57. Gelnar et al. 1994; 58. Koníček and Gelnar 1994; 59. Šebelová et al. 1995; 60. Gelnar et al. 1997; 61. Moravec et al. 1997b; 62. Gelnar et al. 1998; 63. Kadlec et al. 1998.

Piscicola respirans Troschel, 1850
Syn. *Cystobranchus respirans* (Troschel, 1850)
Site: fins and body surface.
Hosts: *Aspius aspius* (19)
Barbus barbus (3, 5, 6, 11, 19)
Barbus petenyi (3, 18, 19)
Chondrostoma nasus (4, 9, 10, 11, 13, 14)
Cobitis taenia (20)
Cyprinus carpio (3)
Esox lucius (5, 6)
Hucho hucho (22)
Leuciscus idus (16)
Lota lota (3)
Rhodeus sericeus (3)
Salmo salar (1)
Salmo trutta m. *fario* (2, 3, 4, 7, 9, 10, 11, 12, 13, 14, 18, 23)
Silurus glanis (5, 6)
Thymallus thymallus (3, 11, 15, 18, 24)

Distribution: **CR:** L – localities not given; D – River Bystřice near Olomouc; O – River Moravice; **SR:** D – Rivers Dunaj (= Danube), Hnilec, Hornád, Hron and Váh; V – River Dunajec.

Records: 1. Poustka 1934; 2. Dyk 1952a; 3. Podubský 1952; 4. Dyk 1954c; 5. Michalovič 1954; 6. Dyk 1955b; 7. Dyk 1955c; 8. Dyk 1956a; 9. Dyk and Lucký 1956b; 10. Dyk and Lucký 1957; 11. Dyk 1961a; 12. Dyk 1961b; 13. Dyk 1964a; 14. Lucký and Dyk 1964; 15. Žitňan 1965c; 16. Ertlová 1967; 17. Ergens and Lom 1970; 18. Žitňan 1970–1971; 19. Žitňan 1971a; 20. Žitňan 1971c; 21. Žitňan 1973b; 22. Žitňan 1976; 23. Žitňan 1980a; 24. Žitňan 1980b.

Fam. Glossiphonidae Vaillant, 1890

Hemiclepsis marginata (O.F. Müller, 1774)
Site: body surface, fins.
Hosts: *Abramis brama* (15)
Barbus barbus (6, 7, 10)
Cyprinus carpio (2, 3, 10, 12, 15)
Esox lucius (15)
Gobio gobio (9, 11, 15)
Gymnocephalus cernuus (4, 15)
Rutilus rutilus (4, 9, 11, 15)
Scardinius erythrophthalmus (4)
Distribution: **CR:** L – South Bohemian ponds; D – River Rokytná, Kníničská water reservoir, South Moravian ponds and pools; O – River Opava; **SR:** D – pools near Dunaj (= Danube) River.
Records: 1. Dyk 1953a; 2. Lucký 1953a; 3. Dyk 1954c; 4. Lucký et al. 1955; 5. Onderiková 1958; 6. Lucký 1959a; 7. Dyk 1961a; 8. Dyk and Lucký 1963; 9. Dyk 1964a; 10. Lucký and Dyk 1964; 11. Rodák et al. 1965; 12. Lucký and Jílek 1967; 13. Ergens and Lom 1970; 14. Koubková and Vojtková 1973; 15. Moravec 1995a.

BIVALVIA

LARVAE

Fam. Unionidae Raphinesque, 1820

Anodonta cygnea (Linnaeus, 1758) glochidia
Site: gills and fins.
Hosts: *Cyprinus carpio* (3)
Esox lucius (2)
Leuciscus cephalus (1)
Leuciscus leuciscus (4)
Phoxinus phoxinus (1)
Distribution: **CR:** L – Lipno water reservoir, Máchovo jezero pond system; O – River Osoblaha.
Records: 1. Ergens 1965a; 2. Ergens 1966a; 3. Moravec 1984e; 4. Moravec 1995a.

Anodonta sp. glochidia
Site: gills and fins.
Hosts: *Chondrostoma nasus* (3)
Esox lucius (1)
Leuciscus cephalus (4)
Perca fluviatilis (5)
Rhodeus sericeus (5)
Rutilus rutilus (5)
Scardinius erythrophthalmus (2)
Distribution: **CR:** L – Máchovo jezero pond system; D – River Morava, fish farm on Jihlava R., South Moravia.
Records: 1. Moravec 1979a; 2. Moravec 1986c; 3. Navrátil 1991; 4. Gelnar et al. 1998; 5. Blažek 2000.

Unio sp. glochidia
Site: gills and fins.
Hosts: *Abramis bjoerkna* (2)
Abramis brama (3)
Alburnoides bipunctatus (2)
Barbus barbus (2)
Ctenopharyngodon idella (4)
Esox lucius (2)
Leuciscus cephalus (1, 2)
Leuciscus leuciscus (1)
Lota lota (2)
Perca fluviatilis (1, 2)

Rhodeus sericeus (1)
Rutilus rutilus (1, 2, 3)
Scardinius erythrophthalmus (1, 2)
Stizostedion lucioperca (2)
Distribution: **CR:** D – River Morava, Kníničská water reservoir; O – River Opava; **SR:** D – basins of Dunaj (= Danube) R. and Latorica R., Nové Zámky.
Records: 1. Dyk 1964a; 2. Lucký and Dyk 1964; 3. Lucký and Jílek 1967; 4. Žitňan 1970b; 5. Blažek 2000.

Unidentified glochidia (*Glochidium* gen. sp.)
Site: gills and fins.
Hosts: *Abramis bjoerkna* (12, 14)
Abramis brama (1, 7, 11, 14)
Abramis sapa (15)
Alburnoides bipunctatus (6, 9)
Alburnus alburnus (6, 7, 10, 14, 15)
Anguilla anguilla (1)
Aspius aspius (15)
Barbatula barbatula (15)
Barbus barbus (9, 12)
Carassius carassius (1)
Cyprinus carpio (17)
Esox lucius (3, 7, 9, 12, 14, 15)
Gobio gobio (6, 15)
Gymnocephalus cernuus (1)
Ictalurus nebulosus (8, 14)
Leuciscus cephalus (6, 9, 12, 13, 15, 16, 18, 19)
Leuciscus idus (2)
Leuciscus leuciscus (13)
Lota lota (12)
Oncorhynchus mykiss (17)
Perca fluviatilis (7, 10, 12, 13, 14, 15, 19)
Phoxinus phoxinus (17)
Proterorhinus marmoratus (21)
Pseudorasbora parva (15)
Rhodeus sericeus (13)
Rutilus rutilus (6, 7, 12, 13, 14, 15, 20)
Scardinius erythrophthalmus (1, 7, 12, 13, 14, 15)
Stizostedion lucioperca (7, 12, 14, 19)
Tinca tinca (1, 7, 15)
Distribution: **CR:** L – River Vltava, pond Malá Kuš; D – Rivers Dyje, Morava, Rokytná and Svratka, Vranovská and Mušov water reservoirs, South-Moravian ponds and pools; O – River Opava; **SR:** D – Rivers Ipeľ, Latorica and Morava.
Records: 1. Frič 1908; 2. Jírovec 1934–1935; 3. Jírovec 1940; 4. Dyk 1953b; 5. Dyk 1954c; 6. Gotthardt 1954; 7. Vojtek et al. 1954; 8. Kašťák 1956d; 9. Lucký 1959a; 10. Vojtková 1959b; 11. Pacák 1962; 12. Dyk and Lucký 1963; 13. Rodák et al. 1965; 14. Ergens et al. 1975; 15. Gelnar et al. 1994;16. Koníček and Gelnar 1994; 17. Moravec 1995a; 18. Šebelová et al. 1995; 19. Moravec et al. 1997b; 20. Kadlec et al. 1998; 21. Koubková and Baruš 2000b.

COPEPODA

Fam. Ergasilidae Edwards, 1840

Ergasilus auritus Markewitsch, 1940
Site: gills and fins.
Host: *Gasterosteus aculeatus* (1)
Distribution: **SR:** D – ponds and open waters in Čunovo, southern Slovakia.
Record: 1. Mészáros 1984.

Ergasilus briani Markewitsch, 1932
Site: gills.
Hosts: *Abramis brama* (3, 4)
Alburnoides bipunctatus (2)
Alburnus alburnus (3, 5)
Leuciscus leuciscus (5)
Phoxinus phoxinus (1)
Vimba vimba (4)
Distribution: **CR:** L – localities not specified; D – Rivers Bečva, Rokytná and Trkmanka, Bystřička water reservoir.
Records: 1. Romanovský 1955c; 2. Lucký 1959a; 3. Dyk and Lucký 1963; 4. Lucký and Dyk 1964; 5. Ergens and Lom 1970.

Ergasilus gibbus Nordmann, 1832
Site: gills.
Hosts: *Anguilla anguilla* (1, 2, 3)
Distribution: **CR:** L – Bohemia (localities not specified); **SR:** D – ponds and open waters in Čunovo, southern Slovakia.
Records: 1. Frič 1873, 2. Šrámek–Hušek 1950, 3. Mészáros 1984.

Ergasilus nasus Beneden, 1871
Site: gills.
Hosts: *Abramis bjoerkna* (1, 2, 3)
Leuciscus cephalus (1)
Distribution: **SR:** D – Čierna Voda and Senné (tributaries of Laborec River).
Records: 1. Kašťák 1956a; 2. Kašťák 1956d; 3. Kašťák 1957b.
Remark: Probably these findings represent misidentifications of *E. sieboldi*.

Ergasilus sieboldi Nordmann, 1832
Syn. *Ergasilus trisetacens* Nordmann, 1832
Site: gills.
Hosts: *Abramis ballerus* (29)
Abramis bjoerkna (3, 4, 11, 12, 13, 21, 23, 29)
Abramis brama (1, 3, 8, 9, 11, 12, 13, 18, 20, 21, 23, 25, 29, 35, 37, 39)
Abramis brama × *Rutilus rutilus* (3)
Acipenser ruthenus (11, 13)
Alburnus alburnus (15, 29, 37)
Anguilla anguilla (18, 34)
Aspius aspius (11, 13, 29, 37, 39)
Barbatula barbatula (17, 21, 23)
Barbus barbus (11, 12, 13)
Carassius carassius (3, 11, 13)
Chondrostoma nasus (22, 29)
Coregonus maraena (38)
Coregonus peled (38)
Ctenopharyngodon idella (18)
Cyprinus carpio (1, 10, 11, 13, 21, 22, 23, 25, 33)
Esox lucius (1, 3, 4, 5, 11, 12, 13, 16, 19, 21, 22, 23, 24, 26, 29, 30, 31, 37, 39)
Gobio gobio (38)
Gymnocephalus cernuus (3, 4, 26, 29)
Gymnocephalus schraetser (29)
Ictalurus nebulosus (29)
Leuciscus cephalus (10, 11, 12, 13, 14, 29, 36, 37)
Leuciscus idus (3, 29, 32)
Leuciscus leuciscus (38)
Perca fluviatilis (3, 8, 9, 12, 18, 21, 23, 26, 29, 37)
Rutilus pigus (29)
Rutilus rutilus (3, 9, 11, 12, 13, 18, 25, 32, 39, 40)
Scardinius erythrophthalmus (3, 11, 12, 13, 21, 23, 29)
Silurus glanis (1, 3, 8, 9, 11, 13, 21, 23, 29)
Stizostedion lucioperca (11, 12, 13, 18, 19, 29, 37, 39)
Thymallus thymallus (19)
Tinca tinca (3, 4, 6, 8, 9, 12, 14, 18, 21, 23, 25, 29, 31, 32, 37)
Vimba vimba (3, 4, 9, 22)
Distribution: **CR:** L – Rivers Labe (= Elbe) and Vltava, Lipno water reservoir, Máchovo jezero pond system, pond Malá Kuš near Lnáře; D – Rivers Morava, Rokytná and Svratka, Morava R. basin, Kníničská and Vranovská water reservoirs, South-Moravian ponds; O – localities not specified; **SR:** D – Rivers Dunaj (= Danube), Bodrog, Ipeľ, Latorica, Morava, Torysa and Váh, ponds near Čunovo, Lučenec, Nové Zámky.
Records: 1. Frič 1873; 2. Šrámek 1900; 3. Frič and Vávra 1903; 4. Frič 1908; 5. Jírovec 1940; 6. Šrámek–Hušek 1950; 7. Dyk 1952b; 8. Dyk 1953c; 9. Dyk 1954c; 10. Gotthardt 1954; 11. Michalovič 1954; 12. Vojtek et al. 1954; 13. Dyk 1955b; 14. Romanovský 1955d; 15. Dyk 1956a; 16. Kašťák 1956d; 17. Lucký 1959a; 18. Vojtková 1959b; 19. Dyk and Štědronský 1960; 20. Pacák 1962; 21. Dyk and Lucký 1963; 22. Hradil 1964; 23. Lucký and Dyk 1964; 24. Ergens 1966a; 25. Lucký and Jílek 1967; 26. Přibyslavský and Lucký 1967; 27. Ergens and Lom 1970; 28. Žitňan 1970b; 29. Ergens et al. 1975; 30. Moravec 1979a; 31. Žitňan 1982b; 32. Mészáros 1984; 33. Moravec 1984e; 34. Moravec 1985b;

35. Moravec 1986c; 36. Koníček and Gelnar 1994; 37. Gelnar et al. 1994; 38. Moravec 1995a; 39. Moravec et al. 1997b; 40. Kadlec et al. 1998.

Ergasilus sp.
Site: gills.
Hosts: *Abramis brama* (1)
Leuciscus idus (1)
Rutilus rutilus (1)
Scardinius erythrophthalmus (1)
Distribution: **CR:** L – Bohemia (localities not specified).
Record: 1. Frič 1908.
Remark: These findings probably concern *E. sieboldi*.

Neoergasilus japonicus (Harada, 1930)
Site: fins, rarely gills.
Hosts: *Abramis bjoerkna* (1)
Alburnus alburnus (1)
Carassius carassius (1)
Leuciscus cephalus (2)
Scardinius erythrophthalmus (1)
Stizostedion lucioperca (1, 2)
Distribution: **SR:** D – River Latorica.
Records: 1. Hanek 1968; 2. Ergens et al. 1975.

Neoergasilus longispinosus Yin, 1956
Site: gills, fins and nasal cavities.
Host: *Ctenopharyngodon idella* (1)
Distribution: **SR:** D – ponds and open waters near Čunovo, southern Slovakia.
Record: 1. Mészáros 1984.

Paraergasilus rylovi Markewitsch, 1937
Site: nasal cavities.
Hosts: *Abramis ballerus* (1, 2)
Abramis bjoerkna (1, 2, 3)
Abramis brama (1, 2, 3)
Alburnus alburnus (1, 2)
Cyprinus carpio (1)
Leuciscus cephalus (1, 2, 3)
Pelecus cultratus (1, 2)
Scardinius erythrophthalmus (3)
Distribution: **SR:** D – River Latorica.
Records: 1. Hanek 1967; 2. Ergens and Lom 1970; 3. Ergens et al. 1975.

Thersitina gasterostei (Pagenstecher, 1861)
Site: gills.
Host: *Gasterosteus aculeatus* (1)
Distribution: **SR:** D – ponds and open waters near Čunovo, southern Slovakia.
Record: Mészáros 1984.

Fam. Lernaeidae Cobbold, 1879

Lamproglena pulchella Nordmann, 1832
Site: gills.
Hosts: *Abramis brama* (19)
Alburnus alburnus (18)
Aspius aspius (1, 4, 9, 18, 19)
Chondrostoma nasus (5, 14)
Cyprinus carpio (6)
Esox lucius (1)
Leuciscus cephalus (1, 3, 4, 5, 6, 7, 8, 10, 11, 12, 13, 14, 15, 16, 17, 18, 21)
Leuciscus idus (3, 12, 19, 20)
Leuciscus leuciscus (3, 4, 18)
Rutilus pigus (19)
Rutilus rutilus (5, 7, 14, 16, 18, 19)
Scardinius erythrophthalmus (3, 11, 19)
Distribution: **CR:** L – Rivers Labe (= Elbe) and Blanice, pond near Prague; D – Rivers Dyje, Morava and Svratka, Vranovská water reservoir, South-Moravian ponds; O – River Osoblaha, **SR:** D – Rivers Hornád, Ipeľ, Latorica and Torysa, Myslava Brook, Oravská water reservoir, localities Čierna Voda and Senné.
Records: 1. Frič 1873; 2. Šrámek 1900; 3. Frič and Vávra 1903; 4. Frič 1908; 5. Dyk 1954c; 6. Gotthardt 1954; 7. Vojtek et al. 1954; 8. Dyk 1955b; 9. Romanovský 1955d; 10. Dyk 1956a; 11. Kašťák 1956d; 12. Kašťák 1957b; 13. Lucký 1959a; 14. Dyk 1961a; 15. Hradil 1964; 16. Lucký and Dyk 1964; 17. Ergens 1965; 18. Ergens and Lom 1970; 19. Ergens et al. 1975; 20. Gelnar et al. 1994; 21. Koníček and Gelnar 1994.

Lernaea ctenopharyngodontis Yin, 1960
Site: skin.
Host: *Ctenopharyngodon idella* (1)
Distribution: **SR:** localities not given.
Record: 1. Žitňan 1982b.

Lernaea cyprinacea Linnaeus, 1758
Site: skin.
Hosts: *Barbus barbus* (11)
Carassius carassius (11)
Cyprinus carpio (5, 9)
Esox lucius (2)
Gobio albipinnatus (8)
Gymnocephalus schraetser (8)
Leuciscus cephalus (11)
Perca fluviatilis (1)
Rhodeus sericeus (10)
Rutilus rutilus (4, 6, 7)
Stizostedion lucioperca (11)
Tinca tinca (9)
Distribution: **CR:** L – Bohemia (Prague); D – River Morava, South-Moravian ponds near Pohořelice; **SR:** D – River Latorica.
Records: 1. Frič 1873; 2. Frič 1908; 3. Šrámek-Hušek 1950; 4. Dyk 1952b; 5. Romanovský 1953; 6. Dyk 1954c; 7. Lucký and Dyk 1964; 8. Ergens et al. 1975; 9. Žitňan 1982b; 10. Gelnar et al. 1994; 11. Moravec 1995a.

Lernaea esocina (Burmeister, 1833)
Site: skin and gills.
Hosts: *Esox lucius* (1, 3, 5, 6, 7)
Phoxinus phoxinus (2)
Distribution: **CR:** L – River Labe (= Elbe), brook Kozákov and brook near Turnov; D – Rivers Morava and Svratka; **SR:** localities not given.
Records: 1. Frič 1873; 2. Vejdovský 1877; 3. Frič and Vávra 1903; 4. Šrámek–Hušek 1950; 5. Dyk 1954c; 6. Lucký and Dyk 1964; 7. Žitňan 1982b; 8. Gelnar et al. 1994.
Remark: According to some authors, this species is a synonym of *L. cyprinacea*.

Fam. Dichelesthiidae Edwards, 1840

Dichelesthium oblongum (Abildgaard, 1794)
Site: gills.
Host: *Acipenser sturio* (1)
Distribution: **CR:** L – Bohemia (localities not given).

Record: 1. Frič 1872 (according to Šrámek–Hušek 1950).
Remark: Šrámek–Hušek's (1950) information is misleading, because in 1873 Frič considered the finding of this parasite in Bohemia as probable, but it was not recorded from there.

Fam. Lernaeopodidae Edwards, 1840

Achtheres percarum Nordmann, 1832
Site: gills, mouth cavity.
Hosts: *Perca fluviatilis* (1, 3, 6)
Stizostedion lucioperca (4, 5, 6, 7, 8)
Stizostedion volgense (7)
Distribution: **CR:** L – Bohemia (localities not given); D – South-Moravia (localities not specified); **SR:** D – Rivers Dunaj (= Danube), Latorica and Váh.
Records: 1. Frič 1873; 2. Šrámek–Hušek 1950; 3. Dyk 1954c; 4. Romanovský 1955d; 5. Kašťák 1956d; 6. Dyk 1961a; 7. Ergens and Lom 1970; 8. Ergens et al. 1975.

Basanistes huchonis (Schrank, 1786)
Site: inner surface of gill-covers.
Host: *Hucho hucho* (1, 2, 3, 4, 5, 6)
Distribution: **SR:** D – Rivers Dunaj (= Danube), Hornád, Orava, Poprad, Turiec and Váh.
Records: 1. Bartoš 1934; 2. Dyk 1955b; 3. Romanovský 1955d; 4. Žitňan 1976; 5. Žitňan 1982b; 6. Žitňan 1985.

Pseudotracheliastes stellifer (Kollar, 1836)
Site: mouth and gill cavities.
Host: *Silurus glanis* (1, 2)
Distribution: **CR** and **SR** (former Czechoslovakia): (localities not given).
Records: 1. Markevich 1956; 2. Ergens and Lom 1970.

Salmincola salmoneus (Linnaeus, 1758)
Syn. *Lernaepoda salmonea* (Blainville, 1822)
Site: gills.
Hosts: *Salmo salar* (1)
Salmo trutta m. *fario* (2)
Distribution: **CR:** L – Bohemia (localities not specified); D – localities not given.
Records: 1. Frič 1873; 2. Ergens and Lom 1970.

Tracheliastes chondrostomi Hanek, 1969
Site: fins.
Host: *Chondrostoma nasus* (1, 2)
Distribution: **SR:** D – River Latorica.
Records: 1. Hanek 1969; 2. Ergens et al. 1975.

Tracheliastes maculatus Kollar, 1836
Site: scales.
Hosts: *Abramis brama* (6)
Chondrostoma nasus (2, 4, 5, 7, 8)
Leuciscus cephalus (9)
Rutilus rutilus (1, 3, 4, 5)
Distribution: **CR:** D – River Svratka; **SR:** D – Rivers Dunaj (= Danube) and Ondava.
Records: 1. Romanovský 1954a; 2. Dyk 1955a; 3. Dyk 1955b; 4. Romanovský 1955d; 5. Dyk 1961a; 6. Pacák 1962; 7. Lucký and Dyk 1964; 8. Kirka et al. 1982; 9. Moravec 1995a.

Tracheliastes polycolpus Nordmann, 1832
Site: fins, less often scales.
Hosts: *Barbus barbus* (4, 5)
Chondrostoma nasus (5)
Cyprinus carpio (5)
Leuciscus cephalus (5)
Phoxinus phoxinus (1, 2)
Distribution: **CR:** L – Brook near Turnov; D – Vranovská water reservoir; O – localities not specified.
Records: 1. Vejdovský 1877; 2. Frič 1908; 3. Šrámek–Hušek 1950; 4. Vojtek et al. 1954; 5. Ergens and Lom 1970.

BRANCHIURA

Fam. Argulidae Müller, 1785

Argulus coregoni Thorell, 1864
Site: skin.
Hosts: *Abramis bjoerkna* (7)
Abramis brama (14, 15, 18, 19)
Alburnus alburnus (7)
Aspius aspius (20)
Barbus barbus (2, 7, 11, 17)
Chondrostoma nasus (5, 7, 9, 10, 16, 18)
Cyprinus carpio (19)
Esox lucius (9, 10, 16, 18)
Hucho hucho (21)
Leuciscus cephalus (5, 7, 9, 10, 15, 18)
Leuciscus idus (15, 18)
Leuciscus leuciscus (8)
Oncorhynchus mykiss (9)
Phoxinus phoxinus (2)
Salmo salar (1)
Salmo trutta (1)
Salmo trutta m. *fario* (3, 4, 5, 6, 7, 9, 10, 16, 18)
Salmo trutta m. *lacustris* (22)
Scardinius erythrophthalmus (15, 18)
Silurus glanis (7)
Stizostedion lucioperca (7)
Thymallus thymallus (5, 9)
Tinca tinca (7)
Vimba vimba (11, 17)
Distribution: **CR:** L – River Vltava, Labe (= Elbe) R. basin, surroundings of Karlovy Vary; D – River Dyje, Knínička water reservoir; O – River Moravice; **SR:** D – Rivers Dunaj (= Danube), Latorica and Torysa; V – River Poprad.
Records: 1. Frič 1908; 2. Kohn 1929; 3. Dyk 1951; 4. Dyk 1954a; 5. Dyk 1954c; 6. Dyk 1955c; 7. Romanovský 1955a; 8. Dyk 1956a; 9. Dyk and Lucký 1956b; 10. Dyk and Lucký 1957; 11. Hradil 1958; 12. Dyk 1961a; 13. Dyk 1961b; 14. Pacák 1962; 15. Dyk and Lucký 1963; 16. Dyk 1964a; 17. Hradil 1964; 18. Lucký and Dyk 1964; 19. Lucký and Jílek 1967; 20. Ergens et al. 1975; 21. Žitňan 1976; 22. Moravec 1995a.

Argulus foliaceus (Linnaeus, 1758)
Site: skin.
Hosts: *Abramis bjoerkna* (18, 38)
Abramis brama (15, 18, 26, 29, 34, 41, 44)

Alburnus alburnus (18, 24)
Barbus barbus (14, 16, 25)
Carassius carassius (12, 14, 16, 18, 25)
Chondrostoma nasus (38)
Coregonus maraena (18)
Ctenopharyngodon idella (36)
Cyprinus carpio (1, 4, 8, 10, 11, 12, 13, 14, 16, 18, 23, 25, 30, 31, 32, 34, 40, 44)
Esox lucius (5, 14, 15, 16, 18, 23, 25, 33, 35, 38, 39)
Gobio gobio (21, 22, 24, 31)
Gymnocephalus cernuus (12)
Ictalurus nebulosus (18)
Leuciscus cephalus (12, 13, 14, 15, 17, 24, 42, 43)
Leuciscus idus (8)
Oncorhynchus mykiss (18)
Perca fluviatilis (12, 15, 18, 26, 30, 32, 35, 38, 42)
Proterorhinus marmoratus (46)
Rhodeus sericeus (18)
Rutilus rutilus (13, 16, 17, 18, 20, 25, 30, 34, 41, 45)
Salmo trutta m. *fario* (18)
Scardinius erythrophthalmus (18, 20, 25, 38, 41)
Silurus glanis (18)
Stizostedion lucioperca (14, 16, 30, 43)
Tinca tinca (18, 25, 30, 32, 34)

Distribution: **CR:** L – Rivers Labe (= Elbe) and Vltava, Lipno water reservoir, Máchovo jezero pond system, pond Rožmberk, pond Malá Kuš near Lnáře; D – River Svratka, Morava R. basin, Kníničská, Mušov and Vranovská water reservoirs, South-Moravian ponds; O – River Moravice, ponds of Odra R. basin; **SR:** D – Rivers Dunaj (= Danube), Bodrog, Hornád, Ipeľ, Latorica, Morava and Váh, ponds near Lučenec.

Recoreds: 1. Frič 1873; 2. Šrámek 1900; 3. Frič and Vávra 1903; 4. Dyk 1939; 5. Jírovec 1940; 6. Hrbáček 1949; 7. Šrámek–Hušek 1950; 8. Dyk 1951; 9. Dyk 1953a; 10. Lucký 1953a; 11. Romanovský 1953; 12. Dyk 1954c; 13. Gotthardt 1954; 14. Michalovič 1954; 15. Vojtek et al. 1954; 16. Dyk 1955b; 17. Lucký et al. 1955; 18. Romanovský 1955a; 19. Romanovský 1955b; 20. Dyk 1956a; 21. Dyk and Lucký 1956b; 22. Dyk and Lucký 1956c; 23. Kašťák 1956d; 24. Dyk and Lucký 1957; 25. Hradil 1958; 26. Vojtková 1959b; 27. Dyk 1961a; 28. Dyk 1961b; 29. Pacák 1962; 30. Dyk and Lucký 1963; 31. Dyk 1964a; 32. Lucký and Dyk 1964; 33. Ergens 1966a; 34. Lucký and Jílek 1967; 35. Přibyslavský et al. 1967; 36. Tesarčík and Ehl 1968; 37. Ergens and Lom 1970; 38. Ergens et al. 1975; 39. Moravec 1979a; 40. Moravec 1984e; 41. Moravec 1986c; 42. Scholz 1987b; 43. Moravec 1995a; 44. Moravec et al. 1997b; 45. Kadlec et al. 1998; 46. Koubková and Baruš 2000b.

Argulus japonicus Thiele, 1900
Syn. *Argulus pellucidus* Wagler, 1935
Site: skin.
Hosts: *Abramis brama* (3, 4, 10, 17, 18)
Barbus barbus (21)
Carassius carassius (11)
Coregonus maraena (10)
Cyprinus carpio (1, 3, 4, 5, 7, 8, 9, 10, 11, 14, 15, 16, 17, 18)
Esox lucius (1, 8, 11, 15, 16, 19)
Gymnocephalus cernuus (10)
Hypophthalmichthys molitrix (20)
Perca fluviatilis (1, 6, 7, 10, 11)
Rutilus rutilus (3, 4, 10, 13, 17, 18)
Scardinius erythrophthalmus (6, 10, 11, 13)
Stizostedion lucioperca (10)
Tinca tinca (7, 11, 15, 16, 17, 18)

Distribution: **CR:** L – South-Bohemian ponds near Blatná, Vodňany and Třeboň, ponds near Lnáře; D – Rivers Dyje, and Svratka, Kníničská and Vranovská water reservoirs, South-Moravian ponds; O – ponds of Odra R. basin; **SR:** D – Rivers Dunaj (= Danube) and Latorica, ponds near Malacky.

Records: 1. Hrbáček 1949; 2. Šrámek–Hušek 1950; 3. Dyk 1953a; 4.

Lucký 1953a; 5. Romanovský 1953; 6. Dyk 1954c; 7. Vojtek et al. 1954; 8. Dyk 1955b; 9. Dyk and Štědronský 1955; 10. Lucký et al. 1955; 11. Romanovský 1955a; 12. Romanovský 1955b; 13. Dyk 1956a; 14. Lucký 1961; 15. Dyk and Lucký 1963; 16. Dyk 1964a; 17. Lucký and Dyk 1964; 18. Lucký and Jílek 1967; 19. Ergens et al. 1975; 20. Lucký 1981b; 21. Moravec 1995a.

Argulus sp.
Site: skin.
Hosts: *Cyprinus carpio* (1)
Esox lucius (1)
Perca fluviatilis (2)
Tinca tinca (1, 2)
Distribution: **CR:** L – localities not given, D – surroundings of Lednice.
Records: 1. Frič 1908; 2. Dyk and Lucký 1954a.

ACARIFORMES

Fam. Eremaeidae Willmann, 1931

Hydrozetes lacustris (Michael, 1882)
Syn. *Notaspis lacustris* Michael, 1882
Site: gills.
Hosts: *Chondrostoma nasus* (5)
Coregonus maraena (2, 4)
Leuciscus cephalus (3)
Distribution: **CR:** L – fish hatchery in Třeboň; D – tanks in Brno, ponds near Pohořelice.
Records: 1. Ginter 1932; 2. Dyk 1954c; 3. Vojtek et al. 1954; 4. Lucký and Dyk 1964; 5. Navrátil 1991.

Hydrozetes sp.
Site: gills.
Host: *Anguilla anguilla* (1)
Distribution: **CR:** L – Máchovo jezero pond system.
Record: 1. Moravec 1985b.

HOST – PARASITE LIST

(fish names designated with an asterisk mean non-indigenous species)

Petromyzontiformes

Fam. Petromyzontidae:

Eudontomyzon danfordi **Regan, 1911**
Monogenea: **Gyrodactylus nemachili*

Lampetra planeri **Bloch, 1784**
Nematoda: *Cucullanus truttae* larv., *Raphidascaris acus* larv.
Nematomorpha: *Gordius aquaticus* larv.

Acipenseriformes

Fam. Acipenseridae:

Acipenser gueldenstaedti **Brandt, 1833**
Trematoda: *Diplostomum spathaceum* larv., *Diplostomum* sp. larv.
Acanthocephala: *Acanthocephalus anguillae*

Acipenser ruthenus **Linnaeus, 1758**
Coelenterata: *Polypodium hydriforme*
Cestoda: *Amphilina foliacea*, *Cyathocephalus truncatus*, *Triaenophorus crassus*
Trematoda: *Crepidostomum auriculatum*, *Skrjabinopsolus semiarmatus*, *Nicolla skrjabini*, *Diplostomum spathaceum* larv., *Diplostomum* sp. larv., *Metagonimus yokogawai* larv.
Nematoda: *Cystoopsis acipenseris*, *Hysterothylacium bidentatum*
Acanthocephala: *Leptorhynchoides plagicephalus*, *Acanthocephalus anguillae*, *Acanthocephalus lucii*, *Pomphorhynchus laevis*
Copepoda: *Ergasilus sieboldi*

Anguilliformes

Fam. Anguillidae:

Anguilla anguilla **(Linnaeus, 1758)**
Monogenea: **Gyrodactylus lucii*, *Pseudodactylogyrus anguillae*, *Pseudodactylogyrus bini*
Cestoda: *Bothriocephalus claviceps*, *Triaenophorus nodulosus*, *Proteocephalus macrocephalus*, *Triaenophorus nodulosus* larv.
Trematoda: *Azygia lucii*, *Metagonimus yokogawai* larv.
Nematoda: *Raphidascaris acus*, *Paraquimperia tenerrima*, *Camallanus lacustris*, *Camallanus truncatus*, *Anguillicola crassus*, *Daniconema anguillae*, **Cystidicoloides ephemeridarum*, *Spinitectus inermis*, *Eustrongylides mergorum* larv., *Raphidascaris acus* larv.
Acanthocephala: *Neoechinorhynchus rutili*, *Acanthocephalus anguillae*, *Acanthocephalus lucii*, *Pomphorhynchus laevis*
Hirudinea: *Piscicola geometra*
Bivalvia: *Glochidium* sp. (larv.)
Copepoda: *Ergasilus gibbus*, *Ergasilus sieboldi*
Acari: *Hydrozetes* sp.

Clupeiformes

Fam. Clupeidae:

Alosa alosa **(Linnaeus, 1758)**
Monogenea: *Mazocraes alosae*
Nematoda: *Anisakis simplex* larv.

Cypriniformes

Fam. Cyprinidae:

***Abramis ballerus* (Linnaeus, 1758)**

Monogenea: *Dactylogyrus cornu*, *Dactylogyrus chranilowi*, *Gyrodactylus elegans*, *Gyrodactylus laevis*, *Gyrodactylus prostae*, *Diplozoon paradoxum*, *Diplozoon* sp., *Paradiplozoon nagibinae*

Cestoda: *Proteocephalus torulosus*, *Neogryporhynchus cheilancristrotus* larv.

Trematoda: *Asymphylodora imitans*, *Palaeorchis incognitus*, *Palaeorchis unicus*, *Phyllodistomum elongatum*, *Phyllodistomum folium*, *Nicolla skrjabini*, *Sphaerostomum globiporum*, *Bucephalus polymorphus* larv., *Diplostomum spathaceum* larv., *Diplostomum* sp. larv., *Tylodelphys clavata* larv., *Hysteromorpha triloba* larv., *Posthodiplostomum cuticola* larv., *Ichthyocotylurus pileatus* larv., *Paracoenogonimus ovatus* larv., *Apophallus muehlingi* larv., *Metagonimus yokogawai* larv.

Nematoda: *Philometra ovata*

Acanthocephala: *Acanthocephalus anguillae*, *Pomphorhynchus laevis*

Hirudinea: *Piscicola geometra*

Copepoda: *Ergasilus sieboldi*, *Paraergasilus rylovi*

***Abramis* (syn. *Blicca*) *bjoerkna* (Linnaeus, 1758)**

Monogenea: *Dactylogyrus alatus, Dactylogyrus cornoides, Dactylogyrus cornu, Dactylogyrus crucifer, Dactylogyrus difformis, Dactylogyrus distinguendus, Dactylogyrus falcatus, Dactylogyrus vistulae, Dactylogyrus nanus, Dactylogyrus similis, Dactylogyrus sphyrna, Dactylogyrus wunderi, Dactylogyrus zandti, Dactylogyrus* sp., *Gyrodactylus carassii, Gyrodactylus laevis, Gyrodactylus prostae, Gyrodactylus vimbi, Diplozoon paradoxum, Diplozoon* sp., *Paradiplozoon bliccae*

Cestoda: *Caryophyllaeides fennica, Archigetes sieboldi, Caryophyllaeus latices, Caryophyllaeus* sp., *Proteocephalus torulosus, Ligula intestinalis* larv., *Neogryporhynchus cheilancristrotus* larv.

Trematoda: *Asymphylodora imitans, Palaeorchis incognitus, Palaeorchis unicus, Phyllodistomum elongatum, Phyllodistomum folium, Allocreadium dogieli, Sphaerostomum bramae, Sphaerostomum globiporum, Bucephalus polymorphus* larv., *Rhipidocotyle campanula* larv., *Diplostomum spathaceum* larv., *Diplostomum* sp. larv., *Tylodelphys clavata* larv., *Hysteromorpha triloba* larv., *Posthodiplostomum cuticola* larv., *Apharyngostrigea cornu* larv., *Ichthyocotylurus pileatus* larv., *Paracoenogonimus ovatus* larv., *Apophallus muehlingi* larv., *Metagonimus yokogawai* larv.

Nematoda: *Schulmanela petruschewskii, Cucullanus dogieli, Philometra ovata, Philometra rischta, Rhabdochona denudata, Raphidascaris acus* larv.

Acanthocephala: *Neoechinorhynchus rutili, Acanthocephalus anguillae, Acanthocephalus lucii, Pomphorhynchus laevis*

Hirudinea: *Piscicola geometra*

Bivalvia: *Unio* sp. larv., *Glochidium* sp. (larv.)

Copepoda: *Ergasilus nasus, Ergasilus sieboldi, Neoergasilus japonicus, Paraergasilus rylovi*

Branchiura: *Argulus coregoni, Argulus foliaceus*

***Abramis brama* (Linnaeus, 1758)**

Monogenea: *Dactylogyrus auriculatus*, *Dactylogyrus cordus*, *Dactylogyrus cornu*, *Dactylogyrus crucifer*, *Dactylogyrus falcatus*, *Dactylogyrus chranilowi*, *Dactylogyrus propinquus*, *Dactylogyrus sphyrna*, *Dactylogyrus suecicus*, *Dactylogyrus wunderi*, *Dactylogyrus zandti*, *Dactylogyrus* sp., *Gyrodactylus cyprini*, *Gyrodactylus elegans*, *Gyrodactylus laevis*, **Gyrodactylus medius*, *Gyrodactylus prostae*, *Gyrodactylus vimbi*, *Gyrodactylus* sp., *Gyrodactylus gracilis* (nomen dubium), *Diplozoon paradoxum*, *Paradiplozoon bliccae*, *Paradiplozoon homoion*

Cestoda: *Caryophyllaeides fennica*, *Archigetes sieboldi*, **Caryophyllaeus fimbriceps*, *Caryophyllaeus laticeps*, *Caryophyllaeus* sp., *Proteocephalus torulosus*, *Ligula intestinalis* larv.

Trematoda: *Asymphylodora imitans*, *Palaeorchis incognitus*, *Phyllodistomum elongatum*, *Phyllodistomum folium*, *Nicolla skrjabini*, *Sphaerostomum bramae*, *Sphaerostomum globiporum*, *Bucephalus polymorphus* larv., *Rhipidocotyle campanula* larv., *Diplostomum spathaceum* larv., *Diplostomum* sp. larv., *Tylodelphys clavata* larv., *Hysteromorpha triloba* larv., *Posthodiplostomum brevicaudatum* larv., *Posthodiplostomum cuticola* larv., *Apharyngostrigea cornu* larv., *Ichthyocotylurus pileatus* larv., *Mesostephanus appendiculatus* larv., *Paracoenogonimus ovatus* larv., *Apophallus muehlingi* larv., *Metagonimus yokogawai* larv.

Nematoda: *Schulmanela petruschewskii*, *Cucullanus dogieli*, *Camallanus truncatus*, *Molnaria intestinalis*, *Philometra ovata*, *Philometra rischta*, *Rhabdochona denudata*, *Raphidascaris acus* larv.

Acanthocephala: *Neoechinorhynchus rutili*, *Acanthocephalus anguillae*, *Acanthocephalus lucii*, *Pomphorhynchus laevis*

Hirudinea: *Piscicola geometra*, *Hemiclepsis marginata*

Bivalvia: *Unio* sp. larv., *Glochidium* sp. (larv.)

Copepoda: *Ergasilus briani*, *Ergasilus sieboldi*, *Ergasilus* sp., *Paraergasilus rylovi*, *Lamproglena pulchella*, *Tracheliastes maculatus*

Branchiura: *Argulus coregoni*, *Argulus foliaceus*, *Argulus japonicus*

Abramis sapa **(Pallas, 1814)**

Monogenea: *Dactylogyrus cornoides*, *Dactylogyrus propinquus*, *Dactylogyrus wunderi*, *Gyrodactylus elegans*, *Gyrodactylus vimbi*, *Diplozoon paradoxum*, *Diplozoon* sp., *Paradiplozoon sapae*

Cestoda: *Caryophyllaeides fennica*, *Caryophyllaeus laticeps*, Caryophyllaeidae gen. sp., *Proteocephalus torulosus*, *Triaenophorus nodulosus* larv., *Ligula intestinalis* larv.

Trematoda: *Bucephalus polymorphus*, *Rhipidocotyle campanula*, *Asymphylodora imitans*, *Palaeorchis incognitus*, *Phyllodistomum elongatum*, *Phyllodistomum folium*, *Nicolla skrjabini*, *Rhipidocotyle campanula* larv., *Diplostomum spathaceum* larv., *Diplostomum* sp. larv., *Tylodelphys clavata* larv., *Hysteromorpha triloba* larv., *Posthodiplostomum cuticola* larv., *Apharyngostrigea cornu* larv., *Ichthyocotylurus platycephalus* larv., *Paracoenogonimus ovatus* larv., *Apophallus muehlingi* larv., *Metagonimus yokogawai* larv.

Nematoda: *Cucullanus dogieli*

Acanthocephala: *Acanthocephalus anguillae*, *Pomphorhynchus laevis*

Hirudinea: *Piscicola geometra*

Bivalvia: *Glochidium* sp. (larv.)

Alburnoides bipunctatus **(Bloch, 1782)**

Monogenea: *Dactylogyrus tissensis*, *Dactylogyrus vistulae*, *Dactylogyrus* sp., *Gyrodactylus gracilihamatus*, *Gyrodactylus hronosus*, *Diplozoon paradoxum*, *Paradiplozoon homoion*, *Octomacrum europaeum*

Cestoda: *Caryophyllaeus laticeps*, *Proteocephalus torulosus*

Trematoda: *Allocreadium isoporum*, *Nicolla skrjabini*, *Diplostomum spathaceum* larv., *Diplostomum* sp. larv., *Tylodelphys clavata* larv., *Paracoenogonimus ovatus* larv., *Apophallus donicus* larv., *Apophallus muehlingi* larv., *Metagonimus yokogawai* larv.

Nematoda: *Pseudocapillaria tomentosa*, *Rhabdochona denudata*

Acanthocephala: *Neoechinorhynchus rutili, Acanthocephalus anguillae, Pomphorhynchus laevis*
Bivalvia: *Unio* sp. larv., *Glochidium* sp. (larv.)
Copepoda: *Ergasilus briani*

Alburnus alburnus (Linnaeus, 1758)
Monogenea: *Dactylogyrus alatus, Dactylogyrus crucifer, Dactylogyrus fallax, Dactylogyrus fraternus, Dactylogyrus minor, Dactylogyrus nanus, Dactylogyrus parvus, Dactylogyrus similis, Dactylogyrus tissensis, Dactylogyrus vistulae, Gyrodactylus carassii, Gyrodactylus glaeseri, Gyrodactylus gracilihamatus, Gyrodactylus hronosus, Gyrodactylus laevis, Gyrodactylus vimbi, Diplozoon paradoxum, Diplozoon* sp., *Paradiplozoon alburni, Paradiplozoon homoion, Paradiplozoon* sp.
Cestoda: *Caryophyllaeides fennica, Caryophyllaeus fimbriceps, Caryophyllaeus laticeps, Caryophyllaeus* sp., Caryophyllaeidae gen. sp., *Proteocephalus torulosus, Ligula intestinalis* larv., *Neogryporhynchus cheilancristrotus* larv.
Trematoda: *Rhipidocotyle campanula, Phyllodistomum dogieli, Phyllodistomum elongatum, Phyllodistomum folium, Phyllodistomum* sp., *Allocreadium isoporum, Nicolla skrjabini, Sphaerostomum bramae, Bucephalus polymorphus* larv., *Rhipidocotyle campanula* larv., *Diplostomum spathaceum* larv., *Diplostomum* sp. larv., *Tylodelphys clavata* larv., *Posthodiplostomum cuticola* larv., *Paracoenogonimus ovatus* larv., *Apophallus donicus* larv., *Apophallus muehlingi* larv., *Metagonimus yokogawai* larv.
Nematoda: *Cucullanus dogieli, Philometra rischta, Rhabdochona denudata, Raphidascaris acus* larv.,
Acanthocephala: *Acanthocephalus anguillae, Pomphorhynchus laevis*
Bivalvia: *Glochidium* sp. (larv.)
Copepoda: *Ergasilus briani, Ergasilus sieboldi, Neoergasilus japonicus, Paraergasilus rylovi, Lamproglena pulchella*
Branchiura: *Argulus coregoni, Argulus foliaceus*

***_Aristichthys nobilis_ (Richardson, 1844)**
Monogenea: *Dactylogyrus nobilis*
Cestoda: *Paradilepis scolecina* larv., *Valipora campylancristrota* larv.
Trematoda: *Rhipidocotyle campanula* larv., *Diplostomum spathaceuam* larv., *Diplostomum* sp. larv.
Hirudinea: *Piscicola geometra*

Aspius aspius (Linnaeus, 1758)
Monogenea: *Dactylogyrus cornu, Dactylogyrus similis, Dactylogyrus tuba, Gyrodactylus* sp., *Diplozoon paradoxum, Paradiplozoon pavlovskii, Paradiplozoon vojteki*
Cestoda: *Caryophyllaeides fennica, Caryophyllaeus laticeps, Triaenophorus crassus, Proteocephalus torulosus, Proteocephalus* sp.
Trematoda: *Phyllodistomum elongatum, Nicolla skrjabini, Sphaerostomum bramae, Sphaerostomum globiporum, Rhipidocotyle campanula* larv., *Diplostomum spathaceum* larv., *Diplostomum* sp. larv., *Tylodelphys clavata* larv., *Posthodiplostomum cuticola* larv., *Ichthyocotylurus pileatus* larv., *Paracoenogonimus ovatus* larv., *Apophallus muehlingi* larv., *Metagonimus yokogawai* larv.
Nematoda: **Paraquimperia tenerrima, Cucullanus dogieli, Camallanus truncatus, Philometra kotlani, Philometra ovata, Philometra rischta, Rhabdochona denudata, Raphidascaris acus* larv., Nematoda gen. sp. larv.
Acanthocephala: *Neoechinorhynchus rutili, Acanthocephalus anguillae, Pomphorhynchus laevis*
Hirudinea: *Piscicola geometra, Piscicola respirans*
Bivalvia: *Glochidium* sp. (larv.)
Copepoda: *Ergasilus sieboldi, Lamproglena pulchella*
Branchiura: *Argulus coregoni*

Barbus barbus (Linnaeus, 1758)

Monogenea: **Dactylogyrus auriculatus, Dactylogyrus carpathicus, Dactylogyrus dyki, Dactylogyrus malaeus, *Dactylogyrus vastator, Gyrodactylus barbi, Gyrodactylus hemibarbi, Gyrodactylus katharineri, Gyrodactylus malmbergi, Gyrodactylus markewitschi, *Gyrodactylus pannonicus, Gyrodactylus* sp., *Diplozoon paradoxum, Diplozoon* sp., *Paradiplozoon homoion, Paradiplozoon tisae*

Cestoda: **Cyathocephalus truncatus, Caryophyllaeides fennica, Caryophyllaeus brachycollis, Caryophyllaeus laticeps, Bathybothrium rectangulum, Proteocephalus torulosus*

Trematoda: *Sanguinicola* sp., *Asymphylodora markewitschi, Allocreadium isoporum, Nicolla skrjabini, Diplostomum spathaceum* larv., *Diplostomum* sp. larv., *Tylodelphys clavata* larv., *Posthodiplostomum cuticola* larv., *Paracoenogonimus ovatus* larv., *Metagonimus yokogawai* larv.

Nematoda: *Pseudocapillaria tomentosa, *Rhabdochona denudata, Rhabdochona hellichi, Raphidascaris acus* larv., *Philometra* sp. larv., *Cystidicoloides ephemeridarum* larv.

Acanthocephala: *Neoechinorhynchus rutili, Acanthocephalus anguillae, Acanthocephalus lucii, Pomphorhynchus laevis*

Hirudinea: *Piscicola geometra, Piscicola respirans, Hemiclepsis marginata*

Bivalvia: *Unio* sp. larv., *Glochidium* sp. (larv.)

Copepoda: *Ergasilus sieboldi, Tracheliastes polycolpus, Lernaea cyprinacea*

Branchiura: *Argulus coregoni, Argulus foliaceus, Argulus japonicus*

Barbus petenyi Heckel, 1847

Monogenea: *Dactylogyrus carpathicus, Dactylogyrus dyki, Dactylogyrus malaeus, Dactylogyrus petenyi, Gyrodactylus malmbergi, Gyrodactylus markewitschi, Gyrodactylus vimbi, Diplozoon paradoxum*

Cestoda: *Caryophyllaeus brachycollis, Caryophyllaeus laticeps, Bathybothrium rectangulum, *Triaenophorus nodulosus, Proteocephalus torulosus*

Trematoda: *Allocreadium isoporum, Nicolla skrjabini, Rhipidocotyle campanula* larv., *Diplostomum* sp. larv., *Metagonimus yokogawai* larv.

Nematoda: **Philometroides sanguinea, *Rhabdochona denudata, Rhabdochona hellichi*

Acanthocephala: *Acanthocephalus anguillae, Pomphorhynchus laevis*

Hirudinea: *Piscicola geometra, Piscicola respirans*

***_Carassius auratus gibelio_ (Bloch, 1783)**

Monogenea: *Dactylogyrus anchoratus, Dactylogyrus crucifer, Dactylogyrus dulkeiti, Dactylogyrus inexpectatus, Dactylogyrus intermedius, Dactylogyrus vastator, Gyrodactylus carassii, Gyrodactylus gurlei, Gyrodactylus longoacuminatus, *Gyrodactylus medius, Gyrodactylus shulmani, Gyrodactylus sprostonae, Paradiplozoon* sp.

Cestoda: *Bothriocephalus acheilognathi, Neogryporhynchus cheilancristrotus* larv.

Trematoda: *Diplostomum spathaceum* larv., *Diplostomum* sp. larv., *Posthodiplostomum cuticola* larv., *Metagonimus yokogawai* larv.

Nematoda: *Pseudocapillaria tomentosa, Schulmanela petruschewskii, Philometroides sanguinea,*

Hirudinea: *Piscicola geometra*

Carassius carassius (Linnaeus, 1758)

Monogenea: *Dactylogyrus amphibothrium, Dactylogyrus anchoratus, Dactylogyrus crassus, Dactylogyrus formosus, Dactylogyrus intermedius, Dactylogyrus vastator, Dactylogyrus wegeneri, Dactylogyrus* sp., *Gyrodactylus carassii, *Gyrodactylus elegans, Gyrodactylus longoacuminatus, *Gyrodactylus medius, Gyrodactylus shulmani, Gyrodactylus vimbi, Gyrodactylus* sp., *Diplozoon paradoxum, Paradiplozoon homoion*

Cestoda: *Khawia rossitensis,*

Caryophyllaeus laticeps, Bothriocephalus acheilognathi, Neogryporhynchus cheilancristrotus larv., *Valipora campylancristrota* larv.
Trematoda: *Asymphylodora markewitschi, Phyllodistomum* sp., *Diplostomum spathaceum* larv., *Diplostomum* sp. larv., *Tylodelphys clavata* larv., *Posthodiplostomum cuticola* larv., *Ichthyocotylurus pileatus* larv., *Paracoenogonimus ovatus* larv., *Metagonimius yokogawai* larv.
Nematoda: *Pseudocapillaria tomentosa, Philometroides sanguinea*
Acanthocephala: *Neoechinorhynchus rutili, Acanthocephalus anguillae, Pomphorhynchus laevis*
Hirudinea: *Piscicola geometra*
Bivalvia: *Glochidium* sp. (larv.)
Copepoda: *Ergasilus sieboldi, Neoergasilus japonicus, Lernaea cyprinacea*
Branchiura: *Argulus foliaceus, Argulus japonicus*

***Chondrostoma nasus* (Linnaeus, 1758)**
Monogenea: *Dactylogyrus dirigerus, Dactylogyrus ergensi, Dactylogyrus chondrostomi, Dactylogyrus nybelini, Dactylogyrus similis, Dactylogyrus sphyrna, Dactylogyrus vistulae, Dactylogyrus* sp., *Tetraonchus monenteron, Gyrodactylus macrocornis, Gyrodactylus paraminimus, Gyrodactylus* sp., *Diplozoon paradoxum, Paradiplozoon* sp.
Cestoda: *Caryophyllaeides fennica, Caryophyllaeus laticeps, Proteocephalus torulosus, Triaenophorus nodulosus* larv.
Trematoda: *Sanguinicola armata, Asymphylodora markewitschi, Allocreadium isoporum, Allocreadium markewitschi, Nicolla skrjabini, Rhipidocotyle campanula* larv., *Diplostomum spathaceum* larv., *Diplostomum* sp. larv., *Tylodelphys clavata* larv., *Posthodiplostomum cuticola* larv., *Ichthyocotylurus pileatus* larv., *Paracoenogonimus ovatus* larv., *Apophallus muehlingi* larv., *Metagonimus yokogawai* larv.
Nematoda: *Philometra rischta*
Acanthocephala: *Neoechinorhynchus rutili, Acanthocephalus anguillae, Pomphorhynchus laevis*
Hirudinea: *Piscicola geometra, Piscicola respirans*
Bivalvia: *Anodonta* sp. larv.
Copepoda: *Ergasilus sieboldi, Lamproglena pulchella, Tracheliastes chondrostomi, Tracheliastes maculatus, Tracheliastes polycolpus*
Branchiura: *Argulus coregoni, Argulus foliaceus*
Acari: *Hydrozetes lacustris*

***_Ctenopharyngodon idella_ (Valenciennes, 1844)**
Monogenea: *Dactylogyrus ctenopharyngodontis, Dactylogyrus lamellatus, Dactylogyrus vistulae, Dactylogyrus* sp., *Gyrodactylus* sp., *Diplozoon paradoxum*
Cestoda: *Caryophyllaeides fennica, Bothriocephalus acheilognathi, Ligula intestinalis* larv., *Neogryporhynchus cheilancristrotus* larv., *Valipora campylancristrota* larv.
Trematoda: *Rhipidocotyle campanula* larv., *Diplostomum spathaceum* larv., *Diplostomum* sp. larv. *Posthodiplostomum cuticola* larv., *Apharyngostrigea cornu* larv., *Apophallus muehlingi* larv., *Metagonimus yokogawai* larv.
Nematoda: *Pseudocapillaria tomentosa*, Nematoda gen. sp.
Acanthocephala: *Neoechinorhynchus rutili, Pomphorhynchus laevis*
Hirudinea: *Piscicola geometra*
Bivalvia: *Unio* sp. larv.
Copepoda: *Ergasilus sieboldi, Neoergasilus longispinosus, Lernaea ctenopharyngodontis*
Branchiura: *Argulus foliaceus*

***Cyprinus carpio* Linnaeus, 1758**
Monogenea: *Acolpenteron* sp., *Dactylogyrus amphibothrium, Dactylogyrus anchoratus, *Dactylogyrus auriculatus,*

Dactylogyrus extensus, Dactylogyrus minutus, Dactylogyrus phoxini, Dactylogyrus vastator, Dactylogyrus sp., *D. achmerowi, Pseudacolpenteron pavlovskii, *Ancyrocephalus paradoxus, Tetraonchus monenteron, Gyrodactylus cyprini, *Gyrodactylus elegans, Gyrodactylus katharineri, Gyrodactylus kherulensis, Gyrodactylus medius, Gyrodactylus shulmani, Gyrodactylus sprostonae, Gyrodactylus stankovici, Gyrodactylus vimbi, Gyrodactylus* sp., *Diplozoon paradoxum, Diplozoon* sp., *Eudiplozoon nipponicum*

Cestoda: *Caryophyllaeides fennica, Khawia sinensis, Archigetes sieboldi, Caryophyllaeus brachycollis, Caryophyllaeus fimbriceps, *Caryophyllaeus laticeps, Caryophyllaeus* sp., *Bothriocephalus acheilognathi, *Bathybothrium rectangulum, Triaenophorus nodulosus* larv., *Ligula intestinalis* larv., *Neogryporhynchus cheilancristrotus* larv., *Paradilepis scolecina* larv., *Valipora campylancristrota* larv.

Trematoda: *Sanguinicola inermis, Phyllodistomum elongatum, Allocreadium carparum, Allocreadium isoporum, Nicolla skrjabini, Sphaerostomum globiporum, Rhipidocotyle campanula* larv., *Diplostomum spathaceum* larv., *Diplostomum sp.* larv., *Tylodelphys clavata* larv., *Posthodiplostomum brevicaudatum* larv., *Posthodiplostomum cuticola* larv., *Apharyngostrigea cornu* larv., *Ichthyocotylurus pileatus* larv., *Metagonimus yokogawai* larv.

Nematoda: *Pseudocapillaria tomentosa, Camallanus lacustris, Contracaecum microcephalum* larv., *Raphidascaris acus* larv.

Acanthocephala: *Neoechinorhynchus rutili, Acanthocephalus anguillae, Acanthocephalus lucii, Acanthocephalus* sp., *Pomphorhynchus laevis*

Hirudinea: *Piscicola geometra, Piscicola respirans, Hemiclepsis marginata*

Bivalvia: *Anodonta cygnea* larv., *Glochidium* sp. (larv.)

Copepoda: *Ergasilus sieboldi, Paraergasilus rylovi, Lamproglena pulchella, Lernaea cyprinacea, Tracheliastes polycolpus*

Branchiura: *Argulus coregoni, Argulus foliaceus, Argulus japonicus, Argulus* sp.

***Gobio albipinnatus* Lukasch, 1933**

Monogenea: *Dactylogyrus cryptomeres, Dactylogyrus finitimus, Gyrodactylus gobii, Gyrodactylus gobioninum, Gyrodactylus markakulensis, Diplozoon* sp., *Paradiplozoon homoion*

Cestoda: *Caryophyllaeides fennica*, Caryophyllaeidae gen. sp., *Bothriocephalus acheilognathi, Ligula intestinalis* larv.

Trematoda: *Allocreadium isoporum, Nicolla skrjabini, Rhipidocotyle campanula* larv., *Diplostomum spathaceum* larv., *Diplostomum* sp. larv., *Metagonimus yokogawai* larv.

Nematoda: *Pseudocapillaria tomentosa, Camallanus truncatus*

Acanthocephala: *Neoechinorhynchus rutili, Pomphorhynchus laevis*

Copepoda: *Lernaea cyprinacea*

***Gobio gobio* (Linnaeus, 1758)**

Monogenea: *Dactylogyrus anchoratus, Dactylogyrus cryptomeres, Tetraonchus monenteron, Gyrodactylus dykovae, *Gyrodactylus elegans, Gyrodactylus gobiensis, Gyrodactylus gobii, Gyrodactylus gobioninum, Gyrodactylus katharineri, Gyrodactylus markakulensis, Gyrodactylus vimbi, Gyrodactylus* sp., *Diplozoon paradoxum, Paradiplozoon homoion*

Cestoda: *Caryophyllaeides fennica, Caryophyllaeus* sp., *Paraglaridacris limnodrili, Bothriocephalus acheilognathi, Proteocephalus torulosus, Ligula intestinalis* larv.

Trematoda: *Allocreadium isoporum, *Allocreadium markewitschi, Nicolla skrjabini, Bucephalus polymorphus* larv., *Diplostomum spathaceum* larv., *Diplostomum* sp. larv., *Tylodelphys clavata* larv., *Posthodiplostomum cuticola* larv., *Ichthyocotylurus pileatus* larv., *Paracoenogonimus ovatus* larv., *Apophallus*

muehlingi larv., *Metagonimus yokogawai* larv.
Nematoda: *Pseudocapillaria tomentosa*, *Philometra abdominalis*, **Philometra ovata*, *Rhabdochona denudata*, *Raphidascaris acus* larv.
Acanthocephala: *Neoechinorhynchus rutili*, *Acanthocephalus anguillae*, *Echinorhynchus truttae*, *Pomphorhynchus laevis*
Hirudinea: *Piscicola geometra*, *Hemiclepsis marginata*
Bivalvia: *Glochidium* sp. (larv.)
Copepoda: *Ergasilus sieboldi*
Branchiura: *Argulus foliaceus*

***Gobio kessleri* Dybowski, 1862**
Monogenea: *Gyrodactylus gobiensis*, *Gyrodactylus gobioninum*, *Diplozoon paradoxum*, *Diplozoon* sp.
Trematoda: *Rhipidocotyle campanula*, *Nicolla skrjabini*, *Rhipidocotyle campanula* larv., *Diplostomum spathaceum* larv., *Metagonimus yokogawai* larv.

***Gobio uranoscopus* (Agassiz, 1828)**
Monogenea: *Gyrodactylus markakulensis*, *Diplozoon* sp.
Trematoda: *Rhipidocotyle campanula*, *Nicolla skrjabini*, *Rhipidocotyle campanula* larv., *Diplostomum* sp. larv., *Paracoenogonimus ovatus* larv., *Metagonimus yokogawai* larv.
Nematoda: *Philometra abdominalis*
Acanthocephala: *Pomphorhynchus laevis*

***Hypophthalmichthys molitrix* (Valenciennes, 1844)**
Monogenea: *Dactylogyrus* sp., *Gyrodactylus* sp., *Paradiplozoon homoion*
Trematoda: *Diplostomum indistinctum* larv., *Diplostomum spathaceum* larv., *Diplostomum volvens* larv.
Hirudinea: *Piscicola geometra*
Branchiura: *Argulus japonicus*

***Leucaspius delineatus* (Heckel, 1843)**
Monogenea: *Dactylogyrus alatus*, *Dactylogyrus anchoratus*, *Dactylogyrus fraternus*, *Gyrodactylus carassii*, *Gyrodactylus laevis*, *Gyrodactylus prostae*, *Gyrodactylus* sp., *Paradiplozoon homoion*
Cestoda: *Neogryporhynchus cheilancristrotus* larv.
Trematoda: *Phyllodistomum folium*, *Rhipidocotyle campanula* larv., *Diplostomum spathaceum* larv., *Tylodelphys clavata* larv., *Posthodiplostomum cuticola* larv., *Ichthyocotylurus pileatus* larv., *Paracoenogonimus ovatus* larv., *Apophallus muehlingi* larv., *Metagonimus yokogawai* larv.
Acanthocephala: *Neoechinorhynchus rutili*, *Pomphorhynchus laevis*
Hirudinea: *Piscicola geometra*

***Leuciscus cephalus* (Linnaeus, 1758)**
Monogenea: *Dactylogyrus borealis*, *Dactylogyrus cordus*, *Dactylogyrus cornu*, **Dactylogyrus crucifer*, *Dactylogyrus difformis*, *Dactylogyrus fallax*, *Dactylogyrus folkmanovae*, *Dactylogyrus micracanthus*, *Dactylogyrus nanoides*, *Dactylogyrus nanus*, *Dactylogyrus navicularis*, *Dactylogyrus naviculoides*, *Dactylogyrus prostae*, *Dactylogyrus rutili*, *Dactylogyrus similis*, *Dactylogyrus sphyrna*, *Dactylogyrus tuba*, *Dactylogyrus vistulae*, *Dactylogyrus vranoviensis*, *Dactylogyrus* sp., **Tetraonchus monenteron*, *Gyrodactylus carassii*, *Gyrodactylus gasterostei*, *Gyrodactylus gracilihamatus*, *Gyrodactylus hemibarbi*, *Gyrodactylus kearni*, *Gyrodactylus laevis*, *Gyrodactylus lamberti*, *Gyrodactylus leucisci*, *Gyrodactylus lomi*, *Gyrodactylus osoblahensis*, *Gyrodactylus prostae*, *Gyrodactylus scardiniensis*, *Gyrodactylus vimbi*, *Gyrodactylus* sp., **Diplozoon paradoxum*, *Diplozoon* sp., *Paradiplozoon bliccae*, *Paradiplozoon ergensi*, *Paradiplozoon homoion*, *Paradiplozoon megan*, *Paradiplozoon rutili*, *Paradiplozoon* sp.
Cestoda: *Caryophyllaeides fennica*, *Caryophyllaeus brachycollis*, *Caryophyllaeus laticeps*, *Caryophyllaeus* sp., Caryophyllaeidae gen. sp., *Bothriocephalus acheilognathi*, **Bathybothrium rectangulum*, *Proteocephalus*

torulosus, *Proteocephalus* sp., *Ligula intestinalis* larv.

Trematoda: *Asymphylodora imitans*, *Palaeorchis incognitus*, *Phyllodistomum dogieli*, *Phyllodistomum elongatum*, *Phyllodistomum folium*, *Allocreadium isoporum*, *Allocreadium* sp., *Nicolla skrjabini*, *Sphaerostomum bramae*, *Sphaerostomum globiporum*, *Bucephalus polymorphus* larv., *Rhipidocotyle campanula* larv., *Diplostomum spathaceum* larv., *Diplostomum* sp. larv., *Tylodelphys clavata* larv., *Posthodiplostomum cuticola* larv., *Apharyngostrigea cornu* larv., *Ichthyocotylurus pileatus* larv., *Paracoenogonimus ovatus* larv., *Apophallus donicus* larv. *Apophallus muehlingi* larv., *Metagonimus yokogawai* larv.

Nematoda: *Pseudocapillaria tomentosa*, *Schulmanela petruschewskii*, *Raphidascaris acus*, **Paraquimperia tenerrima*, *Cucullanus dogieli*, *Camallanus lacustris*, *Molnaria intestinalis*, *Philometra abdominalis*, **Philometra ovata*, *Rhabdochona denudata*, **Rhabdochona hellichi*, **Oswaldocruzia filiformis*, *Raphidascaris acus* larv.

Acanthocephala: *Neoechinorhynchus rutili*, *Acanthocephalus anguillae*, *Acanthocephalus lucii*, *Echinorhynchus truttae*, *Pomphorhynchus laevis*

Hirudinea: *Piscicola geometra*

Bivalvia: *Anodonta cygnea* larv., *Anodonta* sp. larv., *Unio* sp. larv., *Glochidium* sp. (larv.)

Copepoda: *Ergasilus nasus*, *Ergasilus sieboldi*, *Neoergasilus japonicus*, *Paraergasilus rylovi*, *Lamproglena pulchella*, *Tracheliastes maculatus*, *Tracheliastes polycolpus*, *Lernaea cyprinacea*

Branchiura: *Argulus coregoni*, *Argulus foliaceus*, *Argulus japonicus*

Acari: *Hydrozetes lacustris*

***Leuciscus idus* (Linnaeus, 1758)**

Monogenea: *Dactylogyrus crucifer*, *Dactylogyrus similis*, *Dactylogyrus sphyrna*, *Dactylogyrus tuba*, *Dactylogyrus* sp., *Gyrodactylus prostae*, **Diplozoon paradoxum*, *Paradiplozoon megan*

Cestoda: *Caryophyllaeides fennica*, *Caryophyllaeus brachycollis*, *Caryophyllaeus laticeps*, *Caryophyllaeus* sp., *Proteocephalus torulosus*

Trematoda: *Palaeorchis incognitus*, *Nicolla skrjabini*, *Sphaerostomum globiporum*, *Rhipidocotyle campanula* larv., *Diplostomum spathaceum* larv., *Diplostomum* sp. larv., *Tylodelphys clavata* larv., *Posthodiplostomum cuticola* larv., *Apharyngostrigea cornu* larv., *Ichthyocotylurus pileatus* larv., *Paracoenogonimus ovatus* larv., *Prohemistomum* sp. larv., *Apophallus muehlingi* larv., *Metagonimus yokogawai* larv.

Nematoda: *Philometra rischta*, *Spiroxys contortus* larv

Acanthocephala: *Neoechinorhynchus rutili*, *Acanthocephalus anguillae*, *Acanthocephalus lucii*, *Pomphorhynchus laevis*

Hirudinea: *Piscicola geometra*, *Piscicola respirans*

Bivalvia: *Glochidium* sp. (larv.)

Crustacea: *Ergasilus sieboldi*, *Ergasilus* sp., *Lamproglena pulchella*

Branchiura: *Argulus coregoni*, *Argulus foliaceus*

***Leuciscus leuciscus* (Linnaeus, 1758)**

Monogenea: *Dactylogyrus cordus*, *Dactylogyrus crucifer*, *Dactylogyrus fallax*, *Dactylogyrus similis*, *Dactylogyrus sphyrna*, *Dactylogyrus tuba*, *Dactylogyrus vistulae*, **Tetraonchus monenteron*, *Gyrodactylus carassii*, *Gyrodactylus leucisci*, *Gyrodactylus osoblahensis*, *Gyrodactylus prostae*, *Gyrodactylus vimbi*, *Diplozoon paradoxum*, *Paradiplozoon homoion*

Cestoda: *Caryophyllaeides fennica*, **Caryophyllaeus fimbriceps*, *Caryophyllaeus laticeps*, *Caryophyllaeus* sp., *Proteocephalus torulosus*, *Ligula intestinalis* larv.

Trematoda: *Palaeorchis incognitus*,

Allocreadium isoporum, *Nicolla skrjabini*, *Sphaerostomum bramae*, *Sphaerostomum globiporum*, *Rhipidocotyle campanula* larv., *Diplostomum spathaceum* larv., *Diplostomum* sp. larv., *Tylodelphys clavata* larv., *Posthodiplostomum cuticola* larv., *Paracoenogonimus ovatus* larv., *Apophallus muehlingi* larv., *Metagonimus yokogawai* larv.

Nematoda: *Pseudocapillaria tomentosa*, *Philometra abdominalis*, **Philometra ovata*, *Philometra rischta*, *Rhabdochona denudata*, *Raphidascaris acus* larv.

Acanthocephala: *Acanthocephalus anguillae*, *Acanthocephalus lucii*, *Pomphorhynchus laevis*

Bivalvia: *Anodonta cygnea* larv., *Unio* sp. larv., *Glochidium* sp. (larv.)

Copepoda: *Ergasilus briani*, *Ergasilus sieboldi*, *Lamproglena pulchella*

Branchyura: *Argulus coregoni*

***Pelecus cultratus* (Linnaeus, 1758)**

Monogenea: *Dactylogyrus simplicimallaeata*, *Gyrodactylus* sp., **Diplozoon paradoxum*, *Diplozoon* sp., *Paradiplozoon vojteki*

Trematoda: *Nicolla skrjabini*, *Bucephalus polymorphus* larv., *Diplostomum spathaceum* larv., *Diplostomum* sp. larv., *Posthodiplostomum cuticola* larv., *Paracoenogonimus ovatus* larv., *Metagonimus yokogawai* larv.

Nematoda: *Camallanus lacustris*, *Rhabdochona denudata*

Acanthocephala: *Acanthocephalus anguillae*, *Pomphorhynchus laevis*

Copepoda: *Paraergasilus rylovi*

***Phoxinus phoxinus* (Linnaeus, 1758)**

Monogenea: *Dactylogyrus borealis*, *Pellucihaptor merus*, *Gyrodactylus aphyae*, *Gyrodactylus carassii*, *Gyrodactylus laevis*, *Gyrodactylus limneus*, *Gyrodactylus macronychus*, *Gyrodactylus minimus*, *Gyrodactylus pannonicus*, **Gyrodactylus nemachili*, *Diplozoon paradoxum*, *Diplozoon* sp.

Cestoda: *Paracaryophyllaeus gotoi*

Trematoda: *Allocreadium isoporum*, **Allocreadium markewitschi*, *Allocreadium papilligerum*, *Allocreadium* sp., *Rhipidocotyle campanula* larv., *Diplostomum spathaceum* larv., *Diplostomum* sp. larv., *Tylodelphys clavata* larv., *Posthodiplostomum cuticola* larv., **Apatemon cobitidis* larv., *Paracoenogonimus ovatus* larv., *Metorchis xanthosomus* larv., *Metagonimus yokogawai* larv.

Nematoda: *Pseudocapillaria tomentosa*, *Philometra abdominalis*, **Philometra ovata*, **Rhabdochona denudata*, *Rhabdochona phoxini*, **Cystidicoloides ephemeridarum*, *Raphidascaris acus* larv., *Streptocara crassicauda* larv.

Acanthocephala: *Neoechinorhynchus rutili*, *Acanthocephalus anguillae*, *Pomphorhynchus laevis*

Hirudinea: *Piscicola geometra*

Bivalvia: *Anodonta cygnea* larv., *Glochidium* sp. (larv.)

Copepoda: *Ergasilus briani*, *Lernaea esocina*, *Tracheliastes polycolpus*

Branchiura: *Argulus coregoni*

***Pseudorasbora parva* (Schlegel, 1842)**

Nematoda: *Pseudocapillaria tomentosa*

Acanthocephala: *Pomphorhynchus laevis*

Bivalvia: *Glochidium* sp. (larv.)

***Rhodeus sericeus* (Pallas, 1776)**

Monogenea: *Dactylogyrus bicornis*, *Gyrodactylus laevis*, *Gyrodactylus rhodei*, *Diplozoon paradoxum*, *Diplozoon* sp., *Paradiplozoon homoion*

Cestoda: *Neogryporhynchus cheilancristrotus* larv.

Trematoda: *Rhipidocotyle campanula* larv., *Diplostomum spathaceum* larv., *Diplostomum* sp. larv., *Tylodelphys clavata* larv., *Posthodiplostomum cuticola* larv., *Paracoenogonimus ovatus* larv., *Clinostomum complanatum* larv., *Apophallus muehlingi* larv., *Metagonimus yokogawai* larv.

Acanthocephala: *Pomphorhynchus laevis*
Hirudinea: *Piscicola geometra*, *Piscicola respirans*
Bivalvia: *Anodonta* sp. larv., *Unio* sp. larv., *Glochidium* sp. (larv.)
Copepoda: *Lernaea cyprinacea*
Branchiura: *Argulus foliaceus*

***Rutilus pigus* (Lacépède, 1803)**
Monogenea: *Dactylogyrus crucifer*, *Dactylogyrus vistulae*, *Gyrodactylus vimbi*, *Diplozoon paradoxum*, *Paradiplozoon homoion*
Cestoda: *Caryophyllaeides fennica*, *Caryophyllaeus brachycollis*, *Caryophyllaeus laticeps*, *Ligula intestinalis* larv.
Trematoda: *Palaeorchis incognitus*, *Nicolla skrjabini*, *Diplostomum spathaceum* larv., *Diplostomum* sp. larv., *Tylodelphys clavata* larv., *Posthodiplostomum cuticola* larv., *Paracoenogonimus ovatus* larv., *Apophallus muehlingi* larv., *Metagonimus yokogawai* larv.
Nematoda: *Cucullanus dogieli*
Acanthocephala: *Neoechinorhynchus rutili*, *Acanthocephalus anguillae*, *Pomphorhynchus laevis*
Copepoda: *Ergasilus sieboldi*, *Lamproglena pulchella*

***Rutilus rutilus* (Linnaeus, 1758)**
Monogenea: *Dactylogyrus cornu*, *Dactylogyrus crucifer*, **Dactylogyrus difformis*, *Dactylogyrus fallax*, *Dactylogyrus folkmanovae*, **Dactylogyrus macracanthus*, *Dactylogyrus micracanthus*, *Dactylogyrus nanus*, *Dactylogyrus rarissimus*, *Dactylogyrus rutili*, *Dactylogyrus similis*, *Dactylogyrus sphyrna*, *Dactylogyrus suecicus*, *Dactylogyrus vistulae*, *Dactylogyrus zandti*, *Dactylogyrus* sp., *Tetraonchus monenteron*, *Gyrodactylus carassii*, *Gyrodactylus elegans*, *Gyrodactylus gracilihamatus*, *Gyrodactylus laevis*, *Gyrodactylus magnificus*, *Gyrodactylus prostae*, *Gyrodactylus rutilensis*, *Gyrodactylus vimbi*, *Gyrodactylus* sp., *Diplozoon paradoxum*, *Paradiplozoon homoion*
Cestoda: *Caryophyllaeides fennica*, *Caryophyllaeus brachycollis*, *Caryophyllaeus fimbriceps*, *Caryophyllaeus laticeps*, *Caryophyllaeus* sp., *Proteocephalus torulosus*, *Ligula intestinalis* larv.
Trematoda: *Rhipidocotyle campanula*, *Asymphylodora tincae*, *Palaeorchis incognitus*, *Phyllodistomum elongatum*, *Phyllodistomum folium*, *Allocreadium isoporum*, *Sphaerostomum bramae*, *Sphaerostomum globiporum*, *Bucephalus polymorphus* larv., *Rhipidocotyle campanula* larv., *Diplostomum spathaceum* larv., *Diplostomum* sp. larv., *Tylodelphys clavata* larv., *Hysteromorpha triloba* larv., *Posthodiplostomum cuticola* larv., *Apharyngostrigea cornu* larv., *Ichthyocotylurus pileatus* larv., *Paracoenogonimus ovatus* larv., *Apophallus muehlingi* larv., *Metagonimus yokogawai* larv.
Nematoda: *Pseudocapillaria tomentosa*, *Cucullanus dogieli*, *Philometra ovata*, *Philometra rischta*, *Rhabdochona denudata*, *Contracaecum* sp. larv., *Raphidascaris acus* larv., *Streptocara crassicauda* larv.
Acanthocephala: *Neoechinorhynchus rutili*, *Acanthocephalus anguillae*, *Acanthocephalus lucii*, *Polymorphus laevis*
Hirudinea: *Piscicola geometra*, *Hemiclepsis marginata*
Bivalvia: *Anodonta* sp. larv., *Unio* sp. larv., *Glochidium* sp. (larv.)
Copepoda: *Ergasilus sieboldi*, *Ergasilus* sp., *Lamproglena pulchella*, *Lernaea cyprinacea*, *Tracheliastes maculatus*
Branchiura: *Argulus foliaceus*, *Argulus japonicus*

***Scardinius erythrophthalmus* (Linnaeus, 1758)**
Monogenea: *Dactylogyrus crucifer*, *Dactylogyrus difformis*, *Dactylogyrus difformoides*, *Dactylogyrus izjumovae*, *Dactylogyrus micracanthus*, *Dactylogyrus sphyrna*, *Dactylogyrus vistulae*, *Dactylogyrus yinwenyingae*, *Dactylogyrus* sp., **Tetraonchus*

monenteron, *Gyrodactylus carassii*, *Gyrodactylus laevis*, *Gyrodactylus vimbi*, **Diplozoon paradoxum*, *Paradiplozoon bliccae*, *Paradiplozoon homoion*

Cestoda: *Caryophyllaeides fennica*, *Caryophyllaeus laticeps*, *Caryophyllaeus* sp., Caryophyllaeidae gen. sp., *Proteocephalus torulosus*, *Ligula intestinalis* larv., *Neogryporhynchus cheilancristrotus* larv., *Valipora campylancristrota* larv.

Trematoda: *Sanguinicola inermis*, *Asymphylodora imitans*, *Asymphylodora kubanicum*, *Asymphylodora markewitschi*, *Palaeorchis incognitus*, *Phyllodistomum elongatum*, *Phyllodistomum folium*, *Nicolla skrjabini*, *Sphaerostomum globiporum*, *Bucephalus polymorphus* larv., *Rhipidocotyle campanula* larv., *Diplostomum spathaceum* larv., *Diplostomum* sp. larv., *Tylodelphys clavata* larv., *Posthodiplostomum cuticola* larv., *Apharyngostrigea cornu* larv., *Ichthyocotylurus pileatus* larv., *Paracoenogonimus ovatus* larv., *Clinostomum complanatum* larv., *Apophallus muehlingi* larv., *Metagonimus yokogawai* larv.

Nematoda: *Molnaria intestinalis*, *Skrjabillanus scardinii*, *Raphidascaris acus* larv., *Philometra obturans* larv., *Streptocara crassicauda* larv.

Acanthocephala: *Neoechinorhynchus rutili*, *Acanthocephalus anguillae*, *Pomphorhynchus laevis*

Hirudinea: *Piscicola geometra*

Bivalvia: *Anodonta* sp. larv., *Unio* sp. larv., *Glochidium* sp. (larv.)

Copepoda: *Ergasilus sieboldi*, *Ergasilus* sp., *Neoergasilus japonicus*, *Paraergasilus rylovi*, *Lamproglena pulchella*

Branchiura: *Argulus coregoni*, *Argulus foliaceus*, *Argulus japonicus*

***Tinca tinca* (Linnaeus, 1758)**

Monogenea: *Dactylogyrus macracanthus*, *Dactylogyrus tincae*, **Gyrodactylus elegans*, *Gyrodactylus tincae*

Cestoda: *Khawia baltica*, *Archigetes sieboldi*, *Caryophyllaeus fimbriceps*, **Caryophyllaeus laticeps*, *Caryophyllaeus* sp., *Monobothrium wageneri*, Caryophyllaeidae gen. sp., **Proteocephalus percae*, *Neogryporhynchus cheilancristrotus* larv., *Valipora campylancristrota* larv.

Trematoda: *Asymphylodora tincae*, *Phyllodistomum elongatum*, *Phyllodistomum folium*, *Sphaerium globiporum*, *Bucephalus polymorphus* larv., *Diplostomum spathaceum* larv., *Diplostomum* sp. larv., *Tylodelphys clavata* larv., *Ichthyocotylurus pileatus* larv., *Paracoenogonimus ovatus* larv., *Apophallus muehlingi* larv., *Metagonimus yokogawai* larv.

Nematoda: *Pseudocapillaria tomentosa*, *Schulmanela petruschewskii*, *Skrjabillanus tincae*, **Philometra ovata*, *Raphidascaris acus* larv., *Philometra* sp. larv., *Camallanus lacustris* larv., *Syncuaria squamata* larv., *Agamospirura* sp. (larv.)

Acanthocephala: *Neoechinorhynchus rutili*, *Acanthocephalus anguillae*, *Acanthocephalus lucii*, *Pomphorhynchus laevis*

Hirudinea: *Piscicola geometra*

Bivalvia: *Glochidium* sp. (larv.)

Copepoda: *Ergasilus sieboldi*, *Lernaea cyprinacea*

Branchiura: *Argulus coregoni*, *Argulus foliaceus*, *Argulus japonicus*, *Argulus* sp.

***Vimba vimba* (Linnaeus, 1758)**

Monogenea: *Dactylogyrus cornoides*, *Dactylogyrus cornu*, *Dactylogyrus distinguendus*, *Dactylogyrus haplogonoides*, *Dactylogyrus similis*, *Dactylogyrus sphyrna*, *Dactylogyrus tuba*, **Dactylogyrus vastator*, *Dactylogyrus vistulae*, **Gyrodactylus elegans*, *Gyrodactylus prostae*, *Gyrodactylus vimbi*, *Gyrodactylus* sp., *Diplozoon paradoxum*, *Diplozoon* sp.

Cestoda: *Caryophyllaeides fennica*, *Caryophyllaeus laticeps*

Trematoda: *Palaeorchis incognitus*, *Phyllodistomum dogieli*, *Phyllodistomum*

*elongatum, Nicolla skrjabini, *Plagioporus* sp., *Sphaerostomum bramae, Sphaerostomum globiporum, Rhipidocotyle campanula* larv., *Diplostomum spathaceum* larv., *Diplostomum* sp. larv., *Tylodelphys clavata* larv., *Posthodiplostomum cuticola* larv. *Ichthyocotylurus pileatus* larv., *Apharyngostrigea cornu* larv., *Paracoenogonimus ovatus* larv., *Apophallus muehlingi* larv., *Metagonimus yokogawai* larv.
Nematoda: *Philometra ovata*
Acanthocephala: *Acanthocephalus anguillae, Acanthocephalus lucii, Pomphorhynchus laevis*
Hirudinea: *Piscicola geometra*
Copepoda: *Ergasilus briani, Ergasilus sieboldi*
Branchiura: *Argulus coregoni*

Fam. Cobitidae:

***Cobitis elongatoides* Bacescu et Maier, 1969**
Cestoda: *Ligula colymbi* larv.
Trematoda: *Posthodiplostomum cuticola* larv.

***Cobitis taenia* Linnaeus, 1758**
Monogenea: *Gyrodactylus cobitis, Gyrodactylus latus, Gyrodactylus* sp.
Cestoda: *Paracaryophyllaeus gotoi, Ligula colymbi* larv., *Ligula intestinalis* larv.
Trematoda: *Allocreadium transversale, Nicolla skrjabini, Diplostomum* sp. larv., *Tylodelphys clavata* larv., *Posthodiplostomum cuticola* larv., *Holostephanus cobitidis* larv., *Clinostomum complanatum* larv., *Metorchis xanthosomus* larv., *Metagonimus yokogawai* larv.
Nematoda: *Raphidascaris acus* larv.
Acanthocephala: *Neoechinorhynchus rutili, Acanthocephalus anguillae*
Hirudinea: *Piscicola respirans*
Remark: According to recent genetic studies by Ráb et al. (2000), Šlechtová et al. (2000) and Lusk et al. (2000), previous records of *C. taenia* from the Czech Republic may, in fact, represent diploid-polyploid complexes involving *C. taenia*, *C. elongatoides* and *Cobitis* spp.

***Misgurnus fossilis* (Linnaeus, 1758)**
Monogenea: **Dactylogyrus extensus, Ancyrocephalus cruciatus, Gyrodactylus cobitis, *Gyrodactylus elegans, Gyrodactylus fossilis, *Gyrodactylus medius, Gyrodactylus misgurni*
Cestoda: Caryophyllaeidae gen. sp., *Neogryporhynchus cheilancristrotus* larv.
Trematoda: *Sanguinicola armata, Sanguinicola inermis, Sanguinicola intermedia, Allocreadium transversale, Diplostomum spathaceum* larv., *Diplostomum* sp. larv., *Tylodelphys clavata* larv., *Posthodiplostomum cuticola* larv., *Metagonimus yokogawai* larv.
Nematoda: *Raphidascaris acus* larv.

Fam. Balitoridae:

***Barbatula barbatula* (Linnaeus, 1758)** (syn. *Noemacheilus barbatulus* (Linnaeus, 1758))
Monogenea: *Gyrodactylus barbatuli, Gyrodactylus jiroveci, Gyrodactylus menschikowi, Gyrodactylus pavlovskyi, Gyrodactylus sedelnikowi, Gyrodactylus* sp., (?) *Gyrodactylus nemachili*
Cestoda: *Proteocephalus torulosus*
Trematoda: *Crepidostomum farionis, Phyllodistomum* sp., *Allocreadium isoporum, Diplostomum spathaceum* larv., *Diplostomum* sp. larv., *Tylodelphys clavata* larv., *Posthodiplostomum cuticola* larv., *Apatemon cobitidis* larv., *Metagonimus yokogawai* larv.
Nematoda: *Pseudocapillaria tomentosa, *Rhabdochona denudata, Rhabdochona ergensi, *Cystidicoloides ephemeridarum, Raphidascaris acus* larv., *Rhabdochona phoxini* larv., *Cystidicoloides ephemeridarum* larv., *Streptocara crassicauda* larv.
Acanthocephala: *Neoechinorhynchus rutili, Acanthocephalus anguillae, Acanthocephalus lucii, Pomphorhynchus laevis*

Bivalvia: *Glochidium* sp. (larv.)
Copepoda: *Ergasilus sieboldi*

Siluriformes

Fam. Ictaluridae:

***Ictalurus nebulosus* (LeSueur, 1819)**
Monogenea: **Ancyrocephalus cruciatus, Cleidodiscus pricei,* (?) *Gyrodactylus fairporti, Gyrodactylus* sp.
Trematoda: *Diplostomum spathaceum* larv., *Diplostomum* sp. larv., *Metagonimus yokogawai* larv.
Nematoda: *Raphidascaris acus* larv.
Acanthocephala: *Neoechinorhynchus rutili, Acanthocephalus lucii, Pomphorhynchus laevis*
Bivalvia: *Glochidium* sp. (larv.)
Copepoda: *Ergasilus sieboldi*
Branchiura: *Argulus foliaceus*

Fam. Siluridae:

***Silurus glanis* Linnaeus, 1758**
Monogenea: *Silurodiscoides siluri, Silurodiscoides vistulensis*
Cestoda: *Bothriocephalus acheilognathi, Triaenophorus crassus, Proteocephalus osculatus, *Proteocephalus percae, Proteocephalus* sp., *Silurotaenia siluri, Triaenophorus crassus* larv.
Trematoda: *Bucephalus polymorphus, Bunodera luciopercae, Azygia lucii, Orientocreadium siluri, Nicolla skrjabini, Sphaerostomum bramae, Diplostomum spathaceum* larv., *Diplostomum* sp. larv., *Tylodelphys clavata* larv., *Metagonimus yokogawai* larv.
Nematoda: *Schulmanela petruschewskii, Camallanus lacustris, Camallanus truncatus*
Acanthocephala: *Acanthocephalus anguillae, Acanthocephalus lucii, Pomphorhynchus laevis*
Hirudinea: *Piscicola fasciata, Piscicola geometra, Piscicola respirans*
Copepoda: *Ergasilus sieboldi, Pseudotracheliastes stellifer*
Branchiura: *Argulus coregoni, Argulus foliaceus*

Esociformes

Fam. Esocidae:

***Esox lucius* Linnaeus, 1758**
Monogenea: **Ancyrocephalus paradoxus, Tetraonchus monenteron, Gyrodactylus lucii*
Cestoda: **Bothriocephalus acheilognathi, Triaenophorus crassus, Triaenophorus nodulosus, Triaenophorus* sp., **Proteocephalus percae, *Proteocephalus* sp., *Triaenophorus crassus* larv., *Triaenophorus nodulosus* larv., *Ligula intestinalis* larv.
Trematoda: *Bucephalus polymorphus, Rhipidocotyle campanula, Bunodera luciopercae, Phyllodistomum folium, Azygia lucii, *Allocreadium isoporum, Nicolla skrjabini, Diplostomum spathaceum* larv., *Diplostomum* sp. larv., *Tylodelphys clavata* larv., *Ichthyocotylurus pileatus* larv., *Paracoenogonimus ovatus* larv., *Apophallus muehlingi* larv., *Metagonimus yokogowai* larv.
Nematoda: *Raphidascaris acus, Camallanus lacustris, Camallanus truncatus, Esocinema bohemicum, Philometra obturans, *Cystidicoloides ephemeridarum, Raphidascaris acus* larv.
Acanthocephala: *Neoechinorhynchus rutili, Acanthocephalus anguillae, Acanthocephalus lucii, Echinorhynchus truttae, Pomphorhynchus laevis*
Hirudinea: *Piscicola geometra, Piscicola respirans, Hemiclepsis marginatus*
Bivalvia: *Anodonta cygnea* larv., *Anodonta* sp. larv., *Unio* sp. larv., *Glochidium* sp. (larv.)
Copepoda: *Ergasilus sieboldi, Lamproglena pulchella, Lernaea esocina, Lernaea cyprinacea*
Branchiura: *Argulus coregoni, Argulus foliaceus, Argulus japonicus, Argulus* sp.

Fam. Umbridae:

Umbra krameri **Walbaum, 1792**
Monogenea: *Gyrodactylus slovacicus*
Trematoda: Echinostomatidae gen. sp. larv.
Nematoda: *Raphidascaris acus*

Salmoniformes

Fam. Salmonidae:

*****Coregonus maraena*** **(Bloch, 1779)**
Copepoda: *Ergasilus sieboldi*
Branchiura: *Argulus foliaceus*, *Argulus japonicus*
Acari: *Hydrozetes lacustris*

***Coregonus peled (Gmelin, 1788)**
Copepoda: *Ergasilus sieboldi*

Hucho hucho **(Linnaeus, 1758)**
Cestoda: *Cyathocephalus truncatus*, *Triaenophorus nodulosus*, *Triaenophorus nodulosus* larv.
Nematoda: *Raphidascaris acus*, *Cystidicola farionis*, *Cystidicoloides ephemeridarum*, *Raphidascaris acus* larv.
Acanthocephala: *Neoechinorhynchus rutili*
Hirudinea: *Piscicola respirans*
Copepoda: *Basanistes huchonis*
Branchiura: *Argulus coregoni*

*****Oncorhynchus mykiss*** **(Walbaum, 1792)**
Monogenea: *Gyrodactylus bohemicus*, *Gyrodactylus truttae*
Cestoda: *Cyathocephalus truncatus*, *Proteocephalus longicollis*
Trematoda: *Crepidostomum farionis*, *Crepidostomum metoecus*, *Nicolla wisniewskii*, *Diplostomum spathaceum* larv.
Nematoda: *Raphidascaris acus*, **Rhabdochona denudata*, *Camallanus lacustris*, *Cystidicola farionis*, *Cystidicoloides ephemeridarum*, *Raphidascaris acus* larv.
Acanthocephala: *Neoechinorhynchus rutili*, *Acanthocephalus anguillae*, *Acanthocephalus lucii*, *Echinorhynchus truttae*, *Pomphorhynchus laevis*
Bivalvia: *Glochidium* sp. (larv.)
Branchiura: *Argulus coregoni*, *Argulus foliaceus*

Salmo salar **Linnaeus, 1758**
Cestoda: *Eubothrium crassum*, *Tentacularia coryphaenae* larv., *Tetrarhynchus* sp. larv., *Scolex pleuronectis* (larv.)
Trematoda: *Hemiurus appendiculatus*, *Derogenes varicus*
Nematoda: *Hysterothylacium gadi*, *Anisakis simplex* larv. (reported as "*Agamonema capsularia*")
Acanthocephala: *Echinorhynchus salmonis*
Hirudinea: *Piscicola respirans*
Copepoda: *Salmincola salmoneus*
Branchiura: *Argulus coregoni*

Salmo trutta **m.** ***fario*** **Linnaeus, 1758**
Monogenea: *Gyrodactylus truttae*, **Tetraonchus* sp.
Cestoda: *Cyathocephalus truncatus*, *Proteocephalus longicollis*, *Triaenophorus nodulosus* larv.
Trematoda: *Crepidostomium farionis*, *Crepidostomum metoecus*, *Phyllodistomum folium*, *Phyllodistomum megalorchis*, *Phyllodistomum simile*, *Phyllodistomum* sp., *Nicolla skrjabini*, *Nicolla testiobliquum*, *Nicolla wisniewskii*, *Diplostomum spathaceum* larv., *Diplostomum* sp. larv.
Nematoda: *Pseudocapillaria salvelini*, *Raphidascaris acus*, *Cucullanus truttae*, *Camallanus lacustris*, **Rhabdochona denudata*, **Rhabdochona hellichi*, *Cystidicola farionis*, *Cystidicoloides ephemeridarum*, **Oswaldocruzia* sp.,
Nematoda gen. sp., *Raphidascaris acus* larv., **Porrocaecum ensicaudatum* larv., *Streptocara crassicauda* larv.
Acanthocephala: *Neoechinorhynchus rutili*, *Acanthocephalus anguillae*, *Acanthocephalus lucii*, *Echinorhynchus truttae*, *Pomphorhynchus laevis*
Hirudinea: *Piscicola geometra*, *Piscicola respirans*

Copepoda: *Salmincola salmoneus*
Branchiura: *Argulus coregoni, Argulus foliaceus*

***Salmo trutta* m. *lacustris* Linnaeus, 1758**
Monogenea: *Gyrodactylus truttae*
Cestoda: *Cyathocephalus truncatus*
Trematoda: *Crepidostomum farionis, Crepidostomum metoecus*
Nematoda: *Raphidascaris acus, Cystidicola farionis, *Oswaldocruzia* sp., *Raphidascaris acus* larv.
Acanthocephala: *Neoechinorhynchus rutili*
Branchiura: *Argulus coregoni*

****Salvelinus fontinalis* (Mitchill, 1815)**
Monogenea: *Gyrodactylus bohemicus, Gyrodactylus truttae*
Cestoda: *Cyathocephalus truncatus, Proteocephalus longicollis*
Trematoda: *Crepidostomum farionis, Crepidostomum metoecus, Nicolla wisniewskii*
Nematoda: *Raphidascaris acus, *Rhabdochona denudata, Cystidicola farionis*
Acanthocephala: *Neoechinorhynchus rutili*

****Thymallus baicalensis* Dybowski, 1874**
Monogenea: *Tetraonchus borealis*
Cestoda: *Cyathocephalus truncatus*
Trematoda: *Crepidostomum farionis, Crepidostomum metoecus, Nicolla wisniewskii*
Nematoda: *Raphidascaris acus, Cystidicola farionis, Raphidascaris acus* larv.
Acanthocephala: *Neoechinorhynchus rutili, Echinorhynchus truttae*

***Thymallus thymallus* (Linnaeus, 1758)**
Monogenea: *Tetraonchus borealis, Gyrodactylus thymalli, Gyrodactylus truttae*
Cestoda: *Cyathocephalus truncatus*
Trematoda: *Crepidostomum farionis, Crepidostomum metoecus, Nicolla wisniewskii, Diplostomum* sp. larv.
Nematoda: *Raphidascaris acus, *Rhabdochona denudata, Rhabdochona* sp., *Cystidicola farionis, Cystidicoloides ephemeridarum, Raphidascaris acus* larv.
Acanthocephala: *Neoechinorhynchus rutili, Echinorhynchus truttae, Pomphorhynchus laevis*
Hirudinea: *Piscicola geometra, Piscicola respirans*
Copepoda: *Ergasilus sieboldi*
Branchiura: *Argulus coregoni*

Gadiformes

Fam. Lotidae:

***Lota lota* (Linnaeus, 1758)**
Monogenea: *Gyrodactylus lotae*
Cestoda: *Triaenophorus nodulosus, Proteocephalus* sp., *Triaenophorus nodulosus* larv.
Trematoda: *Rhipidocotyle campanula, Bunodera luciopercae, Crepidostomum farionis, Azygia lucii, Nicolla skrjabini, Rhipidocotyle campanula* larv., *Diplostomum spathaceum* larv., *Diplostomum* sp. larv., *Tylodelphys clavata* larv., *Metagonimus yokogawai* larv.
Nematoda: *Raphidascaris acus, Camallanus lacustris, Camallanus truncatus, Raphidascaris acus* larv.
Acanthocephala: *Acanthocephalus anguillae, Acanthocephalus lucii, Echinorhynchus truttae, Pomphorhynchus laevis*
Hirudinea: *Piscicola geometra, Piscicola respirans*
Bivalvia: *Unio* sp. larv., *Glochidium* sp. (larv.)

Gasterosteiformes

Fam. Gasterosteidae:

***Gasterosteus aculeatus* Linnaeus, 1758**
Copepoda: *Ergasilus auritus, Thersitina gasterostei*

Scorpaeniformes

Fam. Cottidae:

***Cottus gobio* Linnaeus, 1758**
Monogenea: *Gyrodactylus cotti,*

Gyrodactylus hrabei
Cestoda: *Cyathocephalus truncatus*
Trematoda: *Crepidostomum farionis, Crepidostomum metoecus, Phyllodistomum simile, Allocreadium sp., Nicolla skrjabini, Diplostomum spathaceum* larv., *Diplostomum* sp. larv., *Apatemon cobitidis* larv.
Nematoda: *Camallanus truncatus, Raphidascaris acus* larv., *Cystidicoloides ephemeridarum* larv.
Acanthocephala: *Neoechinorhynchus rutili, Echinorhynchus truttae, Pomphorhynchus laevis*

***Cottus poecilopus* Heckel, 1836**
Monogenea: *Gyrodactylus cotti, Gyrodactylus hrabei*
Cestoda: *Cyathocephalus truncatus*
Trematoda: *Crepidostomum farionis, Nicolla skrjabini, Nicolla wisniewskii*
Nematoda: *Raphidascaris acus* larv.
Acanthocephala: *Neoechinorhynchus rutili, Echinorhynchus truttae*

Perciformes

Fam. Centrarchidae:

****Lepomis gibbosus* (Linnaeus, 1758)**
Monogenea: *Urocleidus dispar, Urocleidus similis*
Trematoda: *Diplostomum spathaceum* larv., *Diplostomum* sp. larv., *Tylodelphys clavata* larv., *Apophallus donicus* larv., *Metagonimus yokogawai* larv.
Nematoda: *Schulmanela petruschewskii*
Acanthocephala: *Acanthocephalus lucii, Pomphorhynchus laevis*

****Micropterus salmoides* Lacépède, 1802**
Hirudinea: *Piscicola geometra*

Fam. Percidae:

***Gymnocephalus cernuus* (Linnaeus, 1758)**
Monogenea: *Dactylogyrus amphibothrium, *Dactylogyrus haplogonus, Dactylogyrus hemiamphibothrium, Gyrodactylus cernuae, Gyrodactylus longiradix, Diplozoon paradoxum*
Cestoda: *Proteocephalus cernuae, *Proteocephalus osculatus, Proteocephalus percae, *Proteocephalus torulosus, Proteocephalus* sp., *Neogryporhynchus cheilancristrotus* larv.
Trematoda: *Bunodera luciopercae, Phyllodistomum angulatum, Allocreadium* sp., *Nicolla skrjabini, Nicolla testiobliquum,* Echinostomatidae gen. sp. larv., *Diplostomum spathaceum* larv., *Diplostomum volvens* larv., *Diplostomum* sp. larv., *Tylodelphys clavata* larv., *Tylodelphys podicipina* larv., *Ichthyocotylurus pileatus* larv., *Ichthyocotylurus platycephalus* larv., *Apophallus donicus* larv., *Metagonimus yokogawai* larv.
Nematoda: *Raphidascaris acus, Camallanus lacustris, Camallanus truncatus,* (?) *Ichthyobronema* sp., *Porrocaecum crassum* larv.
Acanthocephala: *Neoechinorhynchus rutili, Acanthocephalus anguillae, Acanthocephalus lucii, Pomphorhynchus laevis*
Hirudinea: *Piscicola geometra, Hemiclepsis marginata*
Bivalvia: *Glochidium* sp. (larv.)
Copepoda: *Ergasilus sieboldi*
Branchiura: *Argulus foliaceus, Argulus japonicus*

***Gymnocephalus schraetser* (Linnaeus, 1758)**
Monogenea: *Dactylogyrus amphibothrium*
Cestoda: **Caryophyllaeus laticeps, Proteocephalus cernuae*
Trematoda: *Rhipidocotyle campanula, Bunodera luciopercae, Phyllodistomum angulatum, *Phyllodistomum folium, Nicolla skrjabini, *Nicolla testiobliquum, Diplostomum spathaceum* larv., *Diplostomum* sp. larv., *Tylodelphys clavata* larv., *Ichthyocotylurus pileatus* larv., *Ichthyocotylurus platycephalus* larv.,

Apophallus donicus larv., *Metagonimum yokogawai* larv.
Nematoda: *Camallanus truncatus, Rhabdochona* sp.
Acanthocephala: *Acanthocephalus anguillae, Pomphorhynchus laevis*
Copepoda: *Ergasilus sieboldi, Lernaea cyprinacea*

***Perca fluviatilis* Linnaeus, 1758**
Monogenea: *Dactylogyrus amphibothrium, *Ancyrocephalus cruciatus, Ancyrocephalus paradoxus, Ancyrocephalus percae, Gyrodactylus cernuae, Gyrodactylus longiradix, Gyrodactylus lucii, Gyrodactylus luciopercae,* (?) *Gyrodactylus rarus*
Cestoda: **Caryophyllaeides fennica, Triaenophorus crassus, Triaenophorus nodulosus, Proteocephalus cernuae, *Proteocephalus longicollis, Proteocephalus percae, *Proteocephalus torulosus, Proteocephalus* sp., (?) *Diphyllobothrium latum* larv., *Triaenophorus crassus* larv., *Triaenophorus nodulosus* larv., *Bothriocephalus claviceps* larv., *Neogryporhynchus cheilancristrotus* larv.
Trematoda: *Bucephalus polymorphus, Rhipidocotyle campanula, *Asymphylodora imitans, Bunodera luciopercae, *Allocreadium isoporum, Azygia lucii, Nicolla skrjabini, Nicolla testiobliquum, *Sphaerostomum bramae, Rhipidocotyle campanula* larv., *Diplostomum spathaceum* larv., *Diplostomum* sp. larv., *Tylodelphys clavata* larv., *Tylodelphys podicipina* larv., *Ichthyocotylurus pileatus* larv., *Ichthyocotylurus platycephalus* larv., *Paracoenogonimus ovatus* larv., *Clinostomum complanatum* larv., *Apophallus donicus* larv., *Metagonimus yokogawai* larv.
Nematoda: *Schulmanela petruschewskii, Raphidascaris acus, *Paraquimperia tenerrima, Camallanus lacustris, Camallanus truncatus, *Rhabdochona denudata, *Cystidicoloides ephemeridarum, Contracaecum microcephalum* larv., *Raphidascaris acus* larv., *Porrocaecum semiteres* larv., *Philometra obturans* larv.
Acanthocephala: *Neoechinorhynchus rutili, Acanthocephalus anguillae, Acanthocephalus lucii, Pomphorhynchus laevis*
Hirudinea: *Piscicola geometra*
Bivalvia: *Anodonta* sp. larv., *Unio* sp. larv., *Glochidium* sp. (larv.)
Copepoda: *Ergasilus sieboldi, Lernaea cyprinacea, Achtheres percarum*
Branchiura: *Argulus foliaceus, Argulus japonicus, Argulus* sp.

***Stizostedion lucioperca* (Linnaeus, 1758)**
Monogenea: **Ancyrocephalus cruciatus, Ancyrocephalus paradoxus, Ancyrocephalus percae, Gyrodactylus longiradix, *Gyrodactylus lucii, Gyrodactylus luciopercae*
Cestoda: *Triaenophorus nodulosus, Proteocephalus* sp.
Trematoda: *Bucephalus polymorphus, Rhipidocotyle campanula, Bunodera luciopercae, Azygia lucii, Nicolla skrjabini, Bucephalus polymorphus* larv., *Diplostomum spathaceum* larv., *Diplostomum* sp. larv., *Tylodelphys clavata* larv., *Ichthyocotylurus pileatus* larv., *Ichthyocotylurus platycephalus* larv., *Paracoenogonimus ovatus* larv., *Apophallus donicus* larv., *Metagonimus yokogawai* larv.
Nematoda: *Camallanus lacustris, Camallanus truncatus*
Acanthocephala: *Acanthocephalus anguillae, Acanthocephalus lucii, Pomphorhynchus laevis*
Hirudinea: *Piscicola geometra*
Bivalvia: *Unio* sp. larv., *Glochidium* sp. (larv.)
Copepoda: *Ergasilus sieboldi, Neoergasilus japonicus, Achtheres percarum, Lernaea cyprinacea*
Branchiura: *Argulus coregoni, Argulus foliaceus, Argulus japonicus*

***Stizostedion volgense* (Gmelin, 1788)**
Monogenea: *Ancyrocephalus gussevi,*

Ancyrocephalus paradoxus, Gyrodactylus luciopercae
Trematoda: *Bucephalus polymorphus, Rhipidocotyle campanula, Bunodera luciopercae, Nicolla skrjabini, Rhipidocotyle campanula* larv., *Diplostomum spathaceum* larv., *Diplostomum* sp. larv., *Tylodelphys clavata* larv., *Paracoenogonimus ovatus* larv., *Apophallus donicus* larv., *Metagonimus yokogawai* larv.
Nematoda: *Camallanus truncatus*
Acanthocephala: *Acanthocephalus lucii*
Copepoda: *Achtheres percarum*

***Zingel streber* (Siebold, 1863)**
Monogenea: *Gyrodactylus luciopercae*
Cestoda: *Neogryporhynchus cheilancristrotus* larv.
Trematoda: *Rhipidocotyle campanula, Bunodera luciopercae, Azygia lucii, Nicolla skrjabini, Diplostomum spathaceum* larv., *Diplostomum* sp. larv.
Nematoda: *Camallanus lacustris*
Acanthocephala: *Acanthocephalus anguillae, Pomphorhynchus laevis*

***Zingel zingel* (Linnaeus, 1766)**
Trematoda: *Bucephalus polymorphus, Rhipidocotyle campanula, Bunodera luciopercae, Azygia lucii, Nicolla skrjabini, Nicolla testiobliquum, Diplostomum spathaceum* larv., *Diplostomum* sp. larv., *Apophallus donicus* larv., *Metagonimus yokogawai* larv.
Nematoda: *Camallanus lacustris, Camallanus truncatus, *Rhabdochona hellichi, Rhabdochona* sp., (?) *Ichthyobronema* sp.
Acanthocephala: *Acanthocephalus anguillae, Pomphorhynchus laevis*

Fam. Gobiidae:

***Proterorhinus marmoratus* (Pallas, 1814)**
Monogenea: *Gyrodactylus arcuatus, Gyrodactylus proterorhini*
Cestoda: *Proteocephalus* sp., *Triaenophorus crassus* larv.
Trematoda: *Diplostomum spathaceum* larv., *Diplostomum* sp. larv., *Tylodelphys clavata* larv., *Apatemon cobitidis proterorhini* larv.
Nematoda: *Pseudocapillaria tomentosa, *Philometra ovata, Camallanus lacustris* larv., *Anguinicola crassus* larv., *Raphidascaris acus* larv.
Acanthocephala: *Neoechinorhynchus rutili, Acanthocephalus anguillae, Acanthocephalus lucii, Pomphorhynchus laevis*
Bivalvia: *Glochidium* sp. (larv.)
Branchiura: *Argulus foliaceus*

Fish species presently or earlier occurring in the Czech Republic or the Slovak Republic not yet examined for the presence of metazoan parasites:
Petromyzon marinus Linnaeus, 1758
Lampetra fluviatilis (Linnaeus, 1758)
Eudontomyzon mariae (Berg, 1931)
Huso huso (Linnaeus, 1758)
Acipenser nudiventris Lovetzky, 1828
Acipenser stellatus Pallas, 1771
Acipenser sturio Linnaeus, 1758
Rutilus frisii meidingeri (Nordmann, 1840)
Sabanejewia aurata (Filippi, 1865)
**Poecilia sphenops* Valenciennes, 1846
**Poecilia reticulata* Peters, 1859
**Xiphophorus helleri* Heckel, 1848
Gymnocephalus baloni Holčík et Hensel, 1974
Platichthys flesus (Linnaeus, 1758)

LIST OF METAZOAN PARASITES OF FISHES OF THE CZECH REPUBLIC

(with indication of main river basins; parentheses mean that no localities were reported from the respective basin; parasites designated with an asterisk were not recorded from the Slovak Republic)

Monogenea:

**Acolpenteron* sp.		D	
**Dactylogyrus achmerowi*	L		
Dactylogyrus alatus	L	D	(O)
Dactylogyrus amphibothrium	L	D	(O)
Dactylogyrus anchoratus	L	D	
Dactylogyrus auriculatus	L	D	(O)
Dactylogyrus bicornis	(L)	D	(O)
Dactylogyrus borealis	(L)	D	O
Dactylogyrus carpathicus	L	D	(O)
Dactylogyrus chondrostomi		D	(O)
**Dactylogyrus cordus*	L	D	O
Dactylogyrus cornoides	L	D	
Dactylogyrus cornu	L	D	(O)
**Dactylogyrus crassus*		D	
Dactylogyrus crucifer	L	D	O
Dactylogyrus cryptomeres		D	O
Dactylogyrus difformis	L	D	
**Dactylogyrus dirigerus*		D	
Dactylogyrus distinguendus	(L)	D	
**Dactylogyrus dulkeiti*		D	
Dactylogyrus dyki	(L)	D	(O)
Dactylogyrus ergensi	L	D	(O)
Dactylogyrus extensus	L	D	(O)
Dactylogyrus falcatus		D	
**Dactylogyrus fallax*	L	D	
Dactylogyrus folkmanovae	L	D	O
Dactylogyrus formosus	(L)	D	(O)
Dactylogyrus fraternus	L	D	O
Dactylogyrus haplogonoides	(L)		(O)
**Dactylogyrus haplogonus*		D	
Dactylogyrus hemiamphibothrium	L	D	(O)
Dactylogyrus hypophthalmichthys	no localities		
**Dactylogyrus inexpectatus*		D	
Dactylogyrus intermedius	(L)	D	(O)
Dactylogyrus izjumovae			(O)
Dactylogyrus lamellatus		D	
Dactylogyrus macracanthus	L	D	
Dactylogyrus mallaeus	L	D	(O)
**Dactylogyrus micracanthus*	L	D	
Dactylogyrus minor	L	D	(O)
Dactylogyrus minutus	L	D	(O)
**Dactylogyrus nanoides*	L	D	
Dactylogyrus nanus	L	D	
**Dactylogyrus navicularis*	L	D	
**Dactylogyrus naviculoides*		D	
Dactylogyrus nobilis		D	
Dactylogyrus nybelini	no localities		
Dactylogyrus parvus	L	D	(O)
Dactylogyrus propinquus		D	
Dactylogyrus prostae	L	D	O
**Dactylogyrus rarissimus*	L	D	
**Dactylogyrus rutili*	(L)	D	
Dactylogyrus similis	L	D	
Dactylogyrus simplicimallaeata		D	
Dactylogyrus sphyrna	L	D	(O)
Dactylogyrus suecicus	L	D	
Dactylogyrus tincae	L		
Dactylogyrus tuba	L	D	(O)
Dactylogyrus vastator	L	D	O
Dactylogyrus vistulae	L	D	O
Dactylogyrus vranoviensis	L	D	O
Dactylogyrus wegeneri	(L)	D	(O)
Dactylogyrus wunderi	L	D	(O)
Dactylogyrus zandti	L	D	(O)
Pellucihaptor merus	(L)	(D)	O
Pseudacolpenteron pavlovskii		D	
Ancyrocephalus cruciatus	(L)	D	(O)

Ancyrocephalus paradoxus	L	D	(O)
Ancyrocephalus percae	L	D	(O)
Cleidodiscus pricei	(L)		
**Pseudodactylogyrus anguillae*	L	D	
**Pseudodactylogyrus bini*	L		
Silurodiscoides siluri		D	
Silurodiscoides vistulensis	(L)	D	
Tetraonchus borealis	L	(D)	(O)
Tetraonchus monenteron	L	D	O
Gyrodactylus aphyae	L	(D)	O
**Gyrodactylus barbatuli*		D	
Gyrodactylus barbi	L		(O)
Gyrodactylus bliccensis	no localities		
**Gyrodactylus bohemicus*	L		
Gyrodactylus carassii	L	D	O
Gyrodactylus cernuae	L	D	(O)
Gyrodactylus cobitis	(L)	D	(O)
Gyrodactylus cotti	(L)	(D)	(O)
Gyrodactylus cyprini	L	D	
**Gyrodactylus dykovae*		D	
Gyrodactylus elegans	L	D	O
Gyrodactylus fossilis	(L)	(D)	(O)
**Gyrodactylus gasterostei*		D	
**Gyrodactylus glaeseri*		L	
Gyrodactylus gobiensis	L	D	
Gyrodactylus gobii	(L)	D	(O)
Gyrodactylus gobioninum	L	D	O
Gyrodactylus gracilihamatus		D	(O)
**Gyrodactylus gurlei*		D	
**Gyrodactylus hemibarbi*		D	O
Gyrodactylus hrabei	(L)	(D)	(O)
Gyrodactylus jiroveci	L	D	(O)
Gyrodactylus katharineri	L	D	(O)
**Gyrodactylus kearni*		D	
**Gyrodactylus kherulensis*	L		
Gyrodactylus kobayashii	no localities		
Gyrodactylus laevis	(L)	D	(O)
**Gyrodactylus lamberti*		D	
**Gyrodactylus latus*	L	D	(O)
Gyrodactylus leucisci	L	D	
Gyrodactylus limneus	L	(D)	(O)
**Gyrodactylus lomi*	L	D	
Gyrodactylus longiradix	L	D	(O)
Gyrodactylus longoacuminatus	L	D	O
Gyrodactylus lucii	L	D	O
Gyrodactylus luciopercae	L	D	
Gyrodactylus macrocornis			O
Gyrodactylus macronychus	L	(D)	O
**Gyrodactylus magnificus*		D	O
Gyrodactylus markakulensis	L	(D)	O
Gyrodactylus markewitschi		D	
Gyrodactylus medius	L	D	(O)
Gyrodactylus menschikowi	L	(D)	(O)
Gyrodactylus minimus	(L)	(D)	(O)
Gyrodactylus misgurni		D	
**Gyrodactylus osoblahensis*		D	O
Gyrodactylus pavlovskyi	L		
Gyrodactylus prostae	L	D	(O)
Gyrodactylus rhodei	L	D	(O)
**Gyrodactylus rutilensis*	L	D	
**Gyrodactylus scardiniensis*		D	
Gyrodactylus sedelnikowi	(L)	D	O
Gyrodactylus sprostonae		D	
Gyrodactylus tincae	L	D	
Gyrodactylus truttae	L	(D)	O
Gyrodactylus vimbi	L	D	O
?*Gyrodactylus nemachili*	L	D	O
?**Gyrodactylus rarus*		D	
**Mazocraes alosae*	L		
Diplozoon paradoxum	L	D	O
**Eudiplozoon nipponicum*	L	D	
**Paradiplozoon alburni*		D	
Paradiplozoon bliccae	(L)	D	
Paradiplozoon ergensi	(L)	D	(O)
Paradiplozoon homoion	L	D	(O)
Paradiplozoon megan		D	
Paradiplozoon nagibinae		D	
**Paradiplozoon rutili*		D	
Paradiplozoon sapae		D	
Paradiplozoon tisae	(L)		(O)
Paradiplozoon vojteki		D	
Octomacrum europaeum		D	

Cestoda:

Caryophyllaeides fennica	L	D	O
**Khawia baltica*	L		
**Khawia sinensis*	L	D	O
Archigetes sieboldi	L		
Caryophyllaeus brachycollis	L	D	(O)
Caryophyllaeus fimbriceps	L	D	O
Caryophyllaeus laticeps	L	D	(O)

**Monobothrium wageneri*	L	D	
Paracaryophyllaeus gotoi			O
Bothriocephalus acheilognathi	L	D	
Bothriocephalus claviceps	L	D	(O)
Bathybothrium rectangulum	L	D	O
**Eubothrium crassum*		L	
Triaenophorus crassus	(L)	D	(O)
Triaenophorus nodulosus	L	D	(O)
Proteocephalus cernuae	L	D	
Proteocephalus longicollis	L	D	
**Proteocephalus macrocephalus*	L		
Proteocephalus osculatus	L	D	
Proteocephalus percae	L	D	(O)
Proteocephalus torulosus	L	D	O
Silurotaenia siluri	L	D	
**Tentacularia coryphaenae* larv.	L		
**Tetrarhynchus* sp. larv.	L		
Ligula intestinalis larv.	L	D	O
**Scolex pleuronectis* larv.	L		
Neogryporhynchus cheilancristrotus larv.	L	D	
**Paradilepis scolecina* larv.	L		
Valipora campylancristrota larv.	L	D	

Trematoda:

Bucephalus polymorphus	L	D	
Rhipidocotyle campanula	L	D	(O)
Sanguinicola armata		D	
Sanguinicola inermis	L	D	O
Sanguinicola intermedia	(L)	D	(O)
**Hemiurus appendiculatus*	L		
**Derogenes varicus*	L		
Asymphylodora imitans	(L)	D	(O)
Asymphylodora markewitschi		D	
Asymphylodora tincae	L	D	(O)
Palaeorchis unicus		D	
Bunodera luciopercae	L	D	(O)
Crepidostomum farionis	L	D	O
Crepidostomum metoecus	L	D	O
**Phyllodistomum dogieli*	L	D	
Phyllodistomum elongatum	L	D	(O)
Phyllodistomum folium	L	D	O
Phyllodistomum megalorchis			(O)
Phyllodistomum simile	no localities		
Azygia lucii	L	D	O
**Allocreadium carparum*	L	D	
Allocreadium isoporum	L	D	O
Allocreadium markewitschi	L	D	O
Allocreadium papilligerum	L	(D)	(O)
Allocreadium transversale	(L)	D	(O)
Nicolla skrjabini			O
Sphaerostomum bramae	(L)	D	O
Sphaerostomum globiporum	L		
**Diplostomum indistinctum* larv.		D	
Diplostomum spathaceum larv.	L	D	O
**Diplostomum volvens* larv.	L	D	
Tylodelphys clavata larv.	L	D	(O)
**Tylodelphys podicipina* larv.		L	
**Posthodiplostomum brevicaudatum* larv.	L	D	
Posthodiplostomum cuticola larv.	(L)	D	(O)
Apharyngostrigea cornu larv.	(L)	D	(O)
Ichthyocotylurus pileatus larv.	L	D	
Ichthyocotylurus platycephalus larv.		D	
Apatemon cobitidis larv.		D	O
Mesostephanus appendiculatus larv.	no localities		
Paracoenogonimus ovatus larv.	(L)		
Clinostomum complanatum larv.	(L)	D	
Metorchis xanthosomus larv.			O
Apophallus donicus larv.	(L)	D	(O)
Apophallus muehlingi larv.		D	

Nematoda:

**Pseudocapillaria salvelini*	L		
Pseudocapillaria tomentosa	L	D	(O)
Schulmanela petruschewskii		D	
**Hysterothylacium gadi*	L		
Raphidascaris acus	L	D	O
**Paraquimperia tenerrima*	L	D	
**Cucullanus truttae*	L		
Camallanus lacustris	L	D	O
Camallanus truncatus			D

**Anguillicola crassus*	L	D	
**Esocinema bohemicum*	L		
Molnaria intestinalis	L	D	
Skrjabillanus scardinii	L		
**Skrjabillanus tincae*	L	D	
**Daniconema anguillae*	L		
Philometra abdominalis	L	D	O
Philometra kotlani	L		
Philometra obturans	L		
Philometra ovata	L	D	O
Philometra rischta	L	D	
Philometroides sanguinea	L	D	
Rhabdochona denudata	L	D	O
Rhabdochona ergensi		D	O
Rhabdochona hellichi	L	D	
**Rhabdochona phoxini*		L	D
Cystidicoloides ephemeridarum	L	D	O
**Spinitectus inermis*	L		
**Eustrongylides mergorum* larv.	L		
**Anisakis simplex* larv.		L	
**Contracaecum microcephalum* larv.		L	
**Porrocaecum ensicaudatum* larv.		D	
**Porrocaecum semiteres* larv.		L	
Streptocara crassicauda larv.	L	D	
**Syncuaria squamata* larv.	L		
**Agamospirura* sp. larv.	L		

Nematomorpha:

**Gordius aquaticus* larv.	L		

Acanthocephala:

Neoechinorhynchus rutili	L	D	O
Acanthocephalus anguillae	L	D	O
Acanthocephalus lucii	L	D	O
**Echinorhynchus salmonis*	L		
Echinorhynchus truttae	L	D	(O)
Pomphorhynchus laevis	(L)	D	O

Hirudinea:

Piscicola fasciata		D	
Piscicola geometra	L	D	O
Piscicola respirans	(L)	D	O
Hemiclepsis marginata	L	D	O

Bivalvia:

**Anodonta cygnea* larv.	L		O
**Anodonta* sp. larv.	L	D	
Unio sp. larv.		D	O

Copepoda:

**Ergasilus briani*	(L)	D	
**Ergasilus gibbus*	L		
Ergasilus sieboldi	L	D	(O)
Lamproglena pulchella	L	D	O
Lernaea cyprinacea	L	D	
Lernaea esocina	L	D	
Achtheres percarum	(L)	(D)	
Pseudotracheliastes stellifer	no	localities	
**Salmincola salmoneus*	(L)	(D)	
Tracheliastes maculatus		D	
**Tracheliastes polycolpus*	L	D	(O)

Branchiura:

Argulus coregoni	L	D	O
Argulus foliaceus	L	D	O
Argulus japonicus	L	D	O

Acariformes:

**Hydrozetes lacustris*	L	D	

LIST OF METAZOAN PARASITES OF FISHES OF THE SLOVAK REPUBLIC

(with indication of main river basins; parantheses mean that no localities were reported from respective basins; parasites designated with an asterisk were not recorded from the Czech Republic)

Coelenterata:

*_Polypodium hydriforme_	D	

Monogenea:

Dactylogyrus alatus	D	
Dactylogyrus amphibothrium	D	
Dactylogyrus anchoratus	D	V
Dactylogyrus auriculatus	D	
Dactylogyrus bicornis	D	
Dactylogyrus borealis	D	V
Dactylogyrus carpathicus	D	V
*_Dactylogyrus chranilowi_	D	
Dactylogyrus cornoides	D	
Dactylogyrus cornu	D	
Dactylogyrus crucifer	D	
Dactylogyrus cryptomeres	D	
*_Dactylogyrus ctenopharyngodonis_	D	
Dactylogyrus difformis	D	
*_Dactylogyrus difformoides_	D	
Dactylogyrus distinguendus	D	
Dactylogyrus dyki	D	
Dactylogyrus ergensi	D	
Dactylogyrus extensus	D	V
Dactylogyrus falcatus	D	
*_Dactylogyrus finitimus_	D	
Dactylogyrus folkmanovae	D	
Dactylogyrus formosus	D	
Dactylogyrus fraternus	D	
Dactylogyrus haplogonoides	(D)	
Dactylogyrus hemiamphibothrium	D	
Dactylogyrus hypophthalmichthys	no localities	
Dactylogyrus intermedius	D	
Dactylogyrus izjumovae	D	
Dactylogyrus lamellatus	D	
Dactylogyrus macracanthus	no localities	
Dactylogyrus mallaeus	D	
Dactylogyrus minor	D	
Dactylogyrus minutus	D	
Dactylogyrus nanus	D	
Dactylogyrus nobilis	D	
Dactylogyrus nybelini	no localities	
Dactylogyrus parvus	D	V
*_Dactylogyrus petenyi_	D	
*_Dactylogyrus phoxini_	D	
Dactylogyrus propinquus	D	
Dactylogyrus prostae	D	V
Dactylogyrus similis	D	
Dactylogyrus simplicimallaeata	D	
Dactylogyrus sphyrna	D	
Dactylogyrus suecicus	D	
Dactylogyrus tincae	no localities	
*_Dactylogyrus tissensis_	D	
Dactylogyrus tuba	D	
Dactylogyrus vastator	D	
Dactylogyrus vistulae	D	V
Dactylogyrus vranoviensis	D	
Dactylogyrus wegeneri	D	
Dactylogyrus wunderi	D	
*_Dactylogyrus yinwenyingae_	D	
Dactylogyrus zandti	D	
Pellucihaptor merus	D	
Pseudacolpenteron pavlovskii	D	
Ancyrocephalus cruciatus	D	
*_Ancyrocephalus gussevi_	D	
Ancyrocephalus paradoxus	D	
Ancyrocephalus percae	D	
Cleidodiscus pricei	D	
Silurodiscoides siluri	D	

Silurodiscoides vistulensis	D	
*Urocleidus dispar	D	
*Urocleidus similis	D	
Tetraonchus borealis	D	V
Tetraonchus monenteron	D	
Gyrodactylus aphyae	D	
*Gyrodactylus arcuatus	D	
Gyrodactylus barbi	(D)	
Gyrodactylus bliccensis	no localities	
Gyrodactylus carassii	D	
Gyrodactylus cernuae	D	
Gyrodactylus cobitis	D	
Gyrodactylus cotti	D	
Gyrodactylus cyprini	D	
Gyrodactylus elegans	D	
*Gyrodactylus fairporti	D	
Gyrodactylus fossilis	D	
Gyrodactylus gobiensis	D	
Gyrodactylus gobii	D	
Gyrodactylus gobioninum	D	
Gyrodactylus gracilihamatus	D	
Gyrodactylus hrabei	D	
*Gyrodactylus hronosus	D	
Gyrodactylus jiroveci	D	
Gyrodactylus katharineri	D	
Gyrodactylus kobayashii	no localities	
Gyrodactylus laevis	D	V
Gyrodactylus leucisci	D	
Gyrodactylus limneus	D	V
Gyrodactylus longiradix	D	
Gyrodactylus longoacuminatus	D	
Gyrodactylus lotae	D	
Gyrodactylus lucii	D	
Gyrodactylus luciopercae	D	
Gyrodactylus macrocornis	D	
Gyrodactylus macronychus	D	V
*Gyrodactylus malmbergi	D	V
Gyrodactylus markakulensis	D	
Gyrodactylus markewitschi	D	
Gyrodactylus medius	D	V
Gyrodactylus menschikowi	D	
Gyrodactylus minimus	D	
Gyrodactylus misgurni	D	
*Gyrodactylus pannonicus	D	
*Gyrodactylus paraminimus	D	
Gyrodactylus pavlovskyi	D	
Gyrodactylus prostae	D	
*Gyrodactylus proterorhini	D	
Gyrodactylus rhodei	D	
Gyrodactylus sedelnikowi	D	
Gyrodactylus shulmani	D	
*Gyrodactylus slovacicus	D	
Gyrodactylus sprostonae	D	
*Gyrodactylus stankovici	D	
*Gyrodactylus thymalli	D	
Gyrodactylus tincae	D	
Gyrodactylus truttae	D	V
Gyrodactylus vimbi	D	
*Gyrodactylus gracilis (nomen dubium)	D	
?Gyrodactylus nemachili	D	
Diplozoon paradoxum	D	
Paradiplozoon bliccae	D	
Paradiplozoon ergensi	D	
Paradiplozoon homoion	D	
Paradiplozoon megan	D	
Paradiplozoon nagibinae	D	
*Paradiplozoon pavlovskii	D	
Paradiplozoon sapae	D	
Paradiplozoon tisae	(D)	
Paradiplozoon vojteki	(D)	
Octomacrum europaeum	D	

Cestoda:

*Amphilina foliacea	D	
*Cyathocephalus truncatus	D	V
Caryophyllaeides fennica	D	
*Khawia rossittensis	D	
Archigetes sieboldi	D	
Caryophyllaeus brachycollis	D	V
Caryophyllaeus fimbriceps	D	
Caryophyllaeus laticeps	D	
Paracaryophyllaeus gotoi	(D)	
*Paraglaridacris limnodrili	D	
Bothriocephalus acheilognathi	D	
Bothriocephalus claviceps	D	
Bathybothrium rectangulum	D	V
Triaenophorus crassus	D	
Triaenophorus nodulosus	D	V
Proteocephalus cernuae	D	
Proteocephalus longicollis	D	V
Proteocephalus osculatus	D	
Proteocephalus percae	D	

Proteocephalus torulosus	D	V
Silurotaenia siluri	D	
**Diphyllobothrium latum* larv.	D	
**Ligula colymbi* larv.	D	
Ligula intestinalis larv.	D	
Neogryporhynchus cheilancristrotus larv.	D	
Valipora campylancristrota larv.	D	

Trematoda:

Bucephalus polymorphus	D	
Rhipidocotyle campanula	D	
Sanguinicola armata	D	
Sanguinicola inermis	D	
Sanguinicola intermedia	D	
Asymphylodora imitans	D	
**Asymphylodora kubanicum*	D	
Asymphylodora markewitschi	D	
Asymphylodora tincae	D	
**Palaeorchis incognitus*	D	
Palaeorchis unicus	D	
Bunodera luciopercae	D	
**Crepidostomum auriculatum*	D	
Crepidostomum farionis	D	V
Crepidostomum metoecus	D	V
**Phyllodistomum angulatum*	D	
Phyllodistomum elongatum	D	
Phyllodistomum folium	D	
Phyllodistomum megalorchis	D	
Phyllodistomum simile	no localities	
**Skrjabinopsolus semiarmatus*	D	
Azygia lucii	D	
**Orientocreadium siluri*	D	
**Allocreadium dogieli*	D	
Allocreadium isoporum	D	V
Allocreadium markewitschi	D	
Allocreadium papilligerum	(D)	
Allocreadium transversale	D	
Nicolla skrjabini	D	
**Nicolla testiobliquum*	D	
**Nicolla wisniewskii*	D	V
Sphaerostomum bramae	D	
Sphaerostomum globiporum	D	
*Echinostomatidae gen. sp. larv.	D	
Diplostomum spathaceum larv.	D	
Tylodelphys clavata larv.	D	
**Hysteromorpha triloba* larv.	D	
Posthodiplostomum cuticola larv.	D	
Apharyngostrigea cornu larv.	D	
Ichthyocotylurus pileatus larv.	D	
Ichthyocotylurus platycephalus larv.	D	
Apatemon cobitidis larv.	D	
**Holostephanus cobitidis* larv.	D	
Mesostephanus appendiculatus larv.	D	
Paracoenogonimus ovatus larv.	D	
Prohemistomum sp. larv.	D	
Clinostomum complanatum larv.	D	
Metorchis xanthosomus larv.	D	
Apophallus donicus larv.	D	
Apophallus muehlingi larv.	D	
**Metagonimus yokogawai* larv.	D	

Nematoda:

Pseudocapillaria tomentosa	D	
Schulmanela petruschewskii	D	
**Cystoopsis acipenseris*	D	
**Hysterothylacium bidentatum*	D	
Raphidascaris acus	D	
**Cucullanus dogieli*	D	
Camallanus lacustris	D	
Camallanus truncatus	D	
Molnaria intestinalis	D	
Skrjabillanus scardinii	D	
Philometra abdominalis	D	
Philometra kotlani	D	
Philometra obturans	D	
Philometra ovata	D	
Philometra rischta	D	
Philometroides sanguinea	D	
Rhabdochona denudata	D	V
Rhabdochona ergensi	D	
Rhabdochona hellichi	D	

**Cystidicola farionis*	D	V
Cystidicoloides ephemeridarum	D	V
**Contracaecum* sp. larv.	D	
**Porrocaecum crassum* larv.	D	
**Spiroxys contortus* larv.	D	
Streptocara crassicauda larv.	D	

Acanthocephala:

Neoechinorhynchus rutili	D	V
**Leptorhynchoides plagicephalus*	D	
Acanthocephalus anguillae	D	
Acanthocephalus lucii	D	
Echinorhynchus truttae	D	
Pomphorhynchus laevis	D	V

Hirudinea:

Piscicola fasciata	no localities	
Piscicola geometra	D	
Piscicola respirans	D	V
Hemiclepsis marginata	D	

Bivalvia:

Unio sp. larv.	D	

Copepoda:

**Ergasilus auritus*	D	
Ergasilus gibbus	D	
**Ergasilus nasus*	D	
Ergasilus sieboldi	D	
**Neoergasilus japonicus*	D	
**Neoergasilus longispinosus*	D	
**Paraergasilus rylovi*	D	
**Thersitina gasterostei*	D	
Lamproglena pulchella	D	
Lernaea esocina	no localities	
**Lernaea ctenopharyngodontis*	no localities	
Lernaea cyprinacea	D	
Achtheres percarum	D	
**Basanistes huchonis*	D	
Pseudotracheliastes stellifer	no localities	
**Tracheliastes chondrostomi*	D	
Tracheliastes maculatus	D	

Branchiura:

Argulus coregoni	D	V
Argulus foliaceus	D	
Argulus japonicus	D	

LITERATURE

ADÁMEK Z., BARUŠ V., PROKEŠ M. 1996: Summer diet of roach *(Rutilus rutilus)* infested by *Ligula intestinalis* (Cestoda) plerocercoids in the Dalešice Reservoir (Czech Republic). Folia Zool. 45: 347–354.

ANONYMOUS (? Štěpán V. J.) 1926: K rozšíření některých tasemnic rybích na jihu Čech [Distribution of some fish tapeworms in southern Bohemia]. Čsl. rybář 6 (25): 100. (In Czech.)

ANONYMOUS 1933: Řemenatka [*Ligula*]. Rybář. věstník 13: 154. (In Czech.)

BAKKE T. A., ŽĎÁRSKÁ Z. 1985: Tegumental microtopography and arrangement of papillae in adult *Phyllodistomum folium* (Olfers, 1816) (Digenea: Gorgoderidae) from pikes (*Esox lucius* L.). Folia Parasitol. 32: 43–49.

BARTOŠ E. 1934: Ektoparasit hlavatky *Basanistes huchonis* (Schrank) [*Basanistes huchonis* (Schrank), an ectoparasite of the huchen]. Příroda, Brno, 27: 97–98. (In Czech.)

BARTOŠOVÁ Š., JURAJDA P., ONDRAČKOVÁ M. 1999: Posthodiplostomosis of 0+ juvenile cyprinid fish. (Abstract.) Helminthologia 36: 125.

BARUŠ V. 1994: Úhořům jde o život [Eel's life is at stake]. Rybářství, č. 11, 328–329. (In Czech.)

BARUŠ V. 1995: First record of *Anguillicola crassus* (Nematoda) in the Morava River drainage basin. (Abstract.) Helminthologia 32: 89.

BARUŠ V., KOTULAN A. 1997: Černé mraky nad úhořem [Black clouds over the eel]. Rybářství, č. 11, 476–479. (In Czech.)

BARUŠ V., HALAČKA K., PROKEŠ M. 1996: Presence and distribution of the nematode *Anguillicola crassus* in the European eel in the Morava River drainage basin, Czech Republic. Živočišná výroba 41: 219–224.

BARUŠ V., KRÁČMAR S., TENORA F. 1998a: Amino acid spectra of two nematode species (Nematoda: Dracunculoidea) parasitizing fishes in the Czech Republic. Helminthologia 35: 7–11.

BARUŠ V., KRÁČMAR S., TENORA F. 1998b: Qualitative and quantitative spectrum of amino acids in males and females of three nematode species. Helminthologia 35: 57–63.

BARUŠ V., KRÁČMAR S., TENORA F. 1998c: Amino acid spectrum of *Ligula intestinalis* plerocercoids from the intestine of *Stizostedion lucioperca* (Perciformes). (Abstract.) Helminthologia 35: 167.

BARUŠ V., KRÁČMAR S., TENORA F., PROKEŠ M. 1997a: Amino acid spectrum of *Ligula intestinalis* (Pseudophyllidea) plerocercoids in different life phases. Helminthologia 34: 65–69.

BARUŠ V., KRÁČMAR S., TENORA F., PROKEŠ M. 1998: Amino acid spectrum of uninfected and infected eels (*Anguilla anguilla*) by *Anguillicola crassus* (Nematoda). Folia Zool. 47: 231–234.

BARUŠ V., MORAVEC F., PROKEŠ M. 1999a: Anguillicolosis of the European eel *(Anguilla anguilla)* in the Czech Republic. Czech J. Anim. Sci. 44: 423–431.

BARUŠ V. MORAVEC F., PROKEŠ M. 1999b: Přehled o anguilikolóze úhoře říčního v České republice [Overview of anguillicolosis of the European eel in the Czech Republic]. Bulletin VÚRH Vodňany 35: 34–47. (In Czech.)

BARUŠ V., MORAVEC F., ŠPAKULOVÁ M. 1997b: The Red Data List of helminths parasitizing fishes of the orders Cypriniformes, Siluriformes and

Gadiformes in the Czech Republic and Slovak Republic. Helminthologia 34: 35–44.
BARUŠ V., PROKEŠ M. 1994a: On *Ligula intestinalis* (L.) plerocercoids in fishes from the Nové Mlýny Reservoirs, Czech Republic. (Abstract.) Helminthologia 31: 203.
BARUŠ V., PROKEŠ M. 1994b: Parasite load of *Ligula intestinalis* plerocercoids in adult silver bream, *Blicca bjoerkna*. Helminthologia 31: 91–94.
BARUŠ V., PROKEŠ M. 1995a: First record of the *Ligula intestinalis* plerocercoid in the intergeneric hybrid of cyprinid fishes (Pisces). Helminthologia 32: 49–51.
BARUŠ V., PROKEŠ M. 1995b: Seasonal changes in weight of *Ligula intestinalis* (L.) plerocercoids from different fish intermediate hosts in the Nové Mlýny Reservoirs. (Abstract.) Helminthologia 32: 91–92.
BARUŠ V., PROKEŠ M. 1995c: Length-weight relationship of *Ligula intestinalis* plerocercoids in adult *Blicca bjoerkna*, and discussion on estimation of parasite age. Appl. Parasitol. 36: 192–199.
BARUŠ V., PROKEŠ M. 1996a: Length-weight relations of uninfected and infected eels *(Anguilla anguilla)* by *Anguillicola crassus* (Nematoda). Folia zool. 45: 183–189.
BARUŠ V., PROKEŠ M. 1996b: Parasite load of *Ligula intestinalis* plerocercoids in adult bream, *Abramis brama*. Acta Univ. Carolinae, Biol., 39 (Year 1995): 127–134.
BARUŠ V., PROKEŠ M. 1997: Přehled o ligulóze ryb v České republice [Review of the ligulosis of fishes in the Czech Republic]. Sbor. „Ochrana zdraví ryb“, VÚRH JU Vodňany, str. 61–72.
BARUŠ V., ŠEBELA M., PROKEŠ M. 1997c: On the possible free-living phase of the *Ligula intestinalis* plerocercoids (Ligulidae). Helminthologia 34: 173–174.
BARUŠ V., TENORA F., KRÁČMAR S., DVOŘÁČEK J. 1999: Contents of several inorganic substances in European eel infected and uninfected by *Anguillicola crassus* (Nematoda). Dis. Aquat. Org. 37: 135–137.
BARUŠ V., TENORA F., KRÁČMAR S., PROKEŠ M., DVOŘÁČEK J. 1999: Microelement contents in males and females of *Anguillicola crassus* (Nematoda: Dracunculoidea). Helminthologia 36: 283–285.
BARUŠ V., TENORA F., KRÁČMAR S., PROKEŠ M., DVOŘÁČEK J. 1999: Several inorganic substances in *Ligula intestinalis* plerocercoids (Cestoda) from *Stizostedion lucioperca* (Perciformes). In: Vzaimootnosheniza parazita i khozyaina. Institute of Parasitology RAS, Moscow, pp. 119–121.
BARUŠ V., TENORA F., KRÁČMAR S., ZEMAN L. 1995: Amino acid spectrum of *Ligula intestinalis* (Pseudophyllidea) plerocercoids from different piscean intermediate hosts. Appl. Parasitol. 36: 264–273.
BERNER V. 1948: *Dactylogyrus*. Čsl. rybář 3 (č. 8–9): 142–143. (In Czech.)
BLAŽEK R. 2000: Spatial and seasonal distribution of uniocean glochidia on host fishes. (Abstract.) Helminthologia 37: 174–175.
BLAŽEK R., JURAJDA P., KOUBKOVÁ B., GELNAR M. 1999: Distribution of glochidial larval stages of uniocean mussels (Mollusca: Unionidae) within the populations of the host fishes. (Abstract.) Helminthologia 36: 128.
BRUŇANSKÁ M. 1997: *Proteocephalus exiguus* La Rue, 1911 (Cestoda, Proteocephalidae): ultrastructure of the vitelline cells. Helminthologia 34: 9–13.
BRUŇANSKÁ M. 1999: Ultrastructure of primary embryonic envelopes in *Proteocephalus longicollis* (Cestoda: Proteocephalidea). Helminthologia 36: 83–89.
BRUŇANSKÁ M., FAGERHOLM H.-P., GUWSTAFSSON M. K. S. 2000: Ultrastructure studies of *Proteocephalus longicollis* (Cestoda, Proteocephalidea): transmission electron microscopy of scolex glands. Parasitol. Res. 86: 717–723.
BRUŇANSKÁ M., GUSTAFSSON M. K. S., FAGERHOLM H.-P. 1998: Ultrastructure of presumed sensory receptors in the scolex of adult *Proteocephalus exiguus* (Cestoda, Proteocephalidea). Int. J. Parasitol. 28: 667–677.
ČÁKAY E. 1957a: Nález *Philometra sanguinea* (Rudolphi, 1819) na Slovensku [The finding of

Philometra sanguinea (Rudolphi, 1819) in Slovakia]. Biológia, Bratislava, 12: 909–914. (In Slovak.)
ČÁKAY E. 1957b: *Sterliadochona ssavini* Skrjabin, 1946 nový parazit našich pstruhov [*Sterliadochona ssavini* Skrjabin, 1946, a new parasite of our brown trout]. Biológia, Bratislava, 12: 863–866. (In Slovak.)
ČÁKAY E. 1958: Onemocnenie šťúk pásomnicami [Diseasis of pikes caused by tapeworms]. Čsl. rybářství 13: 98–99. (In Slovak.)
ČATÁR G., SOBOTA K., KVASZ L., HRÚZIK J. 1967: Prvý neimportovaný prípad difylobotroózy v ČSSR [First non-imported case of diphyllobothriosis in Czechoslovakia]. Bratislav. Lékař. listy 47: 241–244. (In Slovak.)
DE N. C., MORAVEC F. 1979: Some new data on the morphology and development of the nematode *Cystidicoloides tenuissima* (Zelder, 1800). Folia parasitol. 26: 231–238.
DIVÍLEK T., VÍCHOVÁ J., JURAJDA P., GELNAR M. 1998: Diversity of metazoan parasites of chub (*Leuciscus cephalus* L.) under the condition of environmental stress. (Abstract.) Helminthologia 35: 164.
DUDIŇÁK V. 2000: Morphological adaptation of isolated population of *Pomphorhynchus laevis* (Acanthocephala) (Abstract). Helminthologia 37: 186.
DUŠEK L., GELNAR M., ŠEBELOVÁ Š. 1998: Biodiversity of parasites in a freshwater environment with respect to pollution: metazoan parasites of chub (*Leuciscus cephalus* L.) as a model for statistical evaluation. Inter. J. Parasitol. 28: 1555–1571.
DVOŘÁKOVÁ P., KOMÁREK O., JURAJDA P., GELNAR M. 1999: Microhabitat specificity of monogeneans parasitizing the gills of chub (*Leuciscus cephalus* L.) from different localities. (Abstract.) Helminthologia 36: 129.
DYK V. 1939: Nejdůležitější nemoci, které ohrožují naše chovy ryb [The most important diseases endangering our fish breedings]. Zvěrolék. obzor 32: 459–471. (In Czech.)
DYK V. 1941: Patologický význam rybí pijavky [Pathological importance of the fish leech]. Zvěrolék. obzor 34: 68–71. (In Czech.)
DYK V. 1944: Více pozornosti cizopasníkům kapřího plůdku [More attention to the parasites of carp fry]. Der böhmish-mährische Fischer – Českomoravský rybář, Wodnian – Vodňany, 23: 29–33. (In Czech.)
DYK V. 1951: Kapřivci na pstruzích [Carp lice on brown trout]. Čsl. rybář 6: 242–243. (In Czech.)
DYK V. 1952a: Nemoci našich ryb [Diseases of our fishes]. Přírodověd. vydavatelství, Praha, 283pp. + 16 tab. (In Czech.)
DYK V. 1952b: Méně známí parasiti jihomoravských ryb. I [Less-known parasites of South Moravian fishes. I]. Čas. Nár. musea, odd. přírodov., 121: 88–91. (In Czech.)
DYK V. 1953a. Méně známí parasiti jihomoravských ryb. II [Less-known parasites of South Moravian fishes. II]. Čas. Nár. musea, odd. přírodověd., 122: 23–27. (In Czech.)
DYK V. 1953b: Málo známí hlísti v zaživadlech pstruhů [Little-known helminths in the digestive tract of trout]. Čsl. rybářství 8: 76. (In Czech.)
DYK V. 1953c: Chlopek v našich rybách [*Ergasilus* in our fishes]. Vesmír 32: 187. (In Czech.)
DYK V. 1953d: Řemenatka v našich rybách [*Ligula* in our fishes]. Vesmír 32: 293. (In Czech.)
DYK V. 1954a: Méně známí parasiti jihomoravských ryb. III [Less-known parasites of South Moravian fishes. III]. Čas. Nár. musea 123: 39–45. (In Czech.)
DYK V. 1954b: Dosavadní výsledky výzkumu cizopasníků moravskoslezských ryb [Hitherto results of investigations into the parasites of Moravian-Silesian fishes]. Zool. a entomol. listy 3 (17): 275–287. (In Czech.)
DYK V. 1954: Také zranění udicí umožňuje napadení ryby cizopasníky [Also the injury by hook enables infection of fish by parasites]. Čsl. rybářství 9: 56. (In Czech.)
DYK V. 1955a: Další nálezy cizopasníků ryb moravskoslezských vod [Additional records of fish parasites in the Moravian-Silesian waters]. Zool. a entomol. listy 4 (18): 165–166. (In Czech.)
DYK V. 1955b: Současný stav průzkumu parasitů slovenských ryb [Present state of investigations of the parasites of Slovak fishes]. Biológia, Bratislava, 10: 162–172. (In Czech.)

DYK V. 1955c: Dva pozoruhodní cizopasníci pstruhů v Moravici [Two remarkable parasites of trout in the Moravice River]. Přírodověd. sborník Ostravského kraje 16: 139–140. (In Czech.)

DYK V. 1955d: Reservoárové druhy našich ryb [Reservoir host species of our fishes]. Sbor. VŠZL v Brně, B, 24: 1–5. (In Czech.)

DYK V. 1956a: Přírodní ohniska parasitárních chorob užitkových ryb [Natural foci of the parasitic diseases of commercial fishes]. In.: Blaškovič D. (ed.): Prírodné ohniská nákaz. Sborník prác o prírodnej ohniskovosti rôznych nákaz človeka, zvierat i rastlín v zmysle učenia akademika J. N. Pavlovského. Vydavatelstvo SAV, Bratislava, pp. 229–239. (In Czech.)

DYK V. 1956b: *Raphidascaris acus,* závažný parasit slovenských ryb [*Raphidascaris acus*, a serious parasite of Slovak fishes]. Biológia, Bratislava, 11: 70–75. (In Czech.)

DYK V. 1956c: Parasitofauna ryb tatranských ples [Parasite fauna of fishes in the Tatras Mountain lakes]. Čs. parasitol. 3: 33–42. (In Czech.)

DYK V. 1957a: Nález vrtejše *Echinorhynchus clavula* Dujardin 1845 a specifičnost našich rybích vrtejšů [The finding of the acanthocephalan *Echinorhynchus clavula* Dujardin 1845 and the host specificity of our fish acanthocephalans]. Sbor. VŠZL v Brně, B, 5 (26): 87–93. (In Czech.)

DYK V. 1957b: Dynamika endoparasitů ryb tatranských jezer [The dynamics of endoparasites of fishes in the Tatras Mountain lakes]. Biológia, Bratislava, 12: 333–351. (In Czech.)

DYK V. 1958a: Dynamika motolice *Crepidostomum farionis* (O. F. Müller 1784) a její vztahy k hostitelům a prostředí [The dynamics of the trematode *Crepidostomum farionis* (O. F. Müller 1784) and its relations to the hosts and the environment]. Čs. parasitol 5: 51–57. (In Czech.)

DYK V. 1958b: Kapřivci a pijavky jsou brzdou rybářského hospodaření [Carp lice and leeches are the brake of fish farming]. Čsl. rybářství 13: 116–117. (In Czech.)

DYK V. 1961a: Nemoci našich ryb. Třetí přepracované vydání [Diseases of our fishes. Third revised edition]. ČSAZV a Stát. zem. nakl., Praha, 404 pp. (In Czech.)

DYK V. 1961b: Ekonomicznie ważne parazytozy ryb w Czechosłowacji [Parasitoses of economic importance in fishes in Czechoslovakia]. Wiadom. parazytol. 7: 802–807. (In Polish.)

DYK V. 1964a: Cizopasníci ryb severomoravských vod [Parasites of fishes in the North Moravian waters]. Zprávy Vlast. úst. v Olomouci, čís. 119, pp. 1–4. (In Czech.)

DYK V. 1964b: Helminthocenoses of the digestive tract of the trout in various types of waters. (Abstract.). In: Ergens R., Ryšavý B. (eds.): Parasitic worms and aquatic conditions. Proceedings of a Symposium held in Prague on October 29th – November 2nd, 1962. Publ. House of the ČSAV, Prague, p. 47.

DYK V. 1965a: Cizopasní červi zaživadel pstruha obecného v potocích brněnského okolí [Parasitic worms of the digestive tract of the brown trout in the brooks of the surroundings of Brno]. Zool. listy 14: 85–87. (In Czech.)

DYK V. 1965b: Invadování zažívadel pstruha motolicí *Crepidostomum farionis* [Infection of the digestive tract of the trout by the trematode *Crepidostomum farionis*]. Biológia, Bratislava, 20: 417–422. (In Czech.)

DYK V. 1965c: Odchylky v intenzitě invazí u hostitelů jedné populace ryb [Deviations in the intensity of infection within the hosts of one population of fishes]. Biológia, Bratislava, 20: 411–416. (In Czech.)

DYK V., DYKOVÁ S. 1957: *Bulbodacnitis globosa* Zeder, 1800, další hlístice našeho pstruha obecného [*Bulbodacnitis globosa* Zeder, 1800, another nematode of our brown trout]. Čsl. parasitol. 4: 113–120. (In Czech.)

DYK V., DYKOVÁ S. 1963: Helminti u pstruhů tatranských bystřin [Helminths of the trout in the Tatras Mountain streams]. Sbor. VŠZ v Brně, B, 11: 487–499. (In Czech.)

DYK V., LUCKÝ Z. 1953a: *Bunodera luciopercae* O. F. Müller, nový parasit našich lipanů a pstruhů [*Bunodera luciopercae* O. F. Müller, a new parasite of our grayling and trout]. Zool. a entomol. listy 2 (16): 298–300. (In Czech.)

DYK V., LUCKÝ Z. 1953b: *Camallanus lacustris* Zoega 1776 a jeho výskyt v jihomoravských candátech [*Camallanus lacustris* Zoega 1777 and its occurrence in South Moravian pikeperches]. Zool. a entomol. listy 2 (16): 89–92 + tab. 4. (In Czech.)

DYK V., LUCKÝ Z. 1954a: Plevelná ryba jako přenašeč cizopasníků [Prey fish as transmitors of parasites]. Čsl. rybářství 9 (č. 1): 13–14. (In Czech.)

DYK V., LUCKÝ Z. 1954b: Revise našich příslušníků čeledi Camallanidae [Revision of our representatives of the family Camallanidae]. Zool. a entomol. listy 3 (17): 222–225. (In Czech.)

DYK V., LUCKÝ Z. 1954c: Příspěvek k poznání ohnisek parasitů plevelných ryb v rybničních soustavách [Contribution to the knowledge of the foci of parasites of unimportant fishes in pond systems]. Čs. parasitol. 1: 77–84. (In Czech.)

DYK V., LUCKÝ Z. 1954d: Endoparasitosy kaprovitých ryb, zvláště ostroretky, v společném životním prostředí [Endoparasitoses of cyprinids, particularly of *Chondrostoma nasus*, in the common life environment]. Věst. Čs. spol. zool. 18: 198–200. (In Czech.)

DYK V., LUCKÝ Z. 1954e: Určování a výskyt hlístice *Raphidascaris acus* (Bloch, 1779) [Identification and occurrence of the nematode *Raphidascaris acus* (Bloch, 1779)]. Sbor. VŠZL v Brně, B, 2: 1–6 + tab. 1. (In Czech.)

DYK V., LUCKÝ Z. 1956a: Zapijavčení plůdku sumce velkého [Leech infection of the catfish fry]. Živoč. výroba, Sbor. ČSAZV, 29: 469–474. (In Czech.)

DYK V., LUCKÝ Z. 1956b: Parasitofauna ryb řeky Moravice [Parasite fauna of fishes of the Moravice River basin]. Přírodověd. sbor. Ostrav. kraje 4: 571–580. (In Czech.)

DYK V., LUCKÝ Z. 1956c: Parazitární hynutí hrouzků obecných (*Gobio gobio* L.) [Parasitic dying of the common gudgeon (*Gobio gobio* L.)]. Veterinární medicina, Sbor. Čsl. akad. zem. věd, 29: 335–338. (In Czech.)

DYK V., LUCKÝ Z. 1957: Průzkum cizopasníků ryb v povodí řeky Moravice [Investigations into fish parasites in the Moravice River basin]. Sbor. VŠZL v Brně, A, č. 1, pp. 71–82. (In Czech.)

DYK V., LUCKÝ Z. 1963: Rezervoáry cizopasníků kapra a jiných užitkových ryb [Reservoir hosts of the parasites of carp and other commercial fishes]. Sbor. VŠZ v Brně, B, 11: 53–69 + 1 tab. (In Czech.)

DYK. V., LUCKÝ Z., ŠTĚDRONSKÝ E. 1955: Hlístice *Raphidascaris acus* v našich mnících [The nematode *Raphidascaris acus* in our burbots]. Živoč. výroba a veter. medicina, Sbor. ČSAZV 28: 887–890. (In Czech.)

DYK. V., LUCKÝ Z., VALENTA Z. 1954: Příspěvek k rozlišení digenetických trematodů z rodu *Bunodera* a *Crepidostomum*, jejich výskyt, hostitelé i pathogenita [Contribution to the differentiation of digenetic trematodes of the genera *Bunodera* and *Crepidostomum*, their occurrence, hosts and pathogenicity]. Sbor. VŠZ a lesn. fak. v Brně, B, 2 (23): 105–115. (In Czech.)

DYK V., ŠTĚDRONSKÝ E. 1955: Kožní změny u kaprů vyvolané kapřivci (*Argulus pellucidus* Wagler) [Skin changes in carps, evoked by carp lice (*Argulus pellucidus* Wagler)]. Sbor. VŠZL v Brně, B, 3 (24) (č. 2): 1–4. (In Czech.)

DYK V., ŠTĚDRONSKÝ E. 1956: Nebezpečí povrchových parasitů pro matečné ryby, přechovávané v sádkách [Danger of superficial parasites for mother fish kept in storage ponds]. Čsl. rybářství 11: 36–37. (In Czech.)

DYK V., ŠTĚDRONSKÝ E. 1958: K poznání helmintů zažívadel pstruha a parmy ve společném životním prostředí [To the recognition of helminths from the digestive tracts of the trout and the barbel in their shared life environment]. Sbor. VŠZL v Brně, B, 6 (27): 107–113. (In Czech.)

DYK V., ŠTĚDRONSKÝ E. 1960: Venujme viac pozornosti zdravotnému stavu rýb [Let us pay more attention to the health condition of fishes]. Poľovníctvo a rybárstvo12 (č. 1): 15. In Slovak.)

DYK V., ŠTĚDRONSKÝ E. 1964: Doplňky k znalosti helmintů jihočeských ryb [Additional knowledge of helminths of South Bohemian

fishes]. Sbor. VŠZ v Brně, B, 12 (33): 175–181. (In Czech.)
DYK V., VALENTA Z. 1955: Nález trematoda z podrodu *Phyllodistomum* v pstruhu obecném [The finding of the trematode of the subgenus *Phyllodistomum* from the brown trout]. Věst. Čs. zool. spol. 19: 190–192. (In Czech.)
ERGENS R. 1956a: Příspěvek k poznání cizopasníků ryb Vranovské přehrady [Contribution to the recognition of fish parasites in the Vranovská water reservoir]. Spisy přírodov. fak. Masaryk. univ. v Brně, č. 372, pp. 1–8. (In Czech.)
ERGENS R. 1956b: Výsledky výzkumu monogenetických motolic rodu *Dactylogyrus* Diesing, 1850 [Results of studies of the monogenetic trematodes of the genus *Dactylogyrus* Diesing, 1850]. Práce brněn. základny ČSAV 28: 346–376. (In Czech.)
ERGENS R. 1957: *Gyrodactylus hraběi* n. sp. s ploutví vranky obecné [*Gyrodactylus hraběi* from the fins of the common bullhead]. Věst. Čs. spol. zool. 21: 143–147. (In Czech.)
ERGENS R. 1958: Nález monogenetické motolice *Pseudacolpenteron pavlovskii* Bychowsky et Gussev, 1955, na žábrech *Cyprinus carpio* L. v ČSR [The finding of the monogenetic trematode *Pseudacolpenteron pavlovskii* Bychowsky et Gussev, 1955 on the gills of *Cyprinus carpio* in Czechoslovakia]. Věst. Čs. spol. zool. 22: 94–96. (In Czech.)
ERGENS R. 1959a: Nálezy dalších druhů rodu *Gyrodactylus* Nordrmann (Monogenoidea) v ČSR [Findings of additional species of the genus *Gyrodactylus* Nordmann (Monogenoidea) in Czechoslovakia]. Čs. parasitol. 6: 73–85. (In Czech.)
ERGENS R. 1959b: Nálezy dalších druhů rodu *Dactylogyrus* Diesing (Monogenoidea) v ČSR a vliv lokalisace na morfologické změny chitinových částí haptoru *Dactylogyrus hemiamphibothrium* [Findings of additional species of the genus *Dactylogyrus* (Monogenoidea) in Czechoslovakia and the influence of localization on the morphological changes of chitinoid parts of the haptor]. Čs. parasitol. 6: 87–92. (In Czech.)
ERGENS R. 1959c: K problému druhové samostatnosti *Dactylogyrus nybelini* Markewitsch, 1933 (Monogenoidea) [On the problem of a species independence of *Dactylogyrus nybelini* Markewitsh, 1933 (Monogenoidea)]. Věst. Čs. spol. zool 23: 156–160. (In Czech.)
ERGENS R. 1960a: *Skrjabillanus* Schigin a Schigina 1958, nový rod cizopasných hlístic (Camallanata Chitwood 1936) pro faunu ČSR [*Skrjabillanus* Schigin and Schigina 1958, a new genus of parasitic nematodes for the fauna of Cyechoslovakia]. Věst. Čs. zool. Spol. 24: 230–231. (In Czech.)
ERGENS R. 1960b: Neue bisher beschriebene *Dactylogyrus*-Arten (Monogenoidea) aus der Tschechoslowakei. Helminthologia 2: 3–8.
ERGENS R. 1961a: Zwei weitere Befunde der *Gyrodactylus*-Art (Monogenoidea) aus der Tschechoslowakei. Věst. Čs. spol. zool. 25: 25–27.
ERGENS R. 1961b: Helmintofauna ryb dvou jihočeských rybničních soustav. I. Cestoidea – tasemnice [Helminth fauna of fishes of two South Bohemian pond systems. I. Cestoidea – tapeworms]. Čs. parasitol. 8: 137–150. (In Czech.)
ERGENS R. 1962: Helmintofauna ryb dvou jihočeských rybničních soustav II. Trematoidea, Monogenoidea, Nematoda, Acanthocephala a Hirudinea [Helminth fauna of fishes of two South Bohemian pond systems. II. Trematoidea, Monogenoidea, Nematoda, Acanthocephala and Hirudinea]. Čs. parasitol. 9: 167–190. (In Czech.)
ERGENS R. 1962: Další nové druhy rodu *Gyrodactylus* Nordmann, 1832 (Monogenoidea) pro faunu Československa [Further new species of the genus *Gyrodactylus* Nordmann, 1832 (Monogenoidea) for the fauna of Czechoslovakia]. Zool. listy 11(25): 327–331. (In Czech.)
ERGENS R. 1962–1963a: Funde von drei von zoogeographischem Standpunkte interessanten Arten des *Gyrodactylus* Nordmann, 1832 (Monogenoidea) in der Tschechoslowakei. Helminthologia 4: 154–159.

ERGENS R. 1962–1963b: Vier neue Arten des Genus *Gyrodactylus* Nordmann 1832 (Monogenoidea) aus Fischen der Tschechoslowakei. Z. Fischerei 11: 735–741.
ERGENS R. 1963: Revise helmintofauny ryb z území Československa I. Rod *Crepidostomum* Braun 1900 (Trematoida: Allocreadiidae) [Revision of the helminth fauna of fishes from the territory of Czechoslovakia I. Genus *Crepidostomum* Braun 1900 (Trematoidea: Allocreadiidae)]. Čs. parasitol. 10: 81–88. (In Czech.)
ERGENS R., 1964: Revise helmintofauny ryb z území Československa II. Rod *Allocreadium* Looss 1900 (Trematoidea: Allocreadiidae) [Revision of the helminth fauna of fishes from the territory of Czechoslovakia II. Genus *Allocreadium* Looss 1900 (Trematoidea: Allocreadiidae)]. Čs. parasitol. 11: 127–140. (In Czech.)
ERGENS R. 1965a: Parasitofauna ryb řeky Osoblahy [Parasite fauna of fishes of the Osoblaha River]. Čs. parasitol. 12: 145–169. (In Czech.)
ERGENS R. 1965b: Die Morphogenese der chitinoiden Teile des Haptors bei *Gyrodactylus decorus* Malmberg, 1965 (Monogenoidea) und ihre morphologish-metrische Variabilität. Z. Parasitenk. 25: 359–370.
ERGENS R. 1965c: Beitrag zum Problem der Arten *Gyrodactylus markewitschi* Kulakowskaja, 1951 und *Gyrodactylus albaniensis* Ergens, 1960 (Monogenoidea). Čs. parasitol. 12: 319–322.
ERGENS R. 1965d: Die Morphogenese der chitinoiden Teile des Haptors bei *Gyrodactylus tincae* (Malmberg, 1956) Malmberg, 1964 (Monogenoidea) und ihre morphologisch--metrische Variabilität. Z. Parasitenk. 26: 173–184.
ERGENS R. 1966a: Results of parasitological investigations on the health of *Esox lucius* L. in the Lipno reservoir. Folia Parasitol.13: 222–236.
ERGENS R. 1966b: Revision of the helminthofauna of fishes in Czechoslovakia III. Genus *Ancyrocephalus* (s.l.) Creplin, 1839 (Monogenoidea: Dactylogynidae). Folia Parasitol. 13: 28–35.
ERGENS R. 1966c: Revision of the helminthofauna of fishes from Czechoslovakia IV. Group of the species *Gyrodactylus elegans* Nordmann, 1832 (Monogenoidea) . Folia Parasitol. 13: 212–221.
ERGENS R. 1966d: The finding of the species *Dactylogyrus nasalis* Strelkow et Ha Ky, 1964 (Dactylogyridae: Monogenoidea) in *Scardinius erythrophthalmus* from the Danube basin. Folia Parasitol. 13: 278–280.
ERGENS R. 1967: New species of the genus *Gyrodactylus* (Monogenoidea) from the Danube basin. Folia Parasitol. 14: 377–379.
ERGENS R. 1970: Synonymy, incidence and distribution of *Gyrodactylus shulmani* Ling-Mo-en, 1962 (Monogenoidea: Gyrodactylidae). Folia Parasitol. 17: 267–270.
ERGENS R. 1969: On the problem of the species *Gyrodactylus nemachili* Achmerow, 1952 (Monogenoidea: Gyrodactylidae). Parazitol. sbor. Zool inst. AN SSSR, Leningrad, 24: 85–90. (In Russian.)
ERGENS R. 1971a: The variability, location and distribution of *Gyrodactylus stankovici* Ergens, 1970 (Gyrodactylidae: Monogenoidea). Folia Parasitol. 18: 377–380.
ERGENS R. 1971b: Über die Synonymik, das Vorkommen und die Verbreitung von *Gyrodactylus carassii* Malmberg 1956 (Monogenoidea: Gyrodactylidae). Zool. Anz. 186: 299–306.
ERGENS R. 1973a: Notes on *Gyrodactylus cobitis* Bychowsky, 1933 (Gyrodactylidae: Monogenoidea). Folia Parasitol. 20: 169–173.
ERGENS R. 1973b: Pereopisanie vidov *Gyrodactylus gobii* Shulman, 1953 i *G. gobioninum* Gussev, 1955 (Monogenoidea: Gyrodactylidae) [Redescription of the species *Gyrodactylus gobii* Shulman, 1953 and *G. gobioninum* Gussev, 1955 (Monogenoidea: Gyrodactylidae)]. Parazitologiya 7: 497–501. (In Russian.)
ERGENS R. 1975a: Contribution to the knowledge of the species *Gyrodactylus aphyae* Malmberg, 1957 and *G. pannonicus* Molnár, 1968 (Monogenoidea: Gyrodactylidae). Věst. Čs. spol. zool. 39: 1–8.

ERGENS R. 1975b: *Gyrodactylus* (Monogenoidea) from some Rhodeinae (Cypriniformes). Folia Parasitol. 22: 33–36.

ERGENS R. 1976a: Variability of hard parts of opisthaptor of two species of *Gyrodactylus* Nordmann, 1832 (Monogenoidea) from *Phoxinus phoxinus* (L.). Folia Parasitol. 23: 111–126.

ERGENS R. 1976b: *Gyrodactylus barbi* sp. n. (Monogenoidea) from the fins of barbels. Věst. Čsl. společ. zool. 40: 161–162.

ERGENS R. 1978a: Redescription of *Gyrodactylus latus* Bychowsky, 1933 and other two new species of the genus *Gyrodactylus* from fishes of the genus *Cobitis* L. Folia Parasitol. 25: 333–337.

ERGENS R. 1978b: Results of reexamination of *Gyrodactylus luciopercae* and *G. longiradix* (Gyrodactylidae: Monogenoidea). Věst. Čs. společ. zool. 42: 249–254.

ERGENS R. 1979: Remarks on the nomenclatoric problem of *Gyrodactylus wageneri* Malmberg, 1957 (Monogenea: Gyrodactylidae). Folia Parasitol. 26: 381–383.

ERGENS R. 1980a: Redescription of *Gyrodactylus longoacuminatus* Žitňan, 1964 (Gyrodactylidae: Monogenea). Folia Parasitol. 27: 123–126.

ERGENS R. 1980b: Six new species of *Gyrodactylus* from freshwater fishes of the Palaearctic Region. Folia Parasitol. 27: 303–307.

ERGENS R. 1980c: On the problem of three species of the genus *Gyrodactylus*, members of "*G. wageneri*-group" (Gyrodactylidae: Monogenea). Helminthologia 17: 257–267.

ERGENS R. 1981: Variability of hard parts of opisthaptor in *Gyrodactylus truttae* Gläser, 1974 (Gyrodactylidae: Monogenea). Folia Parasitol 28: 37–42.

ERGENS R. 1983a: *Gyrodactylus* from Eurasian freshwater Salmonidae and Thymallidae. Folia Parasit. 30: 15–26.

ERGENS R. 1983b: A survey of the results of studies on *Gyrodactylus katharineri* Malmberg, 1964 (Gyrodactylidae: Monogenea). Folia Parasitol. 30: 319–327.

ERGENS R. 1986: Notes on *Gyrodactylus bychowskyi* Sproston, 1946 sensu Lucký 1957 and *G. luckiensis* Prost, 1981 (Monogenea). Folia Parasitol. 33: 380.

ERGENS R. 1988: Two new and three previously described species of the genus *Gyrodactylus* Nordmann, 1832 from the *G. phoxini* group (Monogenea: Gyrodactylidae). Folia Parasitol. 35: 23–30.

ERGENS R. 1990: Two new species of the genus *Gyrodactylus* (Monogenea) from fins of *Leuciscus cephalus* (L.). Folia Parasitol. 37: 337–339.

ERGENS R. 1991a: Variability of the hard parts of opisthaptor of *Gyrodactylus leucisci* Žitňan, 1964 (Monogenea: Gyrodactylidae). Folia Parasitol. 38: 23–28.

ERGENS R. 1991b: A new species of the genus *Gyrodactylus* Nordmann, 1832 (Monogenea) from *Gobio gobio* (L.). Folia Parasitol. 38: 83–85.

ERGENS R. 1992a: Redescription of *Gyrodactylus truttae* Gläser, 1974 (Monogenea: Gyrodactylidae) from *Salmo trutta* m. *fario* L. (Clupeiformes: Salmonidae). Folia Parasitol. 39: 201–206.

ERGENS R. 1992b: *Gyrodactylus bohemicus* sp. n. (Monogenea: Gyrodactylidae) from *Oncorhynchus mykiss* (Walbaum) and *Salvelinus fontinalis* (Mitchill) (Clupeiformes: Salmonidae) in Czechoslovakia. Folia Parasitol. 39: 391–394.

ERGENS R., BYKHOVSKY B. E. 1967: Revision of the species *Gyrodactylus nemachili* Bychowsky, 1936 (Monogenoidea). Folia Parasitol. 14: 225–238.

ERGENS R., ERHARDOVÁ B., KOTRLÁ A., RACZEK V. 1965: Uplatnění pentachorfenolátu sodného v rybářství [Application of sodium pentachlorphenolate in fisheries]. Zprávy TPS Spolana, Praha 7, 27 a 29: 54–56. (In Czech.)

ERGENS R., GELNAR M. 1985: Experimental verification of the effect of temperature on the size of hard parts of opisthaptor of *Gyrodactylus katharineri* Malmberg, 1964 (Monogenea). Folia Parasitol. 32: 377–380.

ERGENS R., GELNAR M. 1988: A new species of the genus *Gyrodactylus* Nordmann, 1832

(Monogenea) from *Leuciscus cephalus* (L.). Folia Parasitol. 35: 375–377.

ERGENS R., GUSSEV A. V. 1965: *Dactylogyrus prostae* Molnár, 1964 (Monogenoidea) aus den Kiemen von *Leuciscus cephalus* (L.) und *Leuciscus cephalus orientalis* Nordmann. Čs. parasitol. 12: 323–325.

ERGENS R., GUSSEV V. A., IZYUMOVA N. A., MOLNÁR K. 1975: Parasite fauna of fishes of the Tisa River basin. Rozpravy ČSAV, řada matem. a přírod. věd, Academia, Praha, 85: 1–117.

ERGENS R., LOM J. 1970: Původci parasitárních nemocí ryb [Agents of parasitic diseases of fishes]. Academia, Praha, 383 pp. (In Czech.)

ERGENS R., LUCKÝ Z. 1959: *Dactylogyrus dyki* n. sp. aus der Kiemen der Barbe (*Barbus barbus* (L.)). Věst. Čs. spol. zool. 23: 351–353.

ERTLOVÁ E. 1967: Výskyt pijavice *Cystobranchus respirans* (Troschel) v československom úseku Dunaja [Occurrence of the leech *Cystobranchus respirans* (Troschel) in the Czechoslovak section of the Danube]. Biológia, Bratislava 22: p. 386. (In Slovak.)

ESCHMEYER W. N. (ed.) 1998. Catalog of fishes. Parts 1, 2, 3. California Academy of Sciences, San Francisco, 1820 pp.

FRIČ A. 1873. O korýších země české [On crustaceans of the Czech Lands]. Archiv pro přírodověd. proskoumání Čech, díl II., 4. odd., pp. 189–250.

FRIČ A. 1893: Losos labský, biologická a anatomická studie [Salmon of the Elbe, its biological and anatomical study]. Vlast. náklad, Praha, 106 pp. (In Czech.)

FRIČ A. 1908: České ryby a jich cizopasníci [Czech fishes and their parasites]. Druhé vydání. Praha, 78 pp. (In Czech.)

FRIČ A., VÁVRA V. 1903: Výzkumy zvířeny ve vodách českých. V. Výzkum Labe a jeho starých ramen [Investigations into the fauna of the Czech waters]. Archiv pro přírodověd. výzkum Čech 11: 1–153.

GELNAR M. 1987a: Experimental verification of the effect of water temperature on micropopulation growth of *Gyrodactylus katharineri* Malmberg, 1964 (Monogenea) parasitizing carp fry (*Cyprinus carpio* L.). Folia Parasitol. 34: 19–23.

GELNAR M. 1987b: Experimental verification of the effect of physical condition of *Gobio gobio* (L.) on the growth rate of micropopulations of *Gyrodactylus gobiensis* Gläser, 1974 (Monogenea). Folia Parasitol. 34: 211–217.

GELNAR M. 1987c: Experimental verification of the effect of the form (scaly or scaleless) of carp host (*C. carpio* L.) on the growth rate of micropopulations of *Gyrodactylus katharineri* Malmberg, 1964 (Monogenea). Folia Parasitol. 34: 305–309.

GELNAR M. 1990: Experimental verification of the water temperature effect on the micropopulation growth of *Gyrodactylus rutilensis* Gläser, 1974 (Monogenea). Folia Parasitol. 37: 113–114.

GELNAR M. 1991: Experimental verification of the effect of constant and changing water temperature on the micropopulation growth in *Gyrodactylus gobiensis* Gläser, 1974 (Monogenea) parasitizing gudgeon (*Gobio gobio* L.). Folia Parasitol. 38: 123–131.

GELNAR M., KOUBKOVÁ B., PLÁŇKOVÁ H., JURAJDA P. 1994: Report on metazoan parasites of fishes of the river Morava with remarks on the effects of water pollution. Helminthologia 31: 47–56.

GELNAR M., LUX E. 1991: On the dissemination of two Far East monogenean parasites (*Dactylogyrus achmerovi* Gussev, 1955 and *Gyrodactylus kherulensis* Ergens, 1974) of carp (*Cyprinus carpio* L.) in Czechoslovakia. Folia Parasitol. 38: 131–132.

GELNAR M., SCHOLZ T., MORAVEC F. 1990: Předběžná zpráva o helmintech parazitujících u ryb řeky Malše [Preliminary report on the helminths parasitizing fishes of the Malše River]. In: Kubečka J. (ed.): Ichtyofauna řeky Malše a nádrže Římov. Jihočeské muzeum v Českých Budějovicích, přírod. vědy. České Budějovice, pp. 142–149. (In Czech.)

GELNAR M., SVOBODOVÁ Z., VYKUSOVÁ B. 1989: *Eudiplozoon nipponicum* (Goto, 1981) – nový cizopasník kapra v našich rybnících

[*Eudiplozoon nipponicum* (Goto, 1981), a new parasite of carp in our ponds]. Čs. rybníkářství, č. 1, 5–12. (In Czech.)

GELNAR M., ŠEBELOVÁ Š., DUŠEK L., KOUBKOVÁ B., JURAJDA P., ZAHRÁDKOVÁ S. 1997: Biodiversity of parasites in freshwater environment in relation to pollution. Parassitologia 39: 189–199.

GELNAR M., ŠPAKULOVÁ M. 1997: A checklist of the monogenean parasites reported from fishes in the Czech and Slovak Republics. (Abstract.) Helminthologia 34: 189.

GINTER O. 1932: *Notaspis lacustris* Michael. jako rybí parasit a jeho hubení [*Notaspis lacustris* Michael. as a parasite of fish and its destroy]. Věst. Čsl. akademie zemědělské 8: 199–201. (In Czech.)

GINTER O. 1933: Příspěvek k biologii roztoče *Notaspis lacustris* Michael. (*Hydrozetes lacustris* Mich.) [Contribution to the biology of the mite *Notaspis lacustris* Michael. (*Hydrozetes lacustris* Mich.)]. Sbor. Klubu přírodověd. v Brně 15: 45–55. (In Czech.)

GOTTHARDT A. 1954: Regionální výskyt parasitů ryb Ipelské kotliny u Lučence [Regional occurrence of parasites of fishes of the Ipeľ River basin near Lučenec]. Sbor. VŠZL v Brně, B, 2 (23): 91–94. (In Czech.)

GUSEV A. V. (ed.) 1985: Opredelitel parazitov presnovodnych ryb fauny SSSR, T. 2. Paraziticheskie mnogokletochnye, pervaja chast [Key to parasites of freshwater fishes of the USSR fauna, Vol. 2. Parasitic metazoans, first part]. Nauka, Moskva, 424 pp. (In Russian.)

GUSTAFFSON M. K. S., FAGERHOLM H.-P., HALTON D. W., HANZELOVÁ V., MAULE A. G., REUTER M., SHAW C. 1995: Neuropeptides and serotonin in the cestode, *Proteocephalus exiguus*: an immunocytochemical study. Int. J. Parasitol. 25: 673–682.

HALAČKA K., LUSKOVÁ V., LUSK S. 2000: Contribution to the occurrence of parasites in *Cobitis elongatoides* and *Cobitis elongatoides* × *C.* sp. Folia Zool. 49 (Suppl. 1): 215–218.

HANEK J. 1967: The finding of *Paraergsilus rylovi* Markewitsch, 1937 (Copepoda: Ergasilidae) on the territory of Czechoslovakia. Folia Parasitol. 14: 369.

HANEK J. 1968: The finding of *Neoergasilus japonicus* (Harada, 1930) (Copepoda: Ergasilidae) in Europe. Folia Parasitol. 15: 277–278.

HANEK G., 1969: *Tracheliastes chondrostomi* n. sp. (Copepoda: Lernaeopodidae) from *Chondrostoma nasus* (L.) in Czechoslovakia. Can. J. Zool. 47: 455–456 + Pl. 1.

HANEL L.: 1988: Vliv ligulózy na růst plotice obecné *(Rutilus rutilus)* ve Slapské údolní nádrži [Influence of ligulosis on the growth of the common roach *(Rutilus rutilus)* in the Slapská valley reservoir]. Živoč. výroba 33: 941–948. (In Czech.)

HANZELOVÁ V. 1992: *Proteocephalus neglectus* as a possible indicator of changes in the ecological balance of aquatic environments. J. Helminthol. 66: 17–24.

HANZELOVÁ V., RYŠAVÝ B. 1999: Synopsis of cestodes in Slovakia V. Dilepididae, Dipylidiidae and Paruterinidae. Helminthologia 36: 111–117.

HANZELOVÁ V., SCHOLZ T. 1992: Redescription of *Proteocephalus neglectus* La Rue, 1911 (Cestoda: Proteocephalidae), a trout parasite, including designation of its lectotype. Folia Parasitol. 39: 317–323.

HANZELOVÁ V., SCHOLZ T. 1993: Systematic status of *Proteocephalus tumidocollis* (Cestoda: Proteocephalidae), a parasite of salmonid fishes in North America. Helminthologia 30: 157–161.

HANZELOVÁ V., ŠNÁBEL V., KRÁĽOVÁ I., SCHOLZ T., D'AMELIO S. 1999: Genetic and morphological variability in cestodes of the genus *Proteocephalus*: geographical variation in *Proteocephalus percae* populations. Can. J. Zool. 77: 1450–1458.

HANZELOVÁ V., ŠNÁBEL V., KRÁĽOVÁ I., SCHOLZ T., DE CHAMBRIER A., VAUCHER C. 1998: Morphological and genetic comparison of different European populations of *Proteocephalus percae* (Cestoda), parasite of perch *(Perca fluviatilis)*. (Abstract) Helminthologia 35: 159–160.

HANZELOVÁ V., ŠNÁBEL V., ŠPAKULOVÁ M. 1996: On the host specificity of fish tapeworm *Proteocephalus exiguus* La Rue, 1911 (Cestoda). Parasite 3: 253–257.

HANZELOVÁ V., ŠNÁBEL V., ŠPAKULOVÁ M., KRÁĽOVÁ I., FAGERHOLM H.-P. 1995: A comparative study of fish parasites *Proteocephalus exiguus* and *P. percae* (Cestoda: Proteocephalidae): morphology, isoenzymes, karyotype. Can. J. Zool. 73: 1191–1198.

HANZELOVÁ V., ŠPAKULOVÁ M. 1992: Biometric variability of *Proteocephalus neglectus* (Cestoda: Proteocephalidae) in two different age groups of the rainbow trout from the Dobšiná water reservoir (East Slovakia). Folia Parasitol. 39: 307–316.

HANZELOVÁ V., ŽITŇAN R. 1982: The seasonal dynamics of the invasion of *Gyrodactylus katharineri* Malmberg, 1964 (Monogenea). Helminthologia 19: 257–265.

HANZELOVÁ V., ŽITŇAN R. 1983: Concurrent invasions with Monogenea and their epizootiological significance. 1st Inter. Symp. Ichthyoparasitol. "Parasites and Parasitic Diseases of Fish", České Budějovice, 8–13 August 1983, Abstracts of papers, pp. 35–36.

HANZELOVÁ V., ŽITŇAN R. 1986: Embryogenesis and development of *Bothriocephalus acheilognathi* Yamaguti, 1934 (Cestoda) in the intermediate host under experimental conditions. Helminthologia 23: 145–155.

HANZELOVÁ V., ŽITŇAN R. 1987: The effect of season on embryogenesis of *Bothriocephalus acheilognathi* Yamaguti, 1934 (Cestoda). Biológia, Bratislava, B – zool. 1, 42: 105–111.

HANZELOVÁ V., ŽITŇAN R., SYSOEV A. V. 1990: The seasonal dynamics of invasion cycle of *Proteocephalus neglectus* (Cestoda). Helminthologia 27: 135–144.

HAVLENA F., PEKÁŘ Č.: 1960: Hubení plovatky mezihostitele krevničky rybí a motolice oční [Destroying of *Lymnaea*, an intermediate host of *Sanguinicola* and fish ocular fluke]. Poľnohospodárstvo, Bratislava, 7: 535–544. (In Czech.)

HRADIL V. 1958: K poznání kapřivců *(Argulus)* ze dvou nových lokalit na východním Slovensku [To the recognition of carp lice *(Argulus)* from two new localities in eastern Slovakia]. Zool. listy 7 (21): 257–360. (In Czech.)

HRADIL V. 1964: Parazitofauna rýb povodia Torysy [Parasite fauna of fishes of the Torysa River drainage system]. Sbor. Východoslov. múzea v Košiciach 5 A: 75–80. (In Slovak.)

HRBÁČEK J. 1949: Naši kapřivci [Our carp lice]. Čsl. rybář 4 (č. 10): 154–155. (In Czech.)

JEDLIČKA L., MATIS D. 1966: Ďalší nález druhu *Cystoopsis acipenseri* Wagner, 1867 (Cystopsidae, Nematoda) v slovenskom úseku Dunaja [An additional finding of the species *Cystoopsis acipenseri* Wagner, 1867 (Cystoopsidae, Nematoda) in the Slovak section of the Danube]. Ac. rer. natur. Mus. Nat. Slov., Bratislava, 12: 139–140. (In Slovak.)

JÍROVEC O. 1934–1935: Larvy škeblí a velevrubů jakožto rybí cizopasníci [Larvae of anodontid and unionid mussels as fish parasites]. Vesmír 13: 49–50. (In Czech.)

JÍROVEC O. 1940: Příspěvek k poznání cizopasníků našich štik [Contribution to the knowledge of the parasites of our pikes]. Čas. Nár. musea, odd. přírodověd., 114: 1–12. (In Czech.)

JURAJDA P., GELNAR M., KOUBKOVÁ B. 1992: Occurrence of ziege *(Pelecus cultratus)* in the River Morava with notes on its parasites. Folia Zool. 41: 187–189.

JURAJDA P., GELNAR M., KOUBKOVÁ B. 1994: Occurrence of zingel *(Zingel zingel)* in the River Morava with notes on its parasites. Folia Zool. 43: 93–96.

JOHN J. 1877: Hlístové českých ryb [Helminths of Czech fishes]. Vesmír 6: 18–21, 54–57, 67–69, 91–94, 116–118. (In Czech.)

KADLEC D., MATĚJUSOVÁ I., ŠIMKOVÁ A., BLAŽEK R., ONDRAČKOVÁ M., JURAJDA P. 1998: The community of metazoan parasites of roach (*Rutilus rutilus* L.) in different biotopes of the lower stretch of the River Morava, Czech Republic. (Abstract.) Helminthologia 35: 163.

KAŠŤÁK V. 1955a: *Echinorhynchus salmonis* Müller 1792 (Acanthocephala), nový parazit rýb na území ČSR [*Echinorhynchus salmonis* Müller 1792 (Acanthocephala), a new parasite

of fishes on the territory of Czechoslovakia]. Biológia, Bratislava, 10: 76–79. (In Slovak.)
KAŠŤÁK V. 1955b: *Azygia lucii* Müller, 1776 (Trematoda), nový parazit rýb na Slovensku [*Azygia lucii* Müller, 1776 (Trematoda), a new parasite of fishes in Slovakia]. Veterinár. časopis, SAV Bratislava, 4: 179–181. (In Slovak.)
KAŠŤÁK V. 1955c: Nález trematóda *Crepidostomum farionis* Müller, 1784 parazita rýb na Slovensku [The finding of the trematode *Crepidostomum farionis* Müller, 1784, a parasite of fishes, in Slovakia]. Biológia, Bratislava, 10: 224–227. (In Slovak.)
KAŠŤÁK V. 1955d: *Dactylogyrus minutus* Kulwieć, 1927 a *Dactylogyrus phoxini* Malewitzkaja, 1949 (Monogenoidea) nové parazity rýb na území ČSR [*Dactylogyrus minutus* Kulwieć, 1927 and *Dactylogyrus phoxini* Malewitzkaja, 1949 (Monogenoidea), new parasites of fishes on the territory of Czechoslovakia]. Biológia, Bratislava, 10: 511–513. (In Slovak.)
KAŠŤÁK V. 1955e: Nová cicavica rodu *Dactylogyrus* Diesing, 1850 [A new trematode of the genus *Dactylogyrus* Diesing, 1850]. Biológia, Bratislava 10: 324–327. In Slovak.)
KAŠŤÁK V. 1956a: Nález *Ergasilus nasus* Beneden (Copepoda) v Československu [The finding of *Ergasilus nasus* Beneden (Copepoda) in Czechoslovakia]. Čs. parasitol. 3: 75–76.
KAŠŤÁK V. 1956b: Výsledky parazitologického prieskumu plotíc *(Rutilus rutilus)* teplého ramena Váhu u Piešťan [Results of parasitological studies of roaches *(Rutilus rutilus)* of the warm branch of the Váh River near Piešťany]. Veterinár. časopis, Bratislava, 5: 123–125. (In Slovak.)
KAŠŤÁK V. 1956c: *Dactylogyrus hovorkai* n. sp., ein neuer Saugwurm der *Dactylogyrus*-Art Diesing, 1850. Z. Fischerei 5: 231–234.
KAŠŤÁK V. 1956d: Predbežná zpráva o nálezoch cudzopasníkov rýb v slovenských vodách [Preliminary report on the finds of fish parasites in Slovak waters]. Biológia, Bratislava, 11: 624–635. (In Slovak.)
KAŠŤÁK V. 1956e: Nález trematóda z čeľade Echinostomatidae Dietz u hrebenačky obyčajnej *(Acerina cernua)* [The finding of a trematode of the family Echinostomatidae Dietz in the ruffe *(Acerina cernua)*]. Biológia, Bratislava, 11: 488–490. (In Slovak.)
KAŠŤÁK V. 1956f: K nálezu *Dactylogyrus anchoratus* Dujardin, 1845 (Monogenoidea) nového parazita rýb na Slovensku [To the finding of *Dactylogyrus anchoratus* Dujardin, 1845 (Monogenoidea), a new parasite of fishes in Slovakia] . Biológia, Bratislava, 12: 299–300. (In Slovak.)
KAŠŤÁK V. 1957a: Nový druh monogenetickej cicavice z rodu *Dactylogyrus* Diesing, 1850 u *Barbus meridionalis petényi* (Heck.) [A new species of the monogenetic trematode of the genus *Dactylogyrus* Diesing, 1850 in *Barbus meridionalis petényi*]. Biológia, Bratislava, 12: 137–139.
KAŠŤÁK V. 1957b: Príspevok k poznaniu fauny cudzopasníkov rýb v inundačnom území lokality Senné [Contribution to the knowledge of the parasites of fishes in the periodically flooded territory of the locality Senné]. Veterinár. časopis SAV 6: 105–110. (In Slovak.)
KAŠŤÁK V. 1957c: Helmintofauna šťúk *(Esox lucius)* Oravskej priehrady [Helminth fauna of pikes *(Esox lucius)* of the Oravská water reservoir). Biológia, Bratislava, 12: 102–112. (In Slovak.)
KAŠŤÁK V. 1957d: Poznatky z doterajšieho prieskumu helmintofauny rýb slovenských vôd [Knowledge from the hitherto investigations into the helminth fauna of fishes in Slovak waters]. Helminthologia 2: 186–222. (In Slovak.)
KAŠŤÁK V., ŽITŇAN R. 1959: K vekovej dynamike niektorých helmintóz rýb [To the age dynamics of some helminthoses of fishes]. Biológia, Bratislava, 14: 940–944. (In Slovak.)
KAŠŤÁK V., ŽITŇAN R. 1960a: The helminth fauna of fishes in Slovak lakes. Helminthologia 2: 318–322.
KAŠŤÁK V., ŽITŇAN R. 1960b: Age dynamics of some fish helminthiases. Helminthologia 2: 77–80.
KIRKA A., MÉSZÁROS J., NAGY Š., ŠPORKA F. 1982: Ichtyofauna a jej súčasné podmienky v rieke Ondave [Ichthyofauna and its present

conditions in the Ondava River]. Poľnohospodárstvo 28: 534–540. (In Slovak.)
KODEDOVÁ I., DOLEŽEL D., BROUČKOVÁ M., JIRKŮ M., HYPŠA V., LUKEŠ J., SCHOLZ T. 2000: On the phylogenetic positions of the Caryophyllidea, Pseudophyllidea and Proteocephalidea (Eucestoda) inferred from 18S rRNA. Inter. J. Parasitol. 30: 1109–1113.
KOHN F. G. 1929: In der Karlsbader Gegend beobachtete tierische Parasiten der Warmblüter und der Fische. Prager Arch. Tiermed. 9: 49–64.
KONÍČEK L., GELNAR M. 1994: Metazoan parasites of chub (*Leuciscus cephalus* L.) – a review. (Abstract.) Helminthologia 31: 207.
KOUBKOVÁ B., BARUŠ V. 2000a: The tubenose goby (*Proterorhinus marmoratus*: Perciformes) as paratenic host of the nematode *Anguillicola crassus* (Dracunculoidea). Helminthologia 37: 43–45.
KOUBKOVÁ B., BARUŠ V. 2000b: Metazoan parasites of the recently established tubenose goby (*Proterorhinus marmoratus*: Gobiidae) population from the South Moravian reservoir, Czech Republic. Helminthologia 37: 89–95.
KOUBKOVÁ B., ŠEBELOVÁ Š., GELNAR M. 1997: Teratologies of attachment clamps of the family Diplozoidae (Palombi, 1949). (Abstract.) Helminthologia 34: 183–184.
KOUBKOVÁ B., VOJTKOVÁ L. 1973: K poznání fauny pijavek (Hirudinea) ČSSR [To the knowledge of the fauna of leeches (Hirudinea) in Czechoslovakia]. Folia Fac. Sci. Nat. Univ. Purk. Brun., Biol. 40, 14: 103–118. (In Czech.)
KRÁČMAR S., KUČERA J., TENORA F., BARUŠ V., MINAŘÍK B. 1998: Statistical evaluation of relations between amino acids values in three nematode species. Acta univ. Agric. et silvic. Mendel. Brun. (Brno) 46: 55–62.
KRÁL J. 1977: Použití piperazinové soli niklosamidu při proteocefalóze pstruha duhového [The use of piperazine niclosamide against the proteocephalosis of rainbow trout]. Veterinaria SPOFA 19: 75–83. (In Czech.)
KRÁĽOVÁ I. 1996: A total DNA characterization in *Proteocephalus exiguus* and *P. percae* (Cestoda: Proteocephalidae): RAPD and hybridization techniques. Parasitol. Res. 82: 668–671.
KRÁĽOVÁ I., ŠPAKULOVÁ M. 1996: Intraspecific variability of *Proteocephalus exiguus* La Rue, 1911 (Cestoda: Proteocephalidae) as studied by the random amplified polymorphic DNA. Parasitol. Res. 82: 542–545.
KRÁĽOVÁ I., VAN DE PEER Y., JIRKŮ M., VAN RANST M., SCHOLZ T., LUKEŠ J. 1997: Phylogenetic analysis of a fish tapeworm, *Proteocephalus exiguus*, based on the small subunit rRNA gene. Mol. Biochem. Parasitol. 84: 263–266.
LUCKÝ Z. 1953a: Průzkum parasitů ryb v povodí řeky Dyje u Podivína [Investigation of the parasites of fishes in the Dyje River basin near Podivín]. Sbor. VŠZL v Brně, Spisy fak. veter., 1 (22): 89–99. (In Czech.)
LUCKÝ Z. 1953b: Nález tasemnice *Eubothrium crassum* (Bloch, 1779) ve střevě třeboňského úhoře [The finding of the cestode *Eubothrium crassum* (Bloch, 1779) in the intestine of an eel from Třeboň]. Čas. Nár. musea, odd. přír., 123: 63–66. (In Czech.)
LUCKÝ Z. 1954: *Tetraonchus monenteron* (Wagener, 1857) Diesing, 1858 nový žaberní cizopasník moravských štik [*Tetraonchus monenteron* (Wagener, 1857) Diesing, 1858, a new gill parasite of Moravian pikes]. Sbor. VŠZL v Brně, B, 2 (23): 141–144. (In Czech.)
LUCKÝ Z. 1955a: První nález motolice *Asymphylodora tincae* (Modeer, 1790) v ČSR [First record of the trematode *Asymphylodora tincae* (Modeer, 1790) in Czechoslovakia]. Věst. Čs. zool. spol. 19: 53–55. (In Czech.)
LUCKÝ Z. 1955b: Příspěvek k bionomii, pathogenitě a rozšíření cizopasných žábrohlístů na jižní Moravě [Contribution to the bionomics, pathogenicity and distribution of parasitic monogeneans in South Moravia]. Sbor. ČSAZV 28: 433–440. (In Czech.)
LUCKÝ Z. 1955c: Nález vajíček hlístice *Hepaticola petruschewskii* (Šulman, 1948) v játrech lína z povodí Dyje [The finding of the eggs of the nematode *Hepaticola petruschewskii* (Shulman, 1948) in the liver of tench from the Dyje River basin]. Čs. parasitol. 2: 99–104. (In Czech.)

LUCKÝ Z. 1965: Metacerkarie *Neascus cuticola* (Nordaman, 1832) cizopasník perlínů na jižní Moravě [Metacercariae *Neascus cuticola* (Nordmann, 1832), parasites of rudds in South Moravia]. Čs. parasitol. 2: 102–104. (In Czech.)

LUCKÝ Z. 1957a: *Apophallus mühlingi* (Jägerskiöld, 1899) Lühe, 1909, metacerkarie cizopasící u hořavek a cejnků malých [*Apophallus mühlingi* (Jägerskiöld, 1899) Lühe, 1909, metacercariae parasitic in bitterlings and white breams]. Čs. parasitol. 4: 223–226. (In Czech.)

LUCKÝ Z. 1957b: Příspěvek k poznání žábrohlístů jižní Moravy [Contribution to the knowledge of the monogeneans of South Moravia]. Sbor. VŠZL v Brně, B, 5 (26): 123–157. (In Czech.)

LUCKÝ Z. 1957c: K rozšíření a specifičnosti příslušníků rodu *Diplostomulum* na jižní Moravě [To the distribution and specificity of members of the genus *Diplostomulum* in South Moravia]. Sbor. VŠZL v Brně, B, 5 (26): 191–192. (In Czech.)

LUCKÝ Z. 1957d: Příspěvek k poznání dynamiky žábrohlístů v inundačním prostoru řeky Dyje [Contribution to the knowledge of the dynamics of monogeneans in the periodically flooded space of the Dyje River]. In: Hovorka J. (ed.): Helminthologia, Práce z 1. konferencie československých helmintológov roku 1956, Vyd. SAV, Bratislava, pp. 223–226. (In Czech.)

LUCKÝ Z. 1958: Příspěvek k poznání tasemnic ryb jižní Moravy [Contribution to the knowledge of the cestodes of fishes in South Moravia]. Sbor. VŠZL v Brně, B, 6 (27): 203–312. (In Czech.)

LUCKÝ Z. 1959a: Výsledky dosavadního výzkumu parazitofauny ryb řeky Rokytné [Results of the hitherto investigation of the parasite fauna of the Rokytná River]. Zool. listy 8 (22): 213–225. (In Czech.)

LUCKÝ Z. 1959b: Příspěvky k poznání žábrohlístů jižní Moravy, II [Contributions to the knowledge of the monogeneans of South Moravia]. Sbor. VŠZL v Brně, B, 7 (28): 353–377. (In Czech.)

LUCKÝ Z. 1961: První cizopasníci u vykuleného kapřího plůdku [First parasites of the hatched carp fry]. Čsl. rybářství 16: 83. (In Czech.)

LUCKÝ Z. 1963: Nález žábrohlísta druhu *Gyrodactylus salaris* Malmberg, 1956 na kůži pstruha amerického duhového (*Trutta gardneri irideus* Gibbons, 1855) [The finding of the monogenean species *Gyrodactylus salaris* Malmberg, 1956 on the skin of the rainbow trout (*Trutta gairdneri irideus* Gibbons, 1855)]. Sbor. VŠZL v Brně, B, 11 (32): 127–130. (In Czech.)

LUCKÝ Z. 1964: Contribution to the biology and pathogenity of *Sanguinicola inermis* in juvenile carp. In: Ergens R., Ryšavý B (eds.): Parasitic worms and aquatic conditions. Proceedings. of a Symposium held in Prague on October 29th – November 2nd, 1962. Publ. House of the ČSAV, Praha, pp. 153–156.

LUCKÝ Z. 1968a: Sanguinikolóza kapřího plůdku a násady [Saguinicolosis of carp fry and stock fish]. Veterinářství 11: 487–489. (In Czech.)

LUCKÝ Z. 1968b: Přehled nemocí býložravých ryb [Review of the diseases of herbivorous fishes]. Buletin VÚR Vodňany 4 (čís. 2): 28–36. (In Czech.)

LUCKÝ Z. 1970: 1970: Pathological changes with posthodiplostomosis of fish fry. Acta Vet. Brno, 39 (Supplemennt 1): 51–66.

LUCKÝ Z. 1981a: *Diplozoon homoion* (Discocotylidae, Monogenoidea), a new parasite in pond-reared *Hypophthalmichthys molitrix*. Acta Vet. Brno 50: 237–244.

LUCKÝ T. 1981b: An investigation of invasive diseases of the fry of herbivorous fishes and their treatment. Proc. Inter. Seminar "Fish, pathogens and environment in European polyculture", Szarvas, Hungary, pp. 259–269.

LUCKÝ Z. 1984: Investigation on invasive diseases of the herbivorous fish fry and their treatment. In: Oláh J. (ed.): Symposia Biologica Hungarica (Proc. Inter. Seminar "Fish, pathogens and environment in European polyculture", 23-27 June 1981, Szarvas, Hungary), Akademiai Kiadó, Budapest, Hungary, 24: 173–180.

LUCKÝ Z., DYK V. 1964: Cizopasníci ryb v řekách a rybnících povodí Odry a Dyje [Parasites of fishes in the rivers and ponds of the Oder and Dyje Rivers basins]. Sbor. VŠZ v Brně, B, 12: 49–73. (In Czech.)

LUCKÝ Z., DYK V., BÁRTÍK M. 1955: Parasitofauna ryb v oblasti reservace Velký Tisý u Lomnice nad Lužnicí [Parasite fauna of fishes in the region of the reserve Velký Tisý near Lomnice nad Lužnicí]. Čas. Nár. musea, odd. přírodověd., 124: 55–64. (In Czech.)

LUCKÝ Z., JÍLEK J. 1967: Parasitofauna der Fische in Stausse von Kníničky I. Sbor. VŠZ v Brně, B, 15: 123–136 + Tab. 1, 2.

LUCKÝ Z., KAZDA V. 1961: Nález *Allocreadium carparum* Odening 1959 a *Allocreadium markewitschi* Kowal 1949 u ryb v ČSSR [The finding of *Allocreadium carparum* Odening 1959 and *Allocreadium markewitschi* Kowal 1949 in fishes in Czechoslovakia]. Čs. parasitol. 8: 253–256. (In Czech.)

LUCKÝ Z., NAVRÁTIL S. 1983: Parasitic diseases of perch *(Perca fluviatilis)* in retention reservoirs of the Morava River basin. 1st. Inter. Symp. Ichthyoparasitol. "Parasites and Parasitic Diseases of Fish", 8–13 August 1983, České Budějovice. Abstracts of papers, p. 55.

LUCKÝ Z., NAVRÁTIL S. 1984: Parasitic diseases of the perch *(Perca fluviatilis)* in retention reservoirs of the Morava River basin. Acta Vet. Brno 53: 81–90.

LUSK S., LUSKOVÁ V., HALAČKA K. 2000: On the occurrence of populations of the genera *Cobitis* and *Sabanejewia* (Pisces, Cobitidae) in the Czech Republic. Folia Zool. 49 (Suppl. 1): 97–106.

MACKIEWICZ J. S. 1994: Order Caryophyllidea van Beneden in Carus, 1863. In: Khalil L. F., Jones A., Bray R. A.: Keys to the cestode parasites of vertebrates. CAB International, Wallingford (UK), pp. 21–43.

MACKO J. K., RYŠAVÝ B., ŠPAKULOVÁ M., KRÁĽOVÁ I. 1993: Synopsis of cestodes in Slovakia I. Cestodaria, Cestoidea: Caryophyllidea, Spathebothriidea, Pseudophyllidea, Proteocephalidea. Helminthologia 30: 85–91.

MARKEVICH A. P. 1956: Paraziticheskie veslonogie ryb SSSR [Parasitic copepods of fishes of the USSR]. Acad. Sci. Ukr. SSR, Kiev, 259 pp. (In Russian.)

MÉSZÁROS J. 1983: A contribution to the knowledge of helminthofauna of fish in flysch belt of East Slovakian rivers. 1st Inter. Symp. Ichthyoparasitol. "Parasites and Parasitic Diseases of Fish", 8–13 August 1983, České Budějovice, Abstracts of papers, p. 69.

MÉSZÁROS J. 1984: Ergasilóza rýb [Ergasilosis of fishes]. Poľovníctvo a rybárstvo 36 (č. 7): 31. (In Slovak.)

MICHALOVIČ M. 1954: Výsledky průzkumu parasitů ryb v podunajské oblasti u Komárna [Results of investigation of the parasites of fishes in the Danube region near Komárno]. Sbor. VŠZL v Brně, B, 2 (23): 67–74 + 1 Plt. (In Czech.)

MITTERPÁK J. 1982: O ekonomicky závažných parazitózách rýb Štátného rybárstva vo Východoslovenskom kraji za obdobie rokov 1976–1981 [Parasitoses of fishes of economic importance in the State Fisheries in the East-Slovak County in the period 1976–1981]. Zprávy Čsl. spol. parasitol. 22 (č. 2): 20–21. (In Slovak.)

MITTERPÁK J., VASIĽ M., HUŇADY J., 1984: K ekonomickej závažnosti *Caryophyllaeus fimbriceps* u kaprovitých rýb [Economic importance of *Caryophyllaeus fimbriceps* in cyprinid fishes]. Veterinářství 34: 125–126. (In Slovak.)

MORAVEC F. 1966a: The finding of the nematode *Paraquimperia tenerrima* (v. Linstow, 1878) from eels *Anguilla anguilla* (L.) in Czechoslovakia. Folia Parasitol. 13: 182–186.

MORAVEC F. 1966b: On the systematics of the genus *Paraquimperia* Baylis,1934 (Nematoda: *Quimperiidae*). Helminthologia 7: 353–364.

MORAVEC F. 1967: The systematic status of the genus *Sterliadochona* Skrjabin, 1946 (Nematoda: Rhabdochonidae). Folia Parasitol. 14: 371–376.

MORAVEC F. 1968a: Species of the genus *Rhabdochona* Railliet, 1916 (Nematoda: Rhabdochonidae) from fishes of Czechoslovakia. Folia Parasitol. 15: 29–40.

MORAVEC F. 1968b: A new nematode genus, *Molnaria* gen. n. (Nematoda: Skrjabillanidae). Folia Parasitol. 15: 322.

MORAVEC F. 1970: On the life history of the nematode *Raphidascaris acus* (Bloch, 1779) in the natural environment of the River Bystřice, Czechoslovakia. J. Fish Biol. 2: 313–322.
MORAVEC F. 1971a: Nematodes of fishes in Czechoslovakia. Přírodověd. práce ústavů ČSAV v Brně 5: 1–49.
MORAVEC F. 1971b: Some notes on the larval stages of *Camallanus truncatus* (Rudolphi, 1814) and *Camallanus lacustris* (Zoega, 1776) (Nematoda; Camallanidae). Helminthologia 10 (Year 1969): 129–135.
MORAVEC F. 1971c: On the poblem of host specificity, reservoir parasitism and secondary invasions of *Camallanus lacustris* (Nematoda; Camallanidae). Helminthologia 10 (Year 1969): 107–144.
MORAVEC F. 1971d: On the life history of the nematode *Cystidicoloides tenuissima* (Zeder, 1800) in the River Bystřice, Czechoslovakia. Folia Parasitol. 18: 107–112.
MORAVEC F. 1972: Studies on the development of the nematode *Rhabdochona (Filochona) ergensi* Moravec, 1968. Folia Parasitol. 19: 321–333.
MORAVEC F. 1974: Some remarks on the development of *Paraquimperia tenerrima* Linstow, 1878 (Nematoda: Quimperiidae). Scripta Fac. Sci. Nat. UJEP Brun., Biol. 5, 4: 135–142.
MORAVEC F. 1975: Reconstruction of the nematode genus *Rhabdochona* Railliet, 1916 with a review of the species parasitic in fishes of Europe and Asia. Studie ČSAV č. 8, Academia, Praha, 1–104.
MORAVEC F. 1976a: Observations on the development of *Rhabdochona phoxini* Moravec, 1968 (Nematoda: Rhabdochonidae). Folia Parasitol. 23: 309–320.
MORAVEC F. 1976b: Occurrence of the encysted larvae of *Cucullanus truttae* (Fabricius, 1794) in the brook lamprey, *Lampetra planeri* (Bl.). Scripta Fac. Sci. Nat. UJEP Brun., Biol. 1, 6: 17–20.
MORAVEC F. 1977a: The development of the nematode *Philometra abdominalis* Nybelin, 1928 in the intermediate host. Folia Parasit. 24: 237–245.
MORAVEC F. 1977b: The life history of the nematode *Philometra abdominalis* in the Rokytka Brook, Czechoslovakia. Věst. Čs. spol. zool. 41: 114–120.
MORAVEC F. 1977c: Life history of the nematode *Rhabdochona phoxini* Moravec, 1968 in the Rokytka Brook, Czechoslovakia. Folia Parasitol. 24: 97–105.
MORAVEC F. 1977d: A new nematode parasite, *Esocinema bohemicum* gen. et sp. nov. (Skrjabillanidae) of the European pike. Folia Parasitol. 24: 86–90.
MORAVEC F. 1977e: First record of the nematode *Spinitectus inermis* (Zeder, 1800) from eels of Czechoslovakia. Folia Parasitol. 24: 85.
MORAVEC F. 1978a: First record of *Molnaria erythropthalmi* larvae in the intermediate host in Czechoslovakia. Folia Parasitol. 25: 141–142.
MORAVEC F. 1978b: Redescription of the nematode *Philometra obturans* (Prenanat, 1886) with a key to the philometrid nematodes parasitic in European freshwater fishes. Folia Parasitol. 25: 115–124.
MORAVEC F. 1978c: The development of the nematode *Philometra obturans* (Prenant, 1886) in the intermediate host. Folia Parasitol. 25: 303–315.
MORAVEC F. 1978d: Přehled endoparazitických červů zjištěných v rybách rybničního systému Máchova jezera [A survey of endoparasitic worms found in fishes of the Mácha Lake fishpond system]. Scripta Fac. Sci. Nat. UJEP Brun., Biol. 2, 8: 77–80.
MORAVEC F. 1978e: O častých parazitech ryb, hlísticích rodu *Rhabdochona* [Nematodes of the genus *Rhabdochona*, frequent parasites of fishes]. Živa 26: 62–63. (In Czech.)
MORAVEC F. 1979a: Occurrence of the endoparasitic helminths in pike (*Esox lucius* L.) from the Mácha Lake fishpond system. Věst. Čs. spol. zool. 43: 174–193.
MORAVEC F. 1979b: Redescription of the nematode *Spinitectus inermis* parasitic in ells, *Anguilla anguilla*, of Europe. Věst. Čs. spol. zool. 43: 35–42.
MORAVEC F. 1979c: Observations on the development of *Cucullanus (Truttaedacnitis)*

truttae (Fabricius, 1794) (Nematoda: Cucullanidae). Folia Parasitol. 26: 295–307.

MORAVEC F. 1980a: Development of the nematode *Philometra ovata* (Zeder, 1803) in the copepod intermediate host. Folia Parasitol. 27: 29–37.

MORAVEC F. 1980b: Biology of *Cucullanus truttae* (Nematoda) in a trout stream. Folia Parasitol. 27: 217–226.

MORAVEC F. 1980c: The lamprey *Lampetra planeri* as a natural intermediate host for the nematode *Raphidascaris acus*. Folia Parasitol. 27: 347–348.

MORAVEC F. 1980d: First record of the nematode *Capillaria salvelini* Polyansky, 1952 from Czechoslovakia. Folia Parasitol. 27: 115–116.

MORAVEC F. 1980e: Revision of nematodes of the genus *Capillaria* from European freshwater fishes. Folia Parasitol. 27: 309–324.

MORAVEC F. 1982a: The finding of the cestode *Proteocephalus neglectus* La Rue, 1911 from brown trout from Czechoslovakia. Folia Parasitol. 29: 189–190.

MORAVEC F. 1982b: A contribution to the binomics of *Crepidostomum metoecus* (Trematoda: Allocreadiidae). Věst. Čs. spol. zool. 46: 15–24.

MORAVEC F. 1983a: Pozoruhodný vývoj hlístice *Cucullanus truttae* – parazita našich pstruhů [A remarkable development of the nematode *Cucullanus truttae*, a parasite of our trouts]. Živa 31 (69): 63–64. (In Czech.)

MORAVEC F. 1983b: Some observations on the binomics of the nematode *Pseudocapillaria brevispicula* (Linstow, 1983). 1st Inter. Symp. Ichthyoparasitol. "Parasites and Parasitic diseases of Fish", 8–13 August 1983, České Budějovice, Abstracts of papers, p. 77.

MORAVEC F. 1984a: Systematic status of *Allocreadium carparum* Odening, 1959 (Digenea) and three other related species from fishes of Czechoslovakia. Folia Parasitol. 31: 215–224.

MORAVEC F. 1984b: First record of the cestode *Khawia baltica* Szidat, 1941 in Czechoslovakia. Folia Parasitol. 31: 373–374.

MORAVEC F. 1984c: Seasonal occurrence and maturatin of *Neoechinorhynchus rutili* (Müller, 1780) (Acanthocephala) in carp (*Cyprinus carpio* L.) of the Mácha Lake fishpond system, Czechoslovakia. Helminthologia 21: 55–65.

MORAVEC F. 1984d: Obecné aspekty binomie parazitických hlístic (Nematoda) sladkovodních ryb [General aspects of the bionomics of parasitic nematodes (Nematoda) of freshwater fishes]. Studie ČSAV č. 4, Academia, Praha, 116 str. (In Czech.)

MORAVEC F. 1984e: Occurrence of endoparasitic helminths in carp *(Cyprinus carpio)* from the Mácha Lake fishpond system. Věst. Čs. spol. zool. 48: 261–278.

MORAVEC F. 1985a: Occurrence of the endoparasitic helminths in tench *(Tinca tinca)* from Mácha Lake fishpond system. Věst. Čs. spol. zool. 49: 32–50.

MORAVEC F. 1985b: Occurrence of endoparasitic helminths in eels (*Anguilla anguilla* (L.)) from the Mácha Lake fishpond system, Czechoslovakia. Folia Parasitol. 32: 113–125.

MORAVEC F. 1986a: The finding of the caryophyllaeid cestode *Archigetes sieboldi* Leuckart, 1878 in carp in Czechoslovakia. Folia Parasitol. 33: 69–70.

MORAVEC F. 1986b: The morphology and systematic status of *Philometra ovata* (Zeder, 1803) (Nematoda: Philometridae). Folia Parasitol. 33: 227–333 + Plts. 1–4.

MORAVEC F. 1986c: Occurrence of endohelminths in three species of cyprinids *(Abramis brama, Rutilus rutilus* and *Scardinius erythrophthalmus)* of the Mácha Lake fishpond system, Czechoslovakia. Věst. Čs. spol. zool. 50: 49–69.

MORAVEC F. 1987: Revision of capillariid nematodes (subfamily Capillariinae) parasitic in fishes. Studie ČSAV č. 3, Academia, Praha, 141 pp.

MORAVEC F. 1989: Observations on the occurrence and maturation of the nematode *Rhabdochona denudata* (Dujardin, 1845) in chub, *Leuciscus cephalus* (L.), of the Rokytná River, Czechoslovakia. Parassitologia 31: 25–35.

MORAVEC F. 1990a: Synopsis of type specimens deposited in the helminthological collection of the Institute of Parasitology, Czechoslovak Academy of Sciences, in České Budějovice. I. Monogenea. Folia Parasitol. 37: 115–129.

MORAVEC F. 1990b: Synopsis of type specimens deposited in the helminthological collection of the Institute of Parasitology, Czechoslovak Academy of Sciences, in České Budějovice. III. Nematoda. Folia Parasitol. 37: 367–383.

MORAVEC F. 1992a: Observations on the bionomy of *Allocreadium isoporum* (Looss, 1894) (Trematoda: Allocreadiidae). Folia Parasitol. 39: 133–144.

MORAVEC F. 1992b: Spreading of the nematode *Anguillicola crassus* (Dracunculoidea) among eel populations in Europe. Folia Parasitol. 39: 247–248.

MORAVEC F. 1994: Parasitic nematodes of freshwater fishes of Europe. Aacademia and Kluwer Acad. Publishers, Praha and Dordrecht, Boston, London, 473 pp.

MORAVEC F. 1995a: Cizopasníci (mnohobuněční cizopasníci) [Parasites (metazoan parasites)]. In: Baruš V., Oliva O. (eds.): Mihulovci – Petromyzontes a ryby – Osteichthyes. Část 1, 2. [Lampreys – Petromyzontes and fishes – Osteichthyes. Parts 1, 2]. Academia, Praha, pp.: Part 1: 71, 78, 391, 410, 437, 459, 474, 494, 512, 523, 535, 554, 558, 576–577, 582; Part 2: 23–24, 28, 38, 45, 51, 57, 63–64, 71, 76–77, 80, 88–89, 100–101, 109, 116, 119, 123, 131, 141, 150, 154, 161, 180, 185–186, 193, 204, 208, 215, 220–221, 234, 261–262, 268, 273, 278, 238, 291, 305, 311, 326–327, 337, 379–380, 387–388, 391, 401–402, 406, 410, 413, 424, 429, 434, 438, 441. (In Czech.)

MORAVEC F. 1995b: Trichopteran larvae (Insecta) as the intermediate hosts of *Rhabdochona hellichi* (Nematoda: Rhabdochonidae), a parasite of *Barbus barbus* (Pisces). Parasitol. Res. 81: 268–270.

MORAVEC F. 1996a: The amphipod *Gammarus fossarum* as a natural true intermediate host of the nematode *Raphidascaris acus*. J. Parasitol. 82: 668–669.

MORAVEC F. 1996b: Aquatic invertebrates (snails) as new paratenic hosts of *Anguillicola crassus* (Nematoda: Dracunculoidea) and the role of paratenic hosts in the life cycle of this parasite. Dis. Aquat. Org. 27: 237–239.

MORAVEC F. 1996c: Some new data on the morphology of *Spinitectus inermis* (Nematoda: Cystidicolidae), a rare parasite of eels, as revealed by SEM. Parasite 3: 169–174.

MORAVEC F. 1997: Současný stav výzkumu anguilikolózy úhořů [Present state of studies on the anguillicolosis of eels]. Sbor. „Ochrana zdraví ryb", VÚRH JU, Vodňany, pp. 56–60. (In Czech.)

MORAVEC F., DE N. C. 1982: Some new data on the binomics of *Cystidicoloides tenuissima* (Nematoda: Cystidicolidae). Věst. Čs. spol. zool. 46: 100–108.

MORAVEC F., DI CAVE D., ORECCHIA P., PAGGI L. 1993: Studies on the development of *Anguillicola crassus* Kuwahara, Niimi et Itagaki, 1974 (Nematoda: Dracunculoidea) in the intermediate host. Folia Parasitol. 40: 39–48.

MORAVEC F., DYKOVÁ J. 1978: On the biology of the nematode *Philometra obturans* (Prenant, 1886) in the fishpond system of Mácha Lake, Czechoslovakia. Folia Parasitolol. 25: 231–240.

MORAVEC F., GELNAR M., ERGENS R., SCHOLZ T. 1997: Metazoan parasites of fishes from the section of the Vltava River supposed to be affected by the operation of the Temelín nuclear electric power-station, Czech Republic. Acta Soc. Zool. Bohem. 61: 65–76.

MORAVEC F., HOLČÍK J., MÉSZÁROS J. 1989: Několik poznámek k helmintofauně jesetera malého (*Acipenser ruthenus* L.) na Slovensku [Several remarks to the helminth fauna of sterlet (*Acipenser ruthenus* L.) in Slovakia]. Biológia, Bratislava, 44: 151–155. (In Czech.)

MORAVEC F. KØIE M. 1987: *Daniconema anguillae* gen. et sp. n., a new nematode of a new family *Daniconematidae* fam. n. parasitic in European eels. Folia Parasitol. 34: 335–340.

MORAVEC F., KONEČNÝ R., BASKA F., RYDLO M., SCHOLZ T., MOLNÁR K., SCHIEMER F. 1997: Endohelminth fauna of barbel, *Barbus barbus* (L.), under ecological conditions of the Danube basin in Central Europe. Academia, Praha, 96 pp.

MORAVEC F., PROKOPIČ J., SHLIKAS A. V. 1987: The biology of nematodes of the family Capillariidae Neveu-Lemaire, 1936. Folia Parasitol. 34: 39–56.

MORAVEC F., SCHOLZ T. 1984: First record of the trematode *Tylodelphys podicipina* Kozicka et Niewiadomska, 1960 in Czechoslovakia. Folia Parasitol. 31: 331–332.

MORAVEC F., SCHOLZ T. 1991a: Observations on the biology of *Pomphorhynchus laevis* (Zoega in Müller, 1776) (Acanthocephala) in the Rokytná River, Czech and Slovak Federative Republic. Helminthologia 28: 23–29.

MORAVEC F., SCHOLZ T. 1991b: Occurrence of endohelminths in chub, *Leuciscus cephalus*, of the Rokytná River, Czechoslovakia. Věst. Čs. Společ. Zool. 55: 12–28.

MORAVEC F., SCHOLZ T. 1994a: Observations on the development of *Syncuaria squamata* (Nematoda: Acuariidae), a parasite of cormorants, in the intermediate and paratenic hosts. Folia Parasitol. 41: 183–192.

MORAVEC F., SCHOLZ T. 1994b: Seasonal occurrence and maturation of *Neoechinorhynchus rutili* (Acanthocephala) in barbel, *Barbus barbus* (Pisces), of the Jihlava River, Czech Republic. Parasite 1: 271–277.

MORAVEC F., SCHOLZ T. 1995: Life history of the nematode *Rhabdochona hellichi*, a parasite of the barbel in the Jihlava River, Czech Republic. J. Helminthol. 69: 59–64.

MORAVEC F., SCHOLZ T., ŠKORÍKOVÁ B. 1996: Metazoan parasites of eels, *Anguilla anguilla*, in the Czech Republic. EMOP VII Abstracts. Parassitologia 38: 43.

MORAVEC F., ŠKORÍKOVÁ B. 1998: Amphibians and larvae of aquatic insects as new paratenic hosts of *Anguillicola crassus* (Nematoda: Dracunculoidea), a swimbladder parasite of eels. Dis. Aquat. Org. 34: 217–222.

MORAVEC F., VOJTKOVÁ L. 1970: The first record of the nematode *Spiroxys contortus* (Rudolphi, 1819) in Czechoslovakia. Folia Parasitol. 17: 298.

MRÁZEK A. 1901a: Ueber das Verhalten der Längsnerven bei *Abothrium rectangulum* (Rud.). Centralblatt Bakt., Parsitenk. Infektionskrank. (= Zbl. Bakt. Abt. 1 Orig.) 29: 569–571.

MRÁZEK A. 1901b: Ueber die Larve von *Caryophyllaeus mutabilis* Rud. Centralblatt. Bakt., Parasitenk. Infektionskrank. (= Zbl. Bakt. Abt. 1 Orig.) 29: 485–491.

NAGY Š. 1985: Potrava miňa (*Lota lota* L.) v rieke Turiec [Food of burbot (*Lota lota* L.) in the Turiec River]. Živoč. výroba 30: 943–952. (In Slovak.)

NAMULENO G., SCHOLZ T. 1994: Biometrical and morphological variability of *Phylodistomum folium* (Olfers, 1816) (Trematoda: Gorgoderidae), a parasite of pike *(Esox lucius)*. Helminthologia 31: 175–184.

NAŠINCOVÁ V., SCHOLZ T. 1994: The life cycle of *Asymphylodora tincae* (Modeer 1790) (Trematoda: Monorchiidae): a unique development in monorchiid trematodes. Parasitol. Res. 80: 192–197.

NAVRÁTIL S. 1991: Parasitoses in the fry of selected freshwater fish species under the conditions of stripping and rearing. Acta Vet. Brno 60: 357–366.

NAVRÁTIL S., PALÍKOVÁ M., KLEMENT R. 1999: Anguillicolosis of eels in water reservoirs of the Morava River basin. (Abstract.) Helminthologia 36: 129.

ONDERIKOVÁ V. 1958: Faunistický prehľad pijavíc (Hirudinea) Žitného ostrova (juhozápadného Slovenska) [Faunistic review of leeches (Hirudinea) of Žitný ostrov (of south-western Slovakia). Biológia, Bratislava, 13: 628–631. (In Slovak.)

ONDRAČKOVÁ M., DYKOVÁ I., JURAJDA P. 1999: Posthodiplostomosis of Cyprinidae. (Abstract.) Helminthologia 36: 125.

ONDRAČKOVÁ M., JURAJDA P. 2000: Metazoan parasite communities of juvenile roach *(Rutilus rutilus)* and Prussian carp *(Carassius auratus gibelio)*. (Abstract.) Helminthologia 37: 174.

OPRAVILOVÁ V. 1968: Zur Kenntnis des Entwicklungszyklus von *Holostephanus cobitidis* sp. n. (Trematoda: Cyathocotylidae). Věst. Čs. společ. zool. 32: 46–65.

PACÁK Š. 1957: Príspevok k štúdiu parazitofauny salmonidov v potoku Demänová [Contribution

to the study of the parasite fauna of salmonids in the Demänová Brook]. Čs. parasitol. 4: 239–247. (In Slovak.)
PACÁK Š. 1958: Príspevok k poznaniu parazitofauny rýb pstruhových vôd so zreteľom na umelý chov salmonidov [Contribution to the knowledge of the parasite fauna of fishes in the trout waters with reference to arteficial breeding of salmonids]. (In Slovak.) Poľnohospodárstvo 5: 569–600.
PACÁK Š. 1960: Nové parazity pstruhov [New parasites of trouts]. Poľovnictvo a rybárstvo 12 (č. 6): 14. (In Slovak.)
PACÁK Š. 1961: Invázne onemocnenia jalca tmavého z hľadiska rozširovania jeho chovu [Invasive diseases of the ide from the viewpoint of the extending of its breeding]. Čsl. rybářství 16: 51–52. (In Slovak.)
PACÁK Š. 1962: Invázne napadnutia pleskáča vysokého *(Albramis brama danubii)* z nášho Podunajska cudzopasníkmi [Invasive attacks in the common bream *(Abramis brama danubii)* of our Danubian region by parasites]. Práce Laboratória rybárstva, Bratislava, 1: 173–198. (In Slovak.)
PÁR O. 1975: Nový cizopasník kapra [A new parasite of carp.]. Veterinářství 12: 268–269. (In Czech.)
PÁR O. 1978: Mezihostitelé *Bothriocephalus gowkongensis* v rybnících na Vodňansku [Intermediate hosts of *Bothriocephalus gowkongensis* in ponds of the Vodňany region]. Buletin VÚRH Vodňany 2: 17–28. (In Czech.)
PÁR O. 1980: Tasemnice rodu *Bothriocephalus* v našich rybách [Tapeworms of the genus *Bothriocephalus* in our fishes]. Veterinářství 30: 178–179. (In Czech).
PÁR O., PÁROVÁ J. 1976: Tasemnice *Bothriocephalus gowkongensis* Yeh, 1955 v rybnících na Vodňansku [The cestode *Bothriocephalus gowkongensis* Yeh, 1955 in ponds of the Vodňanny region]. Buletin VÚRH Vodňany 12: 34–42. (In Czech.)
PÁR O., VOJTEK J. 1972: K poznání metacerkarií ryb z jižního Slovenska [To the knowledge of metacercariae of southern Slovakia]. Folia Fac. Sci. Nat. Univ. Purk. Brun., Biol. 36, 13: 31–78. (In Czech.)
PEJČOCH M. 1968: Beitrag zur Kenntnis der Gattung *Diplozoon* Nordmann, 1832 (Monogenoidea). Spisy přírodov. fak. Univ. J. E. Purkyně v Brně, č. 497: 373–398.
PEJČOCH M. 1974: Eine Übersicht der Arten der Gattung *Diplozoon* Nordmann, 1832 (Discocotylidae, Monogenoidea). Scripta Fac. Sci. Nat. UJEP Brun., Biol. 5, 4: 171–178.
PODUBSKÝ V. 1952: Nepodceňujme škodlivost pijavky rybí [Let us not minimize the danger of the fish leech]. Čsl. rybář 7: 143–144. (In Czech.)
PODUBSKÝ V. 1958: Hrouzci napadení řemenatkou [Gudgeons infested by *Ligula*]. Čsl. rybářství 13: 135. (In Czech.)
PODUBSKÝ V., ŠTĚDRONSKÝ E. 1954: Pozorování k doplnění biologie pijavky rybí (*Piscicola geometra* L.) [Observations on the biology of the fish leech (*Piscicola geometra* L.)]. Sbor. ČSAZV, B, 27: 315–326. (In Czech.)
POUPĚ J. 1996: Hlístice a úhoři [Nematodes and eels]. Rybářství č. 5, 157.
POUSTKA K. 1934: Pijavky parmové na lososech [Barbel leeches on salmon]. Rybářský věstník 14: 111. (In Czech.)
PROUZA A. 1978: Vývojový cyklus tasemnice *Proteocephalus neglectus* La Rue, 1911 [The life cycle of the cestode *Proteocephalus neglectus* La Rue, 1911]. Sbor. věd. prací Ústřed. Stát. veter. ústavu v Praze, č. 8: 56–63. (In Czech.)
PROUZA A. 1982a: Cestodózy hospodářsky významných ryb [Cestodoses of fishes with economic importance]. Zprávy Čsl. Společ. parazitol. 22/2, pp. 21–22. (In Czech.)
PROUZA A. 1982b: Úloha plevelných ryb při rozšiřování botriocefalózy kaprů [Role of unimportant fishes for the distribution of the bothriocephalosis of carps]. Veterinářství 32: 556–557. (In Czech.)
PROUZA A. 1982c: Khawióza kaprů [Khawiosis of carps]. Veterinářství 32: 470–472. (In Czech.)
PROUZA A. 1983: Vývojový cyklus tasemnice *Proteocephalus neglectus* [Life cycle of the cestode Proteocephalus neglectus]. Veterinářství 33: 128–129. (In Czech.)

PROUZA A., TESARČÍK J. 1977: Khawióza kaprů [Khawiosis of carps]. Čs. rybníkářství č. 1: 11–13. (In Czech.)

PŘIBYSLAVSKÝ J., JÍLEK J., LUCKÝ Z. 1965: Die Helminthen der Fische des Stausees Kníničská, neu für die Fauna ČSSR. Věst. Čs. spol. zool. 29: 5–8.

PŘIBYSLAVSKÝ J., LUCKÝ Z. 1967: Parasitenfauna der Fische im Stausee von Kníničky II. Ergebnisse der Forschung von Schmarotzern einiger Raubfische *(Esox lucius, Perca fluviatilis und Acerina ceruna)*. Sbor. VŠZ v Brně, B, 15: 137–145.

RÁB P., RÁBOVÁ M., BOHLEN J., LUSK S. 2000: Genetic differentiation of the two diploid-polyploid complexes of loaches, genus *Cobitis* (Cobitidae) involving *C. taenia*, *C. elongatoides* and *C.* spp. in the Czech Republic: Karyotypes and cytogenetic diversity. Folia Zool. 49 (Suppl. 1): 55–66.

RAŠÍN K. 1928: Příspěvek k biologii druhu *Pomphorhynchus laevis* Müll. (= *Echinorhynchus proteus* Westrumb.) [Contribution to the biology of the species *Pomphorhynchus laevis* Müll. (= *Echinorhynchus proteus* Westrumb.)]. Zpravodaj Morav. zem. rybář. spolku v Brně, pp. 44–64. (In Czech.)

RAŠÍN K. 1930a: Příspěvek k postembryonálnímu vývoji Amphiliny [Contribution to the postembyonal development of *Amphilina*] [Abstrakt]. Zpráva 168. schůze Čs. biol. společ. v Brně dne 11. 12. 1929. Biolog. listy 15: 140. (In Czech.)

RAŠÍN K. 1930b: Nález parasitů *Polypodium hydriforme* Uss. a *Cystoopsis acip.* Wagn. v naší sterledi [The finding of the parasites *Polypodium hydriforme* Uss. and *Cystoopsis acip.* in our sterlet] [Abstract]. Zpráva 168. schůze Čs. biol. společ. v Brně dne 11. 12. 1929. Biolog. listy 15: 140. (In Czech.)

RAŠÍN K. 1930c: Dvě poznámky o *Cystoopsis acipenseri* N. Wagner [Two notes on *Cystoopsis acipenseri* N. Wagner]. Biologické spisy Vysoké školy zvěrolékař., Brno, sv. 9, spis 6: 12 pp. (In Czech.)

RAŠÍN K. 1931: Beitrag sur postembryonalen Entwicklung der *Amphilina foliacea* (Rud.), nebst einer Bemerkung über die Laboratoriumskultur von *Gammarus pulex* (L.). Zeitschr. wissenschaft. Zoologie 138: 555–579.

RAŠÍN K. 1936: Příspěvek k postembryonálnímu vývoji vrtějše *Leptorhynchoides plagicephalus* Westr. 1821 [Contribution to the postembryonal development of *Leptorhynchoides plagicephalus* Westr. 1821]. Věst. 2. sjezdu veterinářů ČSR, Brno, 12.–14. června 1936, (tisk Novina, Brno), p. 20. (In Czech.)

RAŠÍN K. 1949a: Postembryonální vývoj vrtejše *Leptorhynchoides plagicephalus* (Westrumb 1821) [Postembryonal development of the acanthocephalan *Leptorhynchoides plagicephalus*]. Věst. Čsl. zool. spol. 13: 289–294. (In Czech.)

RAŠÍN K. 1949b: *Cocconema šulci* n. sp. (Microsporidia), cizopasník vajíček jesetera malého – (*Acipenser ruthenus* L.) [*Cocconema šulci* n. sp. (Microsporidia), a parasite of the eggs of sterlet (*Acipenser ruthenus* L.)]. Věst. Čsl. zool. spol. 13: 295–298. (In Czech.)

RODÁK L., LUCKÝ Z., DYK V., 1965: Parasitofauna ryb řeky Opavy a jejích postranních tůní [Parasite fauna of fishes of the Opava River and its side pools]. Sbor. VŠZ v Brně, B, 13: 55–62. (In Czech.)

ROMANOVSKÝ A. 1953: Cizopasní korýši na kaprech [Parasitic crustaceans on carps]. Vesmír 32: 186–187. (In Czech.)

ROMANOVSKÝ A. 1954a: *Tracheliastes maculatus* Kollar 1936, nový parasitický Copepod pro území ČSR [*Tracheliastes maculatus* Kollar 1936, a new parasitic copepod for the territory of Czechoslovakia]. Věst. Čs. zool. spol. 18: 71–72. (In Czech.)

ROMANOVSKÝ A. 1954b: Předběžná zpráva o hojném výskytu larev motolice *Diplostomum spathaceum* v našich rybách [Preliminary report on an abundant occurrence of larvae of the trematode *Diplostomum spathaceum* in our fishes]. Zool a entomol. listy 3 (17): 298–300. (In Czech.)

ROMANOVSKÝ A. 1955a: K systematice a rozšíření kapřivců *(Argulus)* v Československu [To the systematics and distribution of carp lice *(Argulus)* in Czechoslovakia]. Věst. Čs. zool. Spol. 19: 27–43. (In Czech.)

ROMANOVSKÝ A. 1955b: Poznámky k bionomii kapřivců [Remarks to the bionomy of carp lice]. Sbor. ČSAZV 28: 941–954. (In Czech.)

ROMANOVSKÝ A. 1955c: Nález *Ergasilus briani* Markewitsch 1932 (Copepoda, Ergasilidae) v Československu [The finding of *Ergasilus briani* Markewitsch 1932 (Copepoda, Ergasilidae) in Czechoslovakia]. Věst. Čs. zool. spol. 19: 187–188. (In Czech.)

ROMANOVSKÝ A. 1955d: Príspevok k poznaniu parazitických veslonôžok Slovenska [Contribution to the knowledge of the parasitic copepods of Slovakia]. Biológia, Bratislava, 10: 227–231. (In Slovak.)

ROMANOVSKÝ A. 1956: Über die Identität der Arten *Argulus japonicus* Thiele und *Argulus pellucidus* Wagler. Zool. Anz. 157: 264–265.

ŘEHULKA J. 1968: Cizopasníci střevle potoční (*Phoxinus phoxinus*) v potoce Hořině [Parasites of the brook minnow *(Phoxinus phoxinus)* in the Hořina Brook]. Sbor. VŠZ v Brně, D, 4: 91–105. (In Czech.)

ŘEHULKA J. 1973: Remarks on the occurrence of *Gyrodactylus salaris* Malmberg, 1957 sensu Ergens, 1961 (Monogenoidea: Gyrodactylidae). Věst. Čs. spol. zool. 37: 293–295.

ŘEHULKA J., TESARČÍK J. 1971: Přehled záchytů *Sanguinicola inermis* u kaprů v průběhu let 1960–1970 [Review of the records of *Sanguinicola inermis* in carps in the years 1960–1970]. Buletin VÚR Vodňany 4: 17–23. (In Czech.)

SCHÄFERNA K. 1925: Parasity způsobený zákal čočky rybího oka [Cataract of the lens of the fish eye caused by parasites]. Sbor. přírodověd., Česká akademie věd a umění, Praha, 1: 43–49. (In Czech.)

SCHÄFERNA K. 1926: Parasity způsobený zákal čočky rybího oka [Cataract of the lens of the fish eye caused by parasites]. Rybář. věstník 6: 18–22. (In Czech.)

SCHÄFERNA K. 1927: Motolice ničící kapří chovy [The fluke destroying carp breedings]. Rybář. věstník 7: 50–53. (In Czech.)

SCHOLZ T. 1986: Observations on the ecology of five species of intestinal helminths in perch (*Perca fluviatilis*) from the Mácha Lake fishpond system, Czechoslovakia. Věst. Čs. spol. zool. 50: 300–320.

SCHOLZ T. 1987a: First record of the cestode *Monobothrium wageneri* Nybelin, 1922 Caryophyllidea) in Czechoslovakia. Folia Parasitol. 34: 128–130.

SCHOLZ T. 1987b: On the occurrence of endoparasitic helminths in perch (*Perca fluviatilis* L.) from the Mácha Lake fishpond system, Czechoslovakia. Acta Univ. Carolinae, Biologica 1985: 301–318.

SCHOLZ T. 1989a: Amphilinida and Cestoda, parasites of fish in Czechoslovakia. Acta Sc. Nat. Brno 23: 1–56.

SCHOLZ T. 1989b: On the ecology of the tapeworm *Proteocephalus torulosus* Batsch, 1786 (Cestoda: Proteocephalidae) in chub (*Leuciscus cephalus* (L.)) from the River Rokytná, Czechoslovakia. Helminthologia 26: 275–287.

SCHOLZ T. 1989c: First record of the larva of *Paradilepis scolecina* (Rudolphi, 1819) (Cyclophyllidea: Dilepididae) in fishes from Czechoslovakia. Folia Parasitol. 36: 65–66.

SCHOLZ T. 1991a: Studies on the development of the cestode *Proteocephalus neglectus* La Rue, 1911 (Cestoda: Proteocephalidae) under experimental conditions. Folia Parasitol. 38: 39–55.

SCHOLZ T. 1991b: Early development of *Khawia sinensis* Hsü, 1935 (Cestoda: Caryophyllidea), a carp parasite. Folia Parasitol. 38: 133–142.

SCHOLZ T. 1991c: Development of *Khawia sinensis* Hsü, 1935 (Cestoda: Caryophyllidea) in the fish host. Folia Parasitol. 38: 225–234.

SCHOLZ T. 1993a: On the development of *Khawia baltica* Szidat, 1942 (Cestoda: Lytocestidae), a parasite of tench, *Tinca tinca* (L.). Folia Parasitol. 40: 99–103.

SCHOLZ T. 1993b: Development of *Proteocephalus torulosus* in the intermediate host under experimental conditions. J. Helminthol. 67: 316–324.

SCHOLZ T. 1997: Life-cycle of *Bothriocephalus claviceps*, a specific parasite of eels. J. Helminthol. 71: 241–248.
SCHOLZ T., DRÁBEK R., HANZELOVÁ V. 1998: Scolex morphology of *Proteocephalus* tapeworms (Cestoda: Proteocephalidae), parasites of freshwater fish in the Palaearctic Region. Folia Parasitol. 45: 27–43.
SCHOLZ T., HANZELOVÁ V. 1998: Tapeworms of the genus *Proteocephalus* Weinland, 1858 (Cestoda: Proteocephalidae), parasites of fishes in Europe. Studie AV ČR 2. Academia, Praha, 118 pp.
SCHOLZ T., HANZELOVÁ V., ŠNÁBEL V. 1995: The taxonomic status of *Proteocephalus dubius* La Rue, 1911 (Cestoda: Proteocephalidae), a puzzling parasite of perch (*Perca fluviatilis* L.). Parasite 2: 231–234.
SCHOLZ T., KEPR T. 1988: The first finding of the tapeworm *Proteocephalus macrocephalus* (Creplin, 1825) (Cestoda: Proteocephalidae) in Czechoslovakia. Folia Parasitol. 35: 111–112.
SCHOLZ T., MORAVEC F. 1990: A record of gravid cestodes *Bothriocephalus claviceps* (Goeze, 1782) from the newts *Triturus vulgaris* (L.) (Amphibia). Folia Parasitol. 37: 283–284.
SCHOLZ T., MORAVEC F. 1993: Finding of *Proteocephalus* sp. larva (Cestoda: Proteocephalidae) in *Sialis lutaria* (Insecta: Megaloptera). Acta Soc. Zool. Bohem. 57: 159–160.
SCHOLZ T., MORAVEC F. 1994: Seasonal dynamics of *Proteocephalus torulosus* (Cestoda: Proteocephalidae) in barbel (*Barbus barbus*) from the Jihlava River, Czech Republic. Folia Parasitol. 41: 253–257.
SCHOLZ T., MORAVEC F. 1996: Seasonal occurrence and maturation of *Bathybothrium rectangulum* (Cestoda: Amphicotylidae) in barbel *Barbus barbus* (Pisces) of the Jihlava River, Czech Republic. Parasite 3: 39–44.
SCHOLZ T., ŠPAKULOVÁ M., ŠNÁBEL V., KRÁĽOVÁ I., HANZELOVÁ V. 1997: A multidisciplinary approach to the systematics of *Proteocephalus macrocephalus* (Creplin, 1825) (Cestoda: Proteocephalidae). Syst. Parasitol. 37: 1–12.
SCHOLZ T., ŠPETA V., ZAJÍČEK J. 1990: Life history of the tapeworm *Khawia sinensis* Hsü, 1935, a carp parasite, in the pond Dražský Skaličany near Blatná, Czechoslovakia, Acta Vet. Brno 59: 51–63.
SCHOLZ T., ŽĎÁRSKÁ Z., DE CHAMBRIER A., DRÁBEK R. 1999: Scolex morphology of the cestode *Silurotaenia siluri* (Batsch, 1786) (Proteocephalidae: Gangesiinae), a parasite of European wels *(Silurus glanis)*. Parasitol. Res. 85: 1–6.
SILES-LUCAS M., TURČEKOVÁ Ľ, CUESTA-BANDERA C. 1995: A genetic comparison of *Proteocephalus exiguus* and *P. percae* by RAPD technique. Helminthologia 32: 201–203.
SVOBODOVÁ Z. 1978: Hodnoty některých exteriérových, kondičních a fyziologických ukazatelů dvouletých kaprů napadených tasemnicí *Bothriocephalus gowkongensis* [Values of some outside, conditional and physiological indicators in two-years old carps infested by the cestode *Bothriocephalus gowkongensis*]. Buletin VÚRH Vodňany 3: 21–25. In Czech.)
SYSOEV A. V., HANZELOVÁ V., YAKUSHEV V. YU., FREZE V. I. 1992: Some peculiarities of the process of infection transmission in cestodes of the genus *Proteocephalus*. Helminthologia 29: 19–23.
ŠEBELOVÁ Š., GELNAR M., KOUBKOVÁ B., KOMÁREK O., JURAJDA P. 2000: Effect of aquatic pollution on a metazoan parasite community of chub (*Leuciscus cephalus* L.). (Abstract.) Helminthologia 37: 173.
ŠEBELOVÁ Š., GELNAR M., KOUBKOVÁ B., ZAHRÁDKOVÁ S. 1995: The effect of aquatic pollution on the composition and structure of metazoan parasite community of chub (*Leuciscus cephalus* L.). (Abstract.) Helminthoplogia 32: 89.
ŠIMKOVÁ A., KOUBKOVÁ B., GELNAR M. 1997: Seasonal changes in the development of reproductive organs of a carp gill parasite *Eudiplozoon nipponicum* (Goto, 1891). (Abstract.) Helminthologia 34: 184.
ŠIMKOVÁ A., SASAL P., KADLEC D., GELNAR M. 2000: Analysis of dactylogyrid (Dactylogyridae, Monogenea) communities within model host

species (*R. rutilus*). (Abstract.) Helminthologia 37: 171–172.

ŠINDLEROVÁ I., GELNAR M., KOUBKOVÁ B., JURAJDA P. 1996: Contribution to the knowledge of metazoan parasites of fishes of the river Dyje, South Moravia, Czech Republic. (Abstract.) Helminthologia 33: 166.

ŠKORÍKOVÁ B., SCHOLZ T., MORAVEC F. 1996a: Occurrence of *Pseudodactylogyrus anguillae* and *P. bini* (Monogenea: Dactylogyridae) in eels, *Anguilla anguilla*, in the Czech Republic. (Abstract.) Helminthologia 33: 168–169.

ŠKORÍKOVÁ B., SCHOLZ T., MORAVEC F. 1996b: Spreading of introduced monogeneans *Pseudodactylogyrus anguillae* and *P. bini* among eel populations in the Czech Republic. Folia Parasitol. 43: 155–156.

ŠLECHTOVÁ V., LUSKOVÁ V., ŠLECHTA V., LUSK S., HALAČKA K., BOHLEN J. 2000: Genetic differentiation of two diploid-polyploid complexes of spined loach, genus *Cobitis* (Cobitidae), in the Czech Republic, involving *C. taenia*, *C. elongatoides*, and *C.* spp.: Allozyme interpopulation and interspecific differences. Folia Zool. (Suppl. 1): 67–78.

ŠNÁBEL V., HANZELOVÁ V., FAGERHOLM H.-P. 1994: Morphological and genetic comparison of two *Proteocephalus* species (Cestoda: Proteocephalidae). Parasitol. Res. 80: 141–146.

ŠNÁBEL V., HANZELOVÁ V., MATTIUCCI S., D'AMELIO S., PAGGI L. 1996: Genetic polymorphism in *Proteocephalus exiguus* shown by enzyme electrophoresis. J. Helminthol. 70: 345–349.

ŠPAKULOVÁ M. 2000: Fish endohelminths of the protected lake Morské oko (the Great Vihorlat Lake, Eastern Slovakia). (Abstract.) Helminthologia 37: 174.

ŠPAKULOVÁ M., HANZELOVÁ V. 1992: The karyotype of *Proteocephalus percae* (Cestoda: Proteocephalidae). Folia Parasitol. 39: 324–326.

ŠPAKULOVÁ M., SCHOLZ T. 1999: A chromosome study of the tapeworm *Bathybothrium rectangulum* (Bloch, 1782) (Cestoda: Pseudophyllidea). Parasitol. Res. 85: 270–273.

ŠRÁMEK A. 1900: O cizopasnících našich ryb [On parasites of our fishes]. Zemský rybář. věstník 4: 519–521. (In Czech.)

ŠRÁMEK A. 1901: Helminthen der an der zoologischen Station in Podiebrad (Böhmen) untersuchten Fischen. In: Frič A., Vávra V.: Arch. Naturw. Landesdurchforschung Böhmens 11: 94–118.

ŠRÁMEK A. 1903: Cizopasní červi ryb na zoologické stanici v Poděbradech zkoumaných [Parasitic worms of fishes examined in the zoological station near Poděbrady]. In: Frič A., Vávra V.: Výzkumy zvířeny ve vodách českých. V. Výzkum Labe a jeho starých ramen [Investigations into the fauna in Czech waters. V. Investigation of the Elbe and its old branches]. Archiv pro přírodov. výzkum Čech 11: 94–118. (In Czech.)

ŠRÁMEK-HUŠEK R. 1950: Klanonožci a kapřivci (Copepoda et Branchyura) Čech a Moravy [Copepods and carp lice (Copepoda and Branchyura) of Bohemia and Moravia]. Entomol. listy, Brno, 13: 82–91. (In Czech.)

ŠTĚDRONSKÝ E., PEKAŘ Č. 1954: Výskyt vrtejšů u kapřího plůdku (K_1) a násady (K_2) [Occurrence of acanthocephalans in carp fry (K_1) and stock fish (K_2)]. Čsl. rybářství 9 (č. 8): 118. (In Czech.)

ŠTĚPÁN V. J. 1918: Jaterní nemoc okouna říčního [Liver disease of perch]. Českomoravský rybář, Vel. Meziříčí, 17: 41–44. (In Czech.)

ŠTĚPÁN V. J. 1925: O parasitech u našich ryb sladkovodních [On parasites of our freshwater fishes]. Věda přírodní 6: 113–114. (In Czech.)

TENORA F., BARUŠ V., KRÁČMAR S., DVOŘÁČEK J. 2000: Concentrations of some heavy metals in *Ligula intestinalis* plerocercoids (Cestoda) and *Philometra ovata* (Nematoda) compared to some their hosts (Osteichthyes). Helminthologia 37: 15–18.

TENORA F., BARUŠ V., KRÁČMAR S., DVOŘÁČEK J., SRNKOVÁ J. 1999: Parallel analysis of some heavy metals concentrations in the *Anguillicola crassus* (Nematoda) and the European eel *Anguilla anguilla* (Osteichthyes). Helminthologia 36: 79–81.

TENORA F., KRÁČMAR S., BARUŠ V., DVOŘÁČEK J. 1997a: A preliminary report on selected chemical elements in plerocercoids of the tapeworm *Ligula intestinalis* (Pseudophyllidea). (Abstract.) Helminthologia 34: 180.

TENORA F., KRÁČMAR S., BARUŠ V., DVOŘÁČEK J. 1997b: Some anorganic substances in plerocwercoids of *Ligula intestinalis* (Pseudophyllidea). Acta univ. agric. et silvic. Mendel. Brun. (Brno) 45: 23–30.

TESARČÍK J. 1970a: Vylučování vajíček vrtejši *Neoechinorhynchus rutili* v podmínkách jihočeské rybniční oblasti [Shedding of eggs by the acanthocephalans *Neoechinorhynchus rutili* under conditions of the South Bohemian pond region]. Živočišná výroba 15 (43): 785–788. (In Czech.)

TESARČÍK J. 1970b: Výzkum nových způsobů prevence a terapie střevních parazitóz kapra [Research of new methods of prevention and therapy of the intestinal parasitoses of carp]. Buletin VÚR Vodňany 6: 33–34. (In Czech.)

TESARČÍK J. F. 1972: O zarazhenii karpa skrebnem *Neoechinorhynchus rutili* [On the infection of carp by the acanthocephalan *Neoechinorhynchus rutili*]. Parazitologiya 6: 190–191. (In Russian.)

TESARČÍK J. 1973: Možnosti preventivních a terapeutických zásahů při cestodozách kaprů [Possibilities of preventive and therapeutic measures in cestodoses of carp]. Čs. rybníkářství, č. 1: 29–31. (In Czech.)

TESARČÍK J., EHL A. 1968: Antiparazitární a protiplísňové koupele a ekonomicky závažná onemocnění a parazitózy nově aklimatizovaných ryb [Antiparasitic and antifungal bathes and diseases and parasitoses of economic importance in newly acclimatized fishes]. Buletin VÚR Vodňany 4: 19–26. (In Czech.)

TESARČÍK J., IVASIK V. 1977: Sravněnie geľmintogauny foreli iz reki Moravica i iz karpatskich prudov [Comparison of the helminth fauna of trout in the Moravice River and in the Carpathian ponds]. Helminthologia 14–15: 773–778. (In Russian.)

TESARČÍK J., PROUZA A. 1974: Úspěšná léčba kaviózy kaprů Taenifuginem [Successful treatment of khawiosis in carps by Taenifungin]. Buletin VÚRH Vodňany 4: 26–31. (In Czech.)

TURČEKOVÁ Ľ., KRÁĽOVÁ I. 1995: Characterization of DNA restriction profiles and rRNA gene restriction fragments length polymorphism of *Proteocephalus exiguus* and *P. neglectus* from geographically distant regions. J. Helminthol. 69: 159–163.

VEJDOVSKÝ F. 1877: Untersuchungen über die Anatomie und Metamorphose von *Tracheliastes polycolpus* Nordm. Zeitschr. wissenschaft. Zoologie, Leipzig, 29: 15–46 + Taf. II–IV.

VEJNAR F. 1956: Příspěvek k helmintofauně našich okounovitých ryb [Conribution to the helminth fauna of our perciform fishes]. Sbor. Vys. školy zemědělské a lesnic. fak. v Brně, B, 4: 161–176. (In Czech.)

VOJTEK J. 1958a: *Dactylogyrus borealis* Nybelin 1937, nový žábrohlíst pro faunu ČSR [*Dactylogyrus borealis* Nybelin, 1934, a new monogenean for the fauna of Czechoslovakia]. Čs. parasitol. 5: 199–202. (In Czech.)

VOJTEK J. 1958b: *Urocleidus* Mueller 1934, nový rod žábrohlístů (Trematoda, Monogenea) pro ČSR [*Urocleidus* Mueller 1934, a new genus of monogeneans (Trematoda, Monogenea) for Czechoslovakia)]. Biológia, Bratislava 13: 612–615. (In Czech.)

VOJTEK J. 1959a: *Metagonimus yokogawai* Katsurada 1912, nový druh matacerkarií pro faunu ČSR [*Metagonimus yokogawai* Katsurada 1912, a new species of metacercariae for the fauna of Czechoslovakia]. Věst. Čs. spol. zool. 23: 70–73. (In Czech.)

VOJTEK J. 1959b: Příspěvek k poznání helmintofauny ryb v okolí Komárna [Contribution to the knowledge of the helminth fauna of fishes in the vicinity of Komárno]. Publ. Fac. Sci. Univ. Brno, No. 407: 437–466. (In Czech.)

VOJTEK J. 1959c: Nález dospělých motolic druhu *Apophallus mühlingi* (Jägerskiöld 1899) Lühe 1909 (Digenea, Heterophyidae) v racku chechtavém [The finding of adult trematodes of the species *Apophallus mühlingi* (Jägerskiöld 1899) Lühe 1909 (Digenea,

Heterophyidae) in the gull *Larus ridibundus*]. Publ. Fac. Univ. Brno, No. 407: 467–469. (In Czech.)

VOJTEK J. 1961: Stadia larwalne robaków pasożytujacych w rybach ČSSR [Helminth larval stages parasitic in fishes in Czechoslovakia]. Wiadom. parazytol. 7: 810–814. (In Polish.)

VOJTEK J. 1964a: The importance of life-history studies for the systematics of the genus *Apatemon* (Trematoda, Strigeidae). In: Ergens R., Ryšavý B.: Parasitic worms and aquatic conditions. Proceedings of a symposium held in Prague on October 29th–November 2nd, 1962. Publ. House of the ČSAV, Prague, pp. 121–130.

VOJTEK J. 1964b: Zur Kenntnis des Entwicklungszyklus von *Apatemon cobitidis* (Linstow, 1890). Z. Parasitenk. 24: 578–599.

VOJTEK J. 1972: Entwicklung der Meatacerkarien *Apatemon cobitidis* (Linstow, 1890) im zweiten Zwischenwirt. Scripta Fac. Sci. Nat. UJEP Brun., Biol. 2, 2: 91–103, + 3 tab.

VOJTEK J. 1974: Metacerkarie z ryb Československa [Metacercariae from fishes in Czechoslovakia]. Folia Fac. Sci. Nat. Univ. Purk. Brun., Biol. 44, 15: 13–51 + 2 tab. (In Czech.)

VOJTEK J. 1981: Fauna motolic ptáků ČSSR [Fauna of the trematodes of birds in Czechoslovakia]. Univerzita J. E. Purkyně v Brně, Brno, 212 pp. (In Czech.)

VOJTEK J. 1989: The present situation of the research into the stages of development of trematodes in Czechoslovakia. Scripta Fac. Sci. Nat. Univ. Purk. Brun., Biol., 19: 339–352.

VOJTEK J., FOLKMANOVÁ B., ERGENS R. 1954: Příspěvek k poznání cizopasníků ryb jižní Moravy [Contribution to the knowledge of the parasites of fishes in South Moravia]. Práce Brněn. základny ČSAV 26: 1–24. (In Czech.)

VOJTEK J., OPRAVILOVÁ V., VOJTKOVÁ L. 1967: The importance of leeches in the life cycle of the order Strigeidida (Trematoda). Folia Parasitol. 14: 107–119.

VOJTEK J., VOJTKOVÁ L. 1971: Life-cycles of trematodes of the order Strigeidida in Czechoslovakia. Scripta Fac. Sci. Nat. UJEP Brun., Biol. 4, 1: 303–312.

VOJTKOVÁ L. 1959a: *Dactylogyrus chranilowi* Bychowsky 1933 a *Dactylogyrus simplicimalleata* Bychowsky 1931, nové druhy žábrohlístů pro faunu ČSR [*Dactylogyrus chranilowi* Bychowsky 1933 and *Dactylogyrus simplicimallaeata* Bychowsky 1931, new species of monogeneans for the fauna of Czechoslovakia]. Biológia, Bratislava, 14: 441–444. (In Czech.)

VOJTKOVÁ L. 1959b: Příspěvek k poznání cizopasníků ryb řeky Svitavy a Svratky. [Contribution to the knowledge of the parasites of fishes of the Rivers Svitava and Svratka]. Publ. Fac. Sci. Univ. Brno, No. 401: 97–123. (In Czech.)

VOJTKOVÁ L. 1971: Beitrag zur Kenntnis der Helminthofauna der Wasserwirbellosen II. Zweiter Teil der Trematodenlarven. Věst. Čs. spol. zool. 35: 136–145.

VOJTKOVÁ L., MORAVEC F. 1973: Die Nematodenlarven (Nematoda) aus den Amphibien in der Tschechoslowakei. Scripta Fac. Sci. Nat. UJEP Brun., Biol. 2, 3: 81–92.

VOJTKOVÁ L., MORAVEC F., KŘIVANEC K. 1972: Příspěvek k poznání larválních stádií hlístic (Nematoda) z obojživelníků v ČSSR [Contribution to the knowledge of larval stages of nematodes (Nematoda) from amphibians in Czechoslovakia]. Folia Fac. Sci. Nat. Univ. Purk. Brun., Biol. 36, 13: 79–92. (In Czech.)

VOLF F. 1946: Přehlížený škůdce kapřího plůdku [An overlooked one who does harm to carp fry]. Čsl. rybář 1: 123. (In Czech.)

VOLF F. 1953: Nejdůležitější nemoci ryb v našem rybářství a ochrana proti nim. [The most important diseases of fishes in our fisheries and a protection against them]. Za socialistické zemědělství, měsíčník ČSAZV pro zemědělskou vědu a praxi, 3: 1156–1162. (In Czech.)

VOLF F., POUSTKA K. 1933: Onemocnění skřelí cejna malého rybožilem (*Ichthyonema* Diesing) [Illness of gill covers of the white bream caused by the nematode (*Ichthyonema* Diesing)]. Rybář. věstník 13: 123–124. (In Czech.)

VOLF F., SMÍŠEK J. 1955: Úhoř v československých vodách [Eel in Czechoslovak waters]. Sbor. ČSAZV 28: 395–410. (In Czech.)

VOSTRADOVSKÝ J. 1998: Budoucnost úhoře nejistá? [Eel's future uncertain?]. Rybářství č. 2, 52–56. (In Czech.)

WILLOMITZER J. 1980a: Seasonal dynamics of parasitoses in grass-carp *(Ctenopharyngodon idella)* fry and fingerlings. Acta veter., Brno, 49: 269–277.

WILLOMITZER J. 1980b: Therapy of major ectoparasitoses in grascarp *(Ctenopharyngodon idella)* fry and fingerlings. Acta veter., Brno, 49: 279–282.

ZUBRICKÝ J. 1981: *Contracaecum (Erschovicaecum) bidentatum* (Linstow, 1899) Skriabin, 1917 nový druh nematóda u jesetera malého *(Acipenser ruthenus* L.) pre Slovensko (ČSSR) [*Contracaecum (Erschovicaecum) bidentatum* (Linstow, 1899) Skriabin, 1917, a new nematode species in sterlet *(Acipenser ruthenus* L.) for Slovakia]. Biológia, Bratislava, 36: 183–186. (In Slovak.)

ŽĎÁRSKÁ Z., NEBESÁŘOVÁ J. 1997: Ultrastructure of three types of scolex gland cells in adult *Bothriocephalus claviceps* (Cestoda: Pseudophyllidea). Folia Parasitol. 44: 139–146.

ŽĎÁRSKÁ Z., NEBESÁŘOVÁ J. 1999a: Regional ultrastructural differences of the scolex and neck tegument of *Proteocephalus macrocephalus* (Eucestoda: Proteocephalidae). Folia Parasitol. 46: 279–283.

ŽĎÁRSKÁ Z., NEBESÁŘOVÁ J. 1999b: Distribution and ultrastructure of two types of scolex gland cells in adult *Proteocephalus macrocephalus* (Cestoda, Proteocephalidae). Parasite 6: 49–56.

ŽITŇAN R. 1959: Beitrag zum Vorkommen von Metacercarien *Metagonimus yokogawai* Katsurada, 1912, im Unterlauf des Flusses Hron. Helminthologia 1: 99–102.

ŽITŇAN R. 1960a: *Gyrodactylus thymalli* sp. nov. aus den Flossen der Äsche (*Thymallus thymallus* L.). Helminthologia 2: 266–269.

ŽITŇAN R. 1960b: Viete, čo je „metagonimóza"? Ochorenie prenosné z rýb [Do you know what is metagonomosis? An illness transferable from fishes]. Poľovníctvo a rybárstvo 12 (č. 2): 16. (In Slovak.)

ŽITŇAN R. 1960c: Poznatky z helmintologického výskumu rýb východného Slovenska [Results of the helminthological research of fishes in eastern Slovakia]. Sbor. Východoslov. múzea, A, 1: 135–147. (In Slovak.)

ŽITŇAN R. 1960d: Vorkommen, Wirte, Lokalisierung und Invasionsextensität der Metazerkarien *Metagonimus yokogawai* (Katsurada, 1912) bei Fischen in den Gewässern slowakischer Flüsse. Helminthologia 2: 295–306.

ŽITŇAN R. 1961: Oslepnutie rýb vyvolané parazitom [Blindness of fishes caused by the parasite]. Poľovnictvo a rybárstvo 13 (č. 10): 15. (In Slovak.)

ŽITŇAN R. 1963: Ligulóza – závažná helmintóza našich rýb [Ligulosis, an important helminthosis of our fishes]. Poľovnictvo a rybárstvo 15 (č.3): 16. (In Slovak.)

ŽITŇAN R. 1964a: Helminty kapra na niektorých rybničných hospodárstvách Slovenska [Helminths in some pond carp cultures in Slovakia]. Poľovníctvo a rybárstvo 16 (č. 4): 15–16. (In Slovak.)

ŽITŇAN R. 1964b: Veková dynamika plerocerkoidov *Ligula intestinalis* (L.) u plotice obyčajnej *(Rutilus rutilus)* a niektoré jej zvláštnosti [Age dynamics of plerocercoids of *Ligula intestinalis* (L.) in roach *(Rutilus rutilus)* and some its peculiarities]. Biológia, Bratislava 19: 107–111. (In Slovak)

ŽITŇAN R. 1964c: On the finding and variability of *Gyrodactylus albaniensis* Ergens, 1960. Studia Helminthologica 1: 105–107.

ŽITŇAN R. 1964d: *Gyrodactylus rhodei* sp. nov. – novyi monogeneticheskiy sosalshchik s kozhi *Rhodeus sericeus amarus* (Bloch) [*Gyrodactylus rhodei* sp. nov., a new monogenetic trematode from the skin of *Rhodeus seriseus amarus* (Bloch)]. Helminthologia 5: 49–51. (In Russian.)

ŽITŇAN R. 1964e: Vier neue Arten monogenetischer Sougwürmer aus der Gattung *Gyrodactylus* Nordmann, 1832. Helminthologia 5: 115–122.

ŽITŇAN R. 1964f: Príspevok k helmintofaune rýb Laborca při Oboríne [Contribution to the helminth fauna of fishes of the Laborec River near Oborín]. Sbor Východoslov. Múzea v Košiciach, 5A: 91–96. (In Slovak.)

ŽITŇAN R. 1964g: Trematódy s rybími doplnkovými hostiteľmi u mačky domácej na južnom

Slovensku [Trematodes with fish complementary hosts in the domestic cat in southern Slovakia]. Studia Helminthologica 1: 109–113. (In Slovak.)

ŽITŇAN R. 1965a: Výsledky doterajšieho výskumu helmintofauny rýb Latorice [Results of the hitherto research of the helminth fauna of fishes of the Latorica River]. Sbor. Východoslov. Múzea v Košiciach, 6 B: 35–44. (In Slovak.)

ŽITŇAN R. 1965b: Monegenoidea der Fische des Flusses Hron (Gran). Helminthologia 6: 29–48.

ŽITŇAN R. 1965c: Parazitné červy lipňa obyčajného [Parasitic worms of the grayling]. Poľovníctvo a rybárstvo 17 (č. 3): 22. (In Slovak.)

ŽITŇAN R. 1966a: Monogenoidea rýb tečúcich vôd Potiskej nížiny (Helminty rýb Potiskej nížiny (ČSSR) I.) [Monogenoids of fishes from the flowing waters of the Tisa River lowlands (Helminths of fishes of the Tisa River lowlands (Czechoslovakia) I)]. Biológia, Bratislava 21: 681–692. (In Slovak.)

ŽITŇAN R. 1966b: Cestoidea a Trematoidea rýb tečúcich vôd Potiskej nížiny (Helminty rýb Potiskej nížiny (ČSSR) II.) [Cestoidea and Trematoidea of fishes from the flowing waters of the Tisa River lowlands (Helminths of fishes of the Tisa River lowlands (Czechoslovakia) II)]. Biológia, Bratislava, 21: 763–770. (In Slovak.)

ŽITŇAN R. 1966c: Parazitické červy rýb potiskej nížiny [Parasitic worms of fishes of the Tisa River lowlands]. Poľovníctvo a rybárstvo 18 (č. 2): 22. (In Slovak.)

ŽITŇAN R. 1966d: Die Trematoden des Flusses Hron. Helminthologia 7: 201–220.

ŽITŇAN R. 1967a: Helmintózne hynutie slížov [Dying of common loaches caused by helminths]. Poľovnictvo a rybárstvo 19 (č. 5): 17. (In Slovak.)

ŽITŇAN R. 1967b: Rybožravé vtáky prenášače helmintóz rýb [Fish-eating birds, transmiters of helminthoses of fishes]. Poľovníctvo a rybárstvo 19 (č. 11): 16–17. (In Slovak.)

ŽITŇAN R. 1967c: Nematoda, Acanthocephala a Hirudinea rýb tečúcich vôd Potiskej nížiny (Helminty rýb Potiskej nížiny (ČSSR) III) [Nematoda, Acanthocephala and Hirudinea of fishes from the flowing waters of the Tisa River lowlands (Helminths of fishes of the Tisa River lowlands (Czechoslovakia) III)]. Biológia, Bratislava, 22: 381–385. (In Slovak.)

ŽITŇAN R. 1967d: K nálezu niektorých zoogeograficky zaujímavých cicavíc z rodu *Dactylogyrus* Diesing, 1850 [To the finding of some zoogeographically interesting trematodes of the genus *Dactylogyrus* Diesing, 1850]. Biológia, Bratislava, 22: 845–851. (In Slovak.)

ŽITŇAN R. 1968a: Die epizootologische Bedeutung der fischfressenden Vögel bei Larval-helminthosen der Fische in der Wässern der Theissebene. Folia Veterinaria, Bratislava, 12 (č. 2): 175.

ŽITŇAN R. 1968b: O náleze a variabilite *Orientocreadium siluri* (Bychowsky et Dubinina, 1954) a *Crepidostomum auriculatum* (Wedl, 1857) v Československu [On the finding and variability of *Orientocreadium siluri* (Bychowsky et Dubinina, 1954) and *Crepidostomum auriculatum* (Wedl, 1857) in Czechoslovakia]. Biológia, Bratislava, 23: 857–862. (In Slovak.)

ŽITŇAN R. 1968c: Helmintofauna rýb československého úseku Tisy [Helminth fauna of fishes of the Czechoslovak section of the Tisa River]. Zbor. Východoslov. múzea, B, 9: 83–89. (In Slovak.)

ŽITŇAN R. 1968d: Cestoidea der Fische in Flusse Hron (ČSSR). Studia Helminthologica 2: 11–20.

ŽITŇAN R. 1969a: Daktylogyróza – častá príčina hynutia mlade kaprov [Dactylogyrosis, a frequent cause of dying of young carps]. Poľovníctvo a rybárstvo 21 (č. 5): 20. (In Slovak.)

ŽITŇAN R. 1969b: Helminty aklimatizovaného *Thymallus arcticus baicalensis* Dyb. na Slovensku a ich epizootologický význam [Helminths of the acclimatized *Thymallus arcticus baicalensis* Dyb. in Slovakia and their epizootological importance]. Biológia, Bratislava, 24: 629–633. (In Slovak.)

ŽITŇAN R. 1970a: K formovaniu parazitofauny rýb našich nových priehrad a viacúčelových

vodných nádrží [To the formation of the parasite fauna of fishes in our new dam water reservoirs and multi-purpose reservoirs]. Poľovníctvo a rybárstvo 22 (č. 4): 21–22. (In Slovak.)
ŽITŇAN R. 1970b: Aklimatizácia bylinožravých rýb a jej tienisté stránky [Acclimatization of herbivorous fishes and its shadows]. Poľovníctvo a rybárstvo 22 (č. 9): 21–22. (In Slovak.)
ŽITŇAN R. 1970–1971: Helminty rýb lipňového pásma rieky Hornád [Helminths of fishes of the grayling zone of the Hornád River]. Zbor. Východoslov. múzea v Košiciach, B, 11–12 B: 17–27. (In Slovak.)
ŽITŇAN R. 1971a: Nematoda, Acanthocephala und Hirudinea bei Fischen im Flusse Hron (ČSSR). Studia helminthologica 2 (Year 1968): 21–32.
ŽITŇAN R. 1971b: Cestoidea der Fische im Flusse Hron (ČSSR). Studia helminthologica 2 (Year 1968): 11–20.
ŽITŇAN R. 1971c: Zur Helminthenfauna der Fische in der Kleinen Donau. Helminthologia 10 (Year 1969): 313–320.
ŽITŇAN R. 1971d: Eshche ne izvestnye monogeneticheskie sosalshchiki roda *Gyrodactylus* Nordmann, 1832 ryb reki Tisy i eyo pritokov [So far unknown monogenetic trematodes of the genus *Gyrodactylus* Nordmann, 1832 from fishes of the Tisa River and its tributaries]. Studia Helminthologica (Year 1968) 2: 213–217. (In Russian.)
ŽITŇAN R. 1972a: Helminty aklimatizovaných rastlinožravých rýb na Slovensku [Helminths of acclimatized herbivorous fishes in Slovakia.]. Sbor. referátů z ichtyologické konference v Brně 16.–17. 3. 1972, Slovenská zool. spol. při SAV, ichtyolog. sekce, pp. 94–103. (In Slovak.)
ŽITŇAN R. 1972b: Verbreitung und epizootologische Bedeutung von *Dactylogyrus extensus* (Monogenoidea) in der Slowakei. Helminthologia 11 (Year 1970): 141–146.
ŽITŇAN R. 1973a: Helminty rýb Dobšinskej (Hnileckej) priehrady a ich epizootologický význam [Helminths of fishes of the Dobšinská (Hnilecká) dam water reservoir] . Biologické práce, Vydavatel. SAV, Bratislava, 87 pp. (In Slovak.)
ŽITŇAN R. 1973b: K otázke vplyvu helmintóz na úspešné obhospodarovanie voľných vôd [On the question of the influence of helminthoses on the succeeful management of the natural waters]. Téze referátů, Symposium k 75. narozeninám prof. B. Kostomarova, VŠZ v Brně, pp. 237–242. (In Slovak.)
ŽITŇAN R. 1973c: Nebezpečná pásomnica ohrozuje naše kaprovité ryby [A dangerous tapeworm endangers our cyprinid fishes]. Poľovnictvo a rybárstvo 25(č. 4): 29–30. (In Slovak.)
ŽITŇAN R. 1974a: Informaciya o predvaritelnykh rezultatach issledovaniya gelmintov rastitelnoyadnykh ryb na Chekhoslovatskom uchastke Dunaja [Information about preliminary results of the investigation of the helminths of herbivorous fishes in the Czechoslovak section of the Danube]. Materialy XVI. sessii smeshannoy komisii po primeneniyu soglasheniya o rybolovstve v vodach Dunaja, Bratislava, pp. 50–54. (In Russian.)
ŽITŇAN R. 1974b: Acclimatization of fish in the Carpathian region of Czechoslovakia and the role of helminths in this process. Ichthyologia, Beograd, 6: 143–155.
ŽITŇAN R. 1974c: Helminty rýb [Helminths of fishes]. Zborník prác o Tatranskom národnom parku 16, Vydavateľ. Osveta, Martin, pp. 39–42. (In Slovak.)
ŽITŇAN R. 1974d: Rozšírenie a epizootologický význam *Dactylogyrus vastator* Nybelin, 1924 (Monogenoidea) na Slovensku [Distribution and epizootological importance of *Dactylogyrus vastator* Nybelin, 1924 (Monogenoidea) in Slovakia]. Biológia (Bratislava) 29: 115–119. (In Slovak.)
ŽITŇAN R. 1976: Príspevok k poznaniu parazitov hlavátky (*Hucho hucho* L.) v tečúčich vodách Slovenska [Contribution to the knowledge of parasites of the hucho (*Hucho hucho* L.) in the flowing waters of Slovakia]. Súbor referátov zo seminára „Súčasný stav, ochrana a perspektíva hlavátky podunajskej (*Hucho hucho* L.) na Slovensku (v ČSSR) a v strednej Európe“,

Žilina 27. februára 1973. Vydavateľ. Príroda, Bratislava, pp. 56–64. (In Slovak.)
Žitňan R. 1977: Najzávažnejšie helmintózy rýb vodnej nádže Zemplínska šírava (The most important helminthoses of fishes of the Zemplínská Šírava water reservoir]. Zbor. referátov (Dodatok) „Biologické problémy vodného hospodárstva IV. Zemplínska šírava". Závodná pobočka SVTS při Povodí Bodrogu a Hormádu, podniku pre správu tokov v Košiciach, Košice 1977, pp. 39–48. (In Slovak.)
Žitňan R. 1978: Parazitické červy kapra [Parasitic worms of carp]. Poľovníctvo a rybárstvo 30 (č. 11): 28–29. (In Slovak.)
Žitňan R. 1979a: Nový parazit kaprov (A new parasite of carps]. Poľovníctvo a rybárstvo 31 (č. 8): 32. (In Slovak.)
Žitňan R. 1979b: Helminty kapra v malých vodných nádržiach [Helminths of carp in small water reservoirs]. Poľovníctvo a rybárstvo 31 (č. 10): 32–33. (In Slovak.)
Žitňan R. 1979c: Helminty kapra v rybníkoch [Helminths of carp in ponds]. Poľovníctvo a rybárstvo 31 (č. 11): 28–29. (In Slovak.)
Žitňan R. 1979d: Prevencia helmintóz kapra [Prevention of helminthoses of carp]. Poľovníctvo a rybárstvo 31 (č. 12): 28–29. (In Slovak.)
Žitňan R. 1979e: Gelmity ryb chekhoslovatskogo uchastka Dunaya i ikh epizootologicheskoe znachenie [Helminths of fishes of the Czechoslovak section of the Danube and their epizootological importance]. Materialy XXI. sesii smeshannoy komissii po primeneniju soglasheniya o rybolovstve v vodach Dunaya, Budapest 1979, pp. 148–162. (In Russian.)
Žitňan R. 1980a: Parazitické červy pstruha potočného [Parasitic worms of the brown trout]. Poľovníctvo a rybárstvo 32 (č. 4): 32–33. (In Slovak.)
Žitňan R. 1980b: Helminty a pijavice lipňa obyčajného na Slovensku [Helminths and leeches of the grayling in Slovakia]. Poľovníctvo a rybárstvo 32 (č. 8): 32–33. (In Slovak.)
Žitňan R. 1981a: Helminty rýb československého úseku Dunaja před výstavbou vodných diel Gabčíkovo – Nagymaros [Helminths of fishes of the Czechoslovak section of the Danube before the construction of the waterworks Gabchikovo-Nagymaros]. Zbor. materiálov z celoštátnej zoolog. konf. „Společenský význam zoologických výskumov při tvorbe a ochrane životného prostredia", Bratislava 24.–28. augusta 1981, Slovenská zool. spoločnosť při SAV, Bratislava. Bratislava, pp. 107–111. (In Slovak.)
Žitňan R. 1981b: Encyklopédia invázných chorôb rýb [Encyclopaedia of invasive diseases of fishes]. Poľovníctvo a rybárstvo 33 (č. 8 a 9): 34 and 35. (In Slovak.)
Žitňan R. 1982a: Aktuálne helmintózy kapra v rybníkoch [Current helminthoses of carp in ponds]. Zprávy Čsl. spol. parazitolog. 22/2: 22–23. (In Slovak.)
Žitňan R. 1982b: Encyklopédia invázných chorôb rýb [Encyclopaedia of invasive diseases of fishes]. Poľovnictvo a rybárstvo 34 (č. 2, 3, 5, 6): 34–35, 34, 35 a 34. (In Slovak.)
Žitňan R. 1983: Some epizootiological aspects of fish bothriocephalosis. 1st Inter. Symp. Ichthyoparasitol. "Parasites and Parasitic Diseases of Fish", 8–13 August 1983, České Budějovice, Abstracts of Papers, p. 124.
Žitňan R. 1985: Naše chránené a ohrozené ryby [Our protected and endangered fishes]. Poľovníctvo a rybárstvo 37 (č. 4, 5, 6, 7, 8, 9, 10, 11, 12): 25, 26–27, 27, 27, 28–29, 27, 28, 27, 27. (In Slovak.)
Žitňan R. 1986: Naše chránené a ohrozené ryby [Our protected and endangered fishes]. Poľovníctvo a rybárstvo 38 (č. 1, 2, 3, 4, 5, 6, 7, 8, 9, 10, 11): 27, 24, 26, 27, 24, 24, 24, 27, 30, 24, 25. (In Slovak.)
Žitňan R., Hanzelová V. 1981: The seasonal dynamics of the invasion cycle of *Dactylogyrus extensus* Mueller et van Cleave, 1932 (Monogenea). Helminthologia 18: 159–167.
Žitňan R., Hanzelová V. 1982a: Negative effect of bothriocephalosis on carp weight gains. Folia Veter. 26: 173–181.
Žitňan R., Hanzelová V. 1982b: Seasonal dynamics of the invasion cycle of *Gyrodactylus shulmani* Ling Mo-en, 1962 (Monogenea). Folia Veter. 26: 183–194.

Žitňan R., Hanzelová V. 1987: Epizootological importance of *Proteocephalus neglectus* La Rue, 1911 (Cestoda). 2nd Inter. Symp. Ichthyoparasitology "Actual Problems in Fish Parasitology", Sep. 27 – Oct. 3, 1987, Tihany, Hungary. Abstracts of papers and posters, p. 112.

Žitňan R., Hanzelová V. 1989a: K rozšíreniu závažnej pásomnice pstruhov *Proteocephalus neglectus* La Rue, 1911 v rôznych typoch vôd Karpatskej oblasti ČSSR [To the distribution of the serious tapeworm, *Proteocephalus neglectus* La Rue, 1911, of trout in different types of water of the Carpathian region of Czechoslovakia]. Poľovníctvo a rybárstvo 41: 380–381. (In Slovak.)

Žitňan R., Hanzelová V. 1989b: Prevencia proti proteocefalóze pstruhov [Prevention against proteocephalosis in trouts]. Poľovníctvo a rybárstvo 41: 421. (In Slovak.)

Žitňan R., Hanzelová V., Příhoda J., Košťan B. 1981: Overenie účinnosti Taenifuginu carp pri liečení botriocefalózy kaprov pri nízkej teplote vody [Verification of the effectivness of Taenifugin for the treatment of the bothriocephalosis of carps in low water temperature]. Biol. chem. Vet. (Praha) 17 (23): 471–477. (In Slovak.)

Žitňan R., Příhoda J., Hanzelová V., Košťan B. 1980: Taenifugin carp – účinné liečivo [Taenifugin carp, an effective medicament]. Poľovníctvo a rybárstvo 32 (č. 6): 32–33. (In Slovak.)

INDEX

Names of parasites
(page numbers of those in Parasite – host list in italics)

Names of fishes

(page numbers of those in Host – parasite list in italics)

Checklist of the Metazoan Parasites of Fishes of the Czech Republic and the Slovak Republic

RNDr. FRANTIŠEK MORAVEC, DrSc.

Vydala Academia
nakladatelství Akademie věd České republiky
Legerova 61, 120 00 Praha 2
s podporou Akademie věd České republiky

Obálku navrhla Michaela Blažejová
Redaktor publikace Pavel Zátka
Technická redaktorka Běla Trpišovská

Vydání 1., Praha 2001
Ediční číslo 4934

Sazba a tisk **SERIFA**®, s. r. o., Jinonická 80, Praha 5

ISBN 80-200-0907-8

František Moravec

TRICHINELLOID NEMATODES PARASITIC IN COLD-BLOODED VERTEBRATES

The present monograph is the first one dealing in detail with all trichinelloid nematodes parasitizing poikilotherm hosts (fishes, amphibians and reptiles), covering the families Trichinellidae, Capillariidae, Trichosomoididae, Cystoopsidae and Dioctowittidae. It is meant to provide basic data on all hitherto known species and to enable their exact determination. It summaries the present knowledge of the taxonomy, morphology, biology and ecology of these parasites, provides keys to identifications, diagnoses of higher taxa and descriptions and illustrations of all species parasitic in cold-blooded hosts. It also gives data on their hosts, localization, distribution and biology. Altogether 113 trichinelloid species and subspecies belonging to 18 genera of 5 families are dealt with in this way. The host – parasite list is provided. The list of references includes about 670 papers.

This publication is intended for parasitologists, biologists, veterinarians, workers in fisheries, university students. It will also be of interest to ichthyologists, batrachologists and herpetologists, museum curators and those engaged in nature conservation.

430 pp. – 140 line figures – in English – laminated hardback